A Unified Grand Tour of Theoretical Physics

A Unified Grand Tour of Theoretical Physics

Ian D Lawrie

Lecturer in Theoretical Physics,
The University of Leeds

Adam Hilger, Bristol and New York

British Library Cataloguing in Publication Data

Lawrie, Ian D.
A unified grand tour of theoretical physics.
1. Theoretical physics
I. Title
530.1

ISBN 0-85274-014-X (hbk)
0-85274-015-8 (pbk)

US Library of Congress Cataloging-in-Publication Data

Lawrie, Ian D.
A unified grand tour of theoretical physics.

Includes bibliographical references.
1. Mathematical physics. I. Title.
QC20.L27 1989 530.1 89-24740
ISBN 0-85274-014-X
ISBN 0-85274-015-8 (pbk.)

Consultant Editor: **E J Squires**

Published under the Adam Hilger imprint by
IOP Publishing Ltd
Techno House, Redcliffe Way, Bristol BS1 6NX, England
335 East 45th Street, New York, NY 10017-3483, USA

Printed in Great Britain by Page Bros (Norwich) Limited

To the memory of RCL and LWN, without both of whom this book would not have been written.

Contents

Preface

A few years ago, I decided to undertake some research having to do with the early history of the universe. It soon became apparent that I should have to improve my understanding of several aspects of theoretical physics, and it was from the ensuing period of self-education that the idea of writing this book emerged. I was particularly struck by two things. The first was the existence of many interrelationships, both physical and mathematical, between branches of theoretical physics that are traditionally treated as autonomous. The second was the lack of any textbook which had the scope to bring out these interrelationships adequately, or which would teach me at least the rudiments of what I needed to know in a relatively short time. It is that gap in the literature which I hope this book will go some way towards filling.

In trying to cover a wide range of topics, I have naturally been unable to give each the more extensive treatment it would receive in a more specialized work. I have tried to bear in mind the needs of three main categories of reader to whom I hope the book will be of use. As an undergraduate, I recall feeling annoying periods of frustration on encountering references to esoteric matters such as field theory and general relativity which were obviously important but said to be 'beyond the scope' of the lectures or recommended textbooks. Things have moved on a little since then, but it is still largely true that undergraduate courses devoted, for example, to gravitation and cosmology or elementary particle physics are required to give a broad view of the phenomenological aspects of their subjects, which leaves little room for exploring deeper aspects of their theoretical foundations. Final-year undergraduates who feel such a deprivation should find some enlightenment in these pages. Courses on 'theoretical physics' are also offered to undergraduates in physics and mathematics, perhaps as an optional alternative to some stint of laboratory work. The purpose of such a course is to illustrate the ways theoretical physicists have of thinking about the world, rather than to explore any of the subfields of physics exhaustively. I hope that this book will be found suitable as a basis for such courses, and I have tried to arrange the material so that lecturers may select topics from it according to their own tastes.

Postgraduate students will no doubt find, as I have done, the need to acquire some familiarity with a wide range of material which is treated adequately only in rather forbidding technical treatises. They, I hope, will find here a palatable introduction to much of what they need and, indeed, a sufficient coverage of those topics which are peripheral to their chosen speciality.

Third, I have tried to provide for professional scientists and engineers who are not theoretical physicists. They, I conceive, may find themselved unsatisfied by semi-popular accounts of advances in the subject but without time for a full-scale assault on the technical literature. For them, this book may perhaps constitute a useful half-way house.

Responsibility for what appears herein is, of course, my own, but I should like to acknowledge the assistance I have received along the way. Much of what I understand of statistical mechanics was imparted some time ago by Michael Fisher. Others who have benefited from his wisdom may recognize his influence in what I have to say, but he naturally bears no responsibility for anything I failed to understand properly. During 1986–7 I spent a sabbatical year at the University of British Columbia, where I had my first opportunity to teach a substantial graduate course on quantum field theory. The discipline of preparing the lectures and the perceptive response of the students who took the course did much to sharpen the somewhat less advanced presentation offered here. Euan Squires was instrumental in securing a contract for the book to be written. I have greatly appreciated his enthusiastic support during the writing and his comments on the first draft of the manuscript. I am also grateful to Gary Gibbons, who read the chapters on relativity and gravitation and saved me from a number of *faux pas*. Professor James Gates reviewed the entire manuscript, and I have greatly appreciated his many detailed comments and suggestions. It is a pleasure to thank Jim Revill, Neil Robertson and Jane Bartholomew at Adam Hilger for their assistance and encouragement during the various stages of production. The greatest thanks, perhaps, are due to my wife Ingrid who encouraged me through the whole venture and patiently allowed herself to be supplanted by textbooks and word processor through more evenings and weekends than either of us cares to remember.

Ian D Lawrie
December 1989

Glossary of Mathematical Symbols

∂_μ	partial derivative ($= \partial/\partial x^\mu$)	27
∇_μ	covariant derivative	34
$\square$	d'Alembertian operator	129
$\vert\,\rangle$ ($\langle\,\vert$)	ket (bra) vector	100–1
$\hat{A}$	operator in Hilbert space	102
$\hat{A}^\dagger$	adjoint operator	103
$\not{a}$	contraction with Dirac matrices ($= \gamma^\mu a_\mu$)	141
$A \overleftrightarrow{\partial_\mu} B$	antisymmetric derivative ($= A\partial_\mu B - (\partial_\mu A)B$)	130
$A_{,\mu}$	partial derivative	35
$A_{;\mu}$	covariant derivative	35
$\{A, B\}_{\mathrm{P}}$	Poisson bracket	57
$a(t)$	Robertson–Walker scale factor	303
A_μ	electromagnetic 4-vector potential	65
α	fine structure constant	204
β	inverse temperature ($= 1/k_{\mathrm{B}}T$)	221
c	fundamental velocity	9
γ^μ	Dirac matrices	135
$\Gamma^\mu{}_{\nu\sigma}$	affine connection coefficients	32
$e^\mu{}_a(x)$	vierbein	149
ε_0	permittivity of free space	9
$F_{\mu\nu}$	field strength tensor	66, 161, 165
G	Newton's gravitational constant	11
G_{F}	Fermi constant	275
$G_{\mathrm{F}}, S_{\mathrm{F}}, D_{\mathrm{F}}$	Feynman propagators	181, 188, 196
$g_{\mu\nu}(x)$	metric tensor field	14, 38
h	Planck's constant ($\hbar = h/2\pi$)	96
H	Hamiltonian	55
$\mathcal{H}$	Liouville operator	57

$\hat{I}$	identity operator	102
j^{μ}	4-vector current density	63
k_{B}	Boltzmann's constant	221
κ	gravitational constant ($= 8\pi G/c^4$)	78
L	Lagrangian	50
$\mathcal{L}$	Lagrangian density	66
Λ	cosmological constant	77
$\Lambda^{\mu'}{}_{\mu}$	coordinate transformation matrix	27
μ_0	permeability of free space	9
p^{μ}	4-momentum	62
$R^{\mu}{}_{\nu\lambda\sigma}$	Riemann curvature tensor	37
$R_{\mu\nu}$	Ricci tensor	37
R	Ricci scalar	41
ρ	phase space probability density	215
$\hat{\rho}$	density operator	229
S	action	50
S	entropy	225
σ	Stefan–Boltzmann radiation constant	236
σ^{i}	Pauli matrices	136
$T^{\mu\nu}$	stress tensor	64
T_{c}	critical temperature	245
τ	proper time	10
τ	imaginary time	232
η	nucleon/photon ratio	317
$\eta_{\mu\nu}$	metric tensor of Minkowski spacetime	15
$\bar{\psi}$	Dirac conjugate wavefunction or field	141
Z_{can}, Z_{gr}	partition functions	221, 223
Ω	grand potential	226
Ω	cosmological density ratio	308

1

Introduction: The Ways of Nature

In the eighteenth century, it became fashionable for wealthy young Englishmen to undertake the Grand Tour, an excursion which may have lasted several years, their principal destinations being Paris and the great cultural centres of Italy—Rome, Venice, Florence and Naples. For many, no doubt, the joys of travelling and occasional revelry were a sufficient inducement. For others, the opportunity to observe at first hand the social, literary and artistic achievements of other nations represented the completion of their liberal education. For a few, perhaps, it was the starting point of an independent intellectual career. It is in somewhat the same spirit that I wish to offer readers of this book a guided grand tour of theoretical physics. The members of my party need be neither wealthy (my publisher permitting), young, English, nor male. I am, however, going to assume that they have a sound knowledge of basic physics, such as a student in his or her final year of undergraduate study ought to possess.

Our itinerary cannot, of course, include everything that is important in theoretical physics. Our principal destinations are those central ideas which form the foundations of our understanding of how the world works—our knowledge, as it now stands, of the ways of nature. In outline, the topics I plan to explore are: the theories of relativity, which concern themselves with the geometrical structure of space and time and from which emerge an account of gravitational phenomena; quantum mechanics and quantum field theory, which attempt to describe the fundamental constitution of matter; and statistical mechanics, which, up to a point, allows us to deduce from this fundamental constitution the properties of the macroscopic systems of which the universe is mainly composed. The universe itself, and especially its early history, form the subject of the final chapter, where many of the ideas we shall have explored must be brought into play.

For some readers, the desire to gain a little insight into our contemporary understanding of the ways of nature will, I hope, be a sufficient inducement to read this book. For others, such as those nearing the end of their undergraduate studies, I hope to provide the opportunity of rounding off that stage of their education by delving a little more deeply into the ways of nature than the core of an undergraduate curriculum normally does. For a few, such as those embarking upon postgraduate study and research in fundamental theoretical physics, I hope to provide a readily digestible introduction to many of the ideas they will need to master.

Before setting out, I should say a few words about the point of view from which the book is written. By and large, I have written only about what I know and what I believe I understand. This, and the limited number of pages at my disposal, have led to the omission of many topics which other writers might consider essential to a theoretical understanding of physics, but that cannot be helped. The topics I have included are those I believe to be fundamental, in the sense I have tried to convey by speaking of the 'ways of nature'. The philosopher Karl Popper would have us believe that scientific theories exist only to be refuted by experimental evidence. If practising scientists really thought in that way, then I doubt that they would consider their expenditure of intellectual effort worthwhile. A good scientific theory is seldom refuted by new experimental evidence for which it cannot properly account. Much more often, it comes to be extended, generalized or reinterpreted as a constituent part of some more comprehensive theory. Every time this happens, we improve our understanding of what the world is really like: we get a clearer picture of the ways of nature.

The way in which such transformations in our understanding come about is not necessarily apparent at the point where a detailed theoretical prediction is confronted with an experimental datum. Take, for example, the transformation of classical Newtonian mechanics into quantum mechanics. We have discovered, amongst other things, that electrons can be diffracted by crystals, a phenomenon for which quantum mechanics can account, but classical mechanics cannot. Therefore, it is often said, classical mechanics must be wrong, or at least no more than an approximation to quantum mechanics with a restricted range of usefulness. It is indeed true that, under appropriate circumstances, the predictions of classical mechanics can be regarded as good approximations to those of quantum mechanics, but that is the less interesting part of the truth. There is, as we shall see, a level of theoretical description (not especially esoteric) at which classical and quantum mechanics are virtually identical, apart from a change of interpretation, and it is the reinterpretation which is vital and profound. It is, I maintain, at such a level of description that an understanding of the ways of nature is to be sought, and it is that level of description which is emphasized in this book.

It would, of course, be absurd to lay claim to any understanding of the ways of nature if our theories could not be tested in detail against experimental observations. Unfortunately, the task of deriving from our fundamental theories precise predictions which can be subjected to stringent experimental tests is often a long and highly technical one. This task, like the devising of the experiments themselves, is essential and intellectually challenging but, for want of the necessary space, I shall not often describe in detail how it can be accomplished. I do not think that this requires any apology. The basic conceptual understanding I hope to provide can, on first acquaintance, be obscured by the technical details of specific applications. The readers will nevertheless want to know by what right the theories I present can claim to describe the ways of nature, and I shall indeed outline, at certain key points, the evidence on which this claim is based. Readers who wish to become professional physicists will, in the end, have to master at least those details which are relevant to their chosen speciality and will find them described in many excellent, specialized textbooks, some of which are mentioned in my bibliography.

Most good scientific theories have been born of the need to understand certain puzzling observations. If, in retrospect, our improved insight into the ways of nature shows us that those observations are no longer puzzling but entirely to be expected, then we feel satisfied that the desired understanding has been achieved. We feel this satisfaction most deeply when the theory we have constructed has a coherent, logical, aesthetically pleasing internal structure, and rests on a few basic assumptions which, though they may not be quite self-evident, have a convincing ring of truth. Almost, though never entirely, we come to feel that things could not really have been any other way. It may be presumptuous to suppose that the ways of nature must necessarily have such a psychological appeal for us. The fact is, though, that the most successful fundamental theories of physics are of this kind, and that, for me and many others, is what makes the enterprise worthwhile.

My desire to bring out this aspect of theoretical physics strongly influences the way this book is written. When discussing, in particular, relativity and the quantum theory, the main part of my treatment begins by describing the theoretical concepts and mathematical structures which lie at the heart of these theories and later develops some of their consequences in particular physical situations. The more traditional method of introducing these subjects is to set out at the beginning the experimental facts which stand in need of explanation and then ask what new theoretical concepts are needed to accommodate them. I realize that, for many readers, the traditional approach is the more easily accessible one. For that reason, I have given, in §§2.0 and 5.0, short summaries of the more traditional development of elementary aspects of the theory. To some extent, these should serve as previews of the more detailed accounts which follow and enable readers to preserve a sense of direction and purpose while the mathematical formalism is developed. Ideally, readers should already be acquainted with special relativity, the wave-mechanical version of quantum theory and their simpler applications. Readers who are thus equipped may prefer to skip these introductory sections or to regard them and the more elementary exercises as a short revision course.

My treatment of mathematical formalism is intended to be complete and explicit. Wherever I have omitted the algebraic details needed to derive an equation, readers should be able to supply them and should usually not be satisfied until they have done so. In some cases, the exercises offer guidance. The exercises should, indeed, be regarded as an integral part of the tour; some of them introduce important ideas which are not dealt with fully in the main text.

There is one other aspect of theoretical physics which I should like readers to be aware of from the start. In recent years especially, it has become apparent that there are many similarities, some of them physical and others mathematical, between areas of physics which, on the face of it, appear to be quite separate. In the course of the book, I emphasize two of these unifying themes particularly. One is that the geometrical ideas we need to describe the structure of space and time also lie at the root of the gauge theories of the fundamental forces, described in chapters 8 and 12, of which the most familiar is electromagnetism. Indeed, once we realize the importance of these ideas, the existence of both gravitational and other forces is seen to be almost inevitable, even if we had not already been aware of them. The other is a basic mathematical similarity between quantum field theory and statistical mechanics which, as I discuss in chapter 10, can appear in several different guises. This is not altogether surprising, since both theories require us to average over uncertainties of one kind or another. The extent of the similarity is, however, quite striking, and becomes particularly apparent in the study of phase transitions, which are discussed in chapter 11. One of my chief ambitions in writing this book is to offer a unified account of theoretical physics in which these interconnections can properly be brought out.

While the connections between different topics will be appreciated only by those who read the book in its entirety, I have tried to arrange the material so that not all of it need be mastered in one go. Readers who are mainly interested in relativity and gravitation may read chapters 2, 3 and 4 and the first three sections of chapter 13 without serious loss of continuity, though the remainder of chapter 13 requires some knowledge of particle physics and statistical mechanics. Similarly, those whose main interest is in particles and field theory may read chapters 3, 5–9 and 12, but should preferably look at §§2.0, 11.4, 11.5 and 11.7 for some background information. They should then be able to follow most of chapter 13. Chapters 3, 5, 6, 10 and 11 may be read as a short course on statistical physics and the theory of phase transitions. Readers who follow one of these schemes may safely ignore occasional references to unfamiliar material, or may like to dip into relevant portions of the chapters they have omitted.

The purpose of this book is entirely pedagogical. I do not aim to describe the history of theoretical physics, nor to give anything approaching a comprehensive survey of the research literature. As far as possible, I have made at least passing mention of the many important ideas which cannot be covered in detail, and the bibliography lists a number of good textbooks and review articles to which interested readers may turn for further information and references to the original literature. I have given some references to the literature where I think that readers will find an original paper particularly enlightening or where it provides a useful historical perspective, but I have by no means listed every paper in these categories. I have certainly not attempted to refer explicitly to the work of every scientist who has made important contributions to the subjects I discuss. To do so would require a book in itself.

It is time for our tour to begin.

2

Geometry

Our tour of theoretical physics begins with geometry, and there are two reasons for this. One is that the framework of space and time provides, as it were, the stage upon which physical events are played out, and it will be helpful to gain a clear idea of what this stage looks like before introducing the cast. As a matter of fact, the geometry of space and time itself plays an active role in those physical processes which involve gravitation (and perhaps, according to some speculative theories, in other processes as well). Thus, our study of geometry will culminate, in chapter 4, in the account of gravity offered by Einstein's general theory of relativity. The other reason for beginning with geometry is that the mathematical notions we develop will reappear in later contexts.

To a large extent, the special and general theories of relativity are negative theories. By this I mean that they consist more in relaxing incorrect, though plausible, assumptions that we are inclined to make about the nature of space and time than in introducing new ones. I propose to explain how this works in the following way. We shall start by introducing a prototype version of space and time, called a 'differentiable manifold', which possesses a bare minimum of geometrical properties—for example, the notion of length is not yet meaningful. (Actually, it may be necessary to abandon even these minimal properties if, for example, we want a geometry which is fully compatible with quantum theory, but that is beyond the scope of this book.) In order to arrive at a structure which more closely resembles space and time as we know them, we then have to endow the manifold with additional properties, known as an 'affine connection' and a 'metric'. Two points then emerge: first, the common-sense notions of Euclidean geometry correspond to very special choices for these affine and metric properties; second, other possible choices lead to geometrical states of affairs which have a natural interpretation in terms of gravitational effects. Stretching the point slightly, it may be said that, merely by *avoiding* unnecessary assumptions, we are able to see gravitation as something entirely to be expected, rather than as a phenomenon in need of explanation.

To me, this insight into the ways of nature is immensely satisfying, and it is in the hope of communicating this satisfaction to readers that I have chosen to approach the subject in this way. Unfortunately, the assumptions we are to avoid are, by and large, *simplifying* assumptions, so by avoiding them we let ourselves in for some degree of complication in the mathematical formalism. Therefore, to help readers preserve a sense of direction, I will, as promised in chapter 1, provide an introductory section outlining a more traditional approach to relativity and gravitation, in which we ask how our naïve geometrical ideas must be modified to embrace certain observed phenomena.

2.0 The Special and General Theories of Relativity

2.0.1 The special theory

The special theory of relativity is concerned in part with the relation between observations of some set of physical events in two inertial frames of reference which are in relative motion. By an inertial frame, we mean one in which Newton's first law of motion holds:

> Every body continues in its state of rest, or of uniform motion in a right line, unless it is compelled to change that state by forces impressed on it. (Newton 1686)

It is worth noting that this definition by itself is in danger of being a mere tautology, since a 'force' is in effect defined by Newton's second law in terms of the acceleration it produces:

> The change of motion is proportional to the motive force impressed; and is made in the direction of the right line in which that force is impressed. (Newton 1686)

So, from these definitions alone, we have no way of deciding whether some observed acceleration of a body relative to a given frame should be attributed, on the one hand, to the action of a force or, on the other hand, to an acceleration of the frame of reference. Eddington (1929) has made this point by a facetious rerendering of the first law:

> Every body tends to move in the track in which it actually does move, except insofar as it is compelled by material impacts to follow some other track than that in which it would otherwise move.

The extra assumption we need, of course, is that forces can arise only from the influence of one body on another. An inertial frame is one relative to which any body sufficiently well isolated from all other matter for these influences to be negligible does not accelerate. In practice, needless to say, this isolation cannot be achieved. The successful application of Newtonian mechanics depends on our being able systematically to identify, and take proper account of, all those forces which cannot be eliminated. To proceed, we must take it to be established that, in principle, frames of reference can be constructed, relative to which any isolated body will, as a matter of fact, always refuse to accelerate. These frames we call inertial.

Obviously, any two inertial frames must either be relatively at rest or have a uniform relative velocity. Consider, then, two inertial frames, S and S′ (standing for *S*ystems of coordinates) with Cartesian axes so arranged that the x and x' axes lie in the same line, and suppose that S′ moves in the positive x direction with speed v relative to S. Taking y' parallel to y and z' parallel to z, we have the arrangement shown in figure 2.1. We assume that the sets of apparatus used to measure distances and times in the two frames are identical and, for simplicity, that both clocks are adjusted to read zero at the moment the two origins coincide.

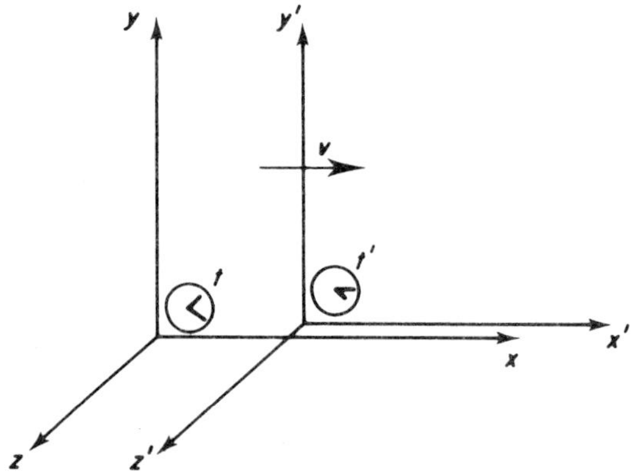

Figure 2.1 Two systems of Cartesian coordinates in relative motion.

Suppose that an event at the coordinates (x, y, z, t) relative to S is observed at (x', y', z', t') relative to S′. According to the Galilean, or common-sense, view of space and time, these two sets of coordinates must be related by

$$x' = x - vt \quad y' = y \quad z' = z \quad t' = t. \tag{2.1}$$

Since the path of a moving particle is just a sequence of events, we easily find that its velocity relative to S, in vector notation $\boldsymbol{u} = \mathrm{d}\boldsymbol{x}/\mathrm{d}t$, is related to its velocity $\boldsymbol{u}' = \mathrm{d}\boldsymbol{x}'/\mathrm{d}t'$ relative to S′ by $\boldsymbol{u}' = \boldsymbol{u} - \boldsymbol{v}$, with $\boldsymbol{v} = (v, 0, 0)$, and that its acceleration is the same in both frames, $\boldsymbol{a}' = \boldsymbol{a}$.

Despite its intuitive plausibility, the common-sense view turns out to be mistaken in several respects. The special theory of relativity hinges on the fact that the relation $\boldsymbol{u}' = \boldsymbol{u} - \boldsymbol{v}$ is not true. That is to say, this relation disagrees with experimental evidence, although discrepancies are detectable only when the magnitude of $\boldsymbol{u}'$ is an appreciable fraction of a fundamental speed, c, whose value is approximately $2.998 \times 10^8\ \mathrm{m\,s^{-1}}$. So far as is known, electromagnetic radiation always travels through a vacuum at this speed which is, of course, generally called the speed of light. Indeed, the speed of light is predicted by Maxwell's electromagnetic theory to be $(\varepsilon_0\mu_0)^{-1/2}$ (in SI units, where ε_0 and μ_0 are known as the permittivity and permeability of free space, respectively) but the theory does not single out any special frame relative to which this velocity should be measured. For quite some time after the appearance of Maxwell's theory (published in its final form in 1864; see also Maxwell (1873)), it was thought that electromagnetic radiation consisted of vibrations of a medium, the 'luminiferous ether', and had the speed c relative to the rest frame of the ether. However, a number of experiments cast doubt on this interpretation. The most celebrated, that of Michelson and Morley (1887), showed that the speed of the Earth relative to the ether must, at any time of year, be considerably smaller than that of its orbit around the Sun. Had the ether theory been correct, of course, the speed of the Earth relative to it should have changed by an amount equal to twice its orbital speed over a period of 6 months. The experiment seems to imply, then, that light always travels with the same speed, c, relative to the apparatus used to observe it.

In his paper of 1905, Einstein makes the fundamental assumption (though he expresses things a little differently) that *light travels with exactly the same speed, c, relative to any inertial frame.* Since this is clearly incompatible with the Galilean transformation law given in (2.1), he takes the remarkable step of modifying this law to read

$$
\begin{aligned}
x' &= \frac{x - vt}{(1 - v^2/c^2)^{1/2}} \qquad y' = y \\
z' &= z \qquad t' = \frac{t - vx/c^2}{(1 - v^2/c^2)^{1/2}}.
\end{aligned}
\tag{2.2}
$$

These equations are known as the *Lorentz transformation*, because a set of equations having essentially this form was first written down by H A Lorentz (1904) in the course of his attempt to explain the results of Michelson and Morley. However, Lorentz believed that his equations described a mechanical effect of the ether upon bodies moving through it, which he attributed to a modification of intermolecular forces. He does not appear to have interpreted them as Einstein did, namely as a general law relating coordinate systems in relative motion. The assumptions which lead to this transformation law are set out in exercise 2.1, where readers are invited to complete its derivation. Here, let us note that (2.2) does indeed embody the assumption that light travels with speed c relative to any inertial frame. For example, if a pulse of light is emitted from the common origin of S and S′ at time $t = t' = 0$, then the equation of the resulting spherical wavefront at time t relative to S is $x^2 + y^2 + z^2 = c^2t^2$. Using (2.2), we easily find that its equation at time t' relative to S′ is $x'^2 + y'^2 + z'^2 = c^2t'^2$.

Many of the elementary consequences of special relativity follow directly from the Lorentz transformations, and we shall meet some of them in later chapters. What particularly concerns us at present—and what makes Einstein's interpretation of the transformation equations so remarkable—is the change that these equations require us to make in our view of space and time. On the face of it, equations (2.1) or (2.2) simply tell us how to relate observations made in two different frames of reference. At a deeper level, however, they contain information about the structure of space and time which is independent of any frame of reference. Consider two events with spacetime coordinates $(\boldsymbol{x}_1, t_1)$ and $(\boldsymbol{x}_2, t_2)$ relative to S. According to the Galilean transformation law (2.1), the time interval $t'_2 - t'_1$ between them relative to S′ is equal to the interval $t_2 - t_1$ relative to S. Likewise, the spatial distances $|\boldsymbol{x}'_2 - \boldsymbol{x}'_1|$ and $|\boldsymbol{x}_2 - \boldsymbol{x}_1|$ are equal. This means that the time interval and the distance between the two events each have the same value in *every* inertial frame, and hence have real physical meanings which are independent of any system of coordinates. According to the Lorentz transformation (2.2), however, both the time interval and the distance between the two events have different values in different inertial frames. Since these frames are arbitrarily chosen by us, neither the time interval nor the distance has any definite independent meaning. The one quantity which does have a definite, frame-independent meaning is the *proper time interval*, $\Delta\tau$, defined by

$$c^2(\Delta\tau)^2 = c^2(\Delta t)^2 - (\Delta \boldsymbol{x})^2 \tag{2.3}$$

where $\Delta t = t_2 - t_1$ and $\Delta\boldsymbol{x} = \boldsymbol{x}_2 - \boldsymbol{x}_1$.

We see, therefore, that the Galilean transformation can be correct only in a *Galilean spacetime*, that is a spacetime in which both time intervals and spatial distances have well defined meanings. For the Lorentz transformation to be correct, the structure of space and time must be such that only proper time intervals are well defined. There are, as we shall see, many such structures. The one in which Lorentz transformations are valid is called *Minkowski spacetime*, after H Minkowski who first clearly described its geometrical properties (Minkowski 1908). These properties are summarized by the definition (2.3) of proper time intervals. In this definition, the constant c does not refer to the velocity of anything. Although it has dimensions of velocity, its role is really no more than that of a conversion factor between units of length and time. Thus, although the special theory of relativity arose from attempts to understand the propagation of light, it has nothing to do with electromagnetic radiation as such. Indeed, it is not in essence about relativity either! Its essential feature is the structure of space and time expressed by (2.3), and the law for transforming between frames in relative motion serves only as a clue to what this structure is. With this in mind, Minkowski (1908) says of the name 'relativity' that it '... seems to me very feeble'.

The geometrical structure of space and time restricts the laws of motion which may govern the dynamical behaviour of objects which live there. This is true, at least, if one accepts the *principle of relativity*, expressed by Einstein (1905) as follows:

> The laws by which the states of physical systems undergo change are not affected, whether these changes of state be referred to the one or the other of two systems of coordinates in uniform translatory motion.

Any inertial frame, that is to say, should be as good as any other as far as the laws of physics are concerned. Mathematically, this means that the equations expressing these laws should be *covariant*—they should have the same form in any inertial frame. Consider, for example, two objects, with masses m_1 and m_2 situated at x_1 and x_2 on the x axis of S. According to Newtonian mechanics and the Newtonian theory of gravity, the equation of motion for particle 1 is

$$m_1 \frac{\mathrm{d}^2 x_1}{\mathrm{d}t^2} = (Gm_1 m_2) \frac{x_2 - x_1}{|x_2 - x_1|^3} \tag{2.4}$$

where $G = 6.67 \times 10^{-11}\,\mathrm{N\,m^2\,kg^{-2}}$ is Newton's gravitational constant. If spacetime is Galilean and the transformation law (2.1) is valid, then $\mathrm{d}^2x'/\mathrm{d}t'^2 = \mathrm{d}^2x/\mathrm{d}t^2$ and $(x_2' - x_1') = (x_2 - x_1)$, so in S′ the equation has exactly the same form and Einstein's principle is satisfied. In Minkowski spacetime, we must use the Lorentz transformation. The acceleration relative to S′ is not equal to the acceleration relative to S (see exercise 2.2), but worse is to come! On the right-hand side, x_1 and x_2 refer to two events, namely the objects reaching these two positions, which occur simultaneously as viewed from S. As viewed from S′, however, these two events are separated by a time interval $(t_2' - t_1') = (x_1' - x_2')v/c^2$, as readers may easily verify from (2.2). In Minkowskian space, therefore, (2.4) does not satisfy the principle of relativity. It is unsatisfactory as a law of motion because it implies that there is a preferred inertial frame, namely S, relative to which the force depends only on the instantaneous separation of the two objects, while relative to any other frame it depends on the distance between their positions at different times. Actually, we do not know *a priori* that there is no such preferred frame. In the end, we trust the principle of relativity because the theories that stem from it explain a number of observed phenomena for which Newtonian mechanics cannot account.

We might imagine that electrical forces would present a similar problem, since we obtain Coulomb's law for particles with charges q_1 and q_2 merely by replacing the constant in parentheses in (2.4) with $-q_1q_2/4\pi\varepsilon_0$. In fact, Maxwell's theory of electromagnetism is not covariant under Galilean transformations but can be made covariant under Lorentz transformations with only minor modifications. We shall deal with electromagnetism in some detail later on, and I do not want to enter into the technicalities at this point. We may note, however, the features which favour Lorentz covariance. In Maxwell's theory, the forces between charged particles are transmitted by electric and magnetic fields. We know that the fields due to a charged particle do indeed appear different in different inertial frames: in a frame in which the particle is at rest, we see only an electric field, while in a frame in which the particle is moving, we also see a magnetic field. Moreover, disturbances in these fields are transmitted at the speed of light. The problem of simultaneity is avoided because a second particle responds not directly to the first one, but rather to the electromagnetic field at its own position. The expression analogous to the right-hand side of (2.4) for the Coulomb force is valid only when the two particles are held fixed relative to each other. In such a case, there is no problem, and readers are encouraged to satisfy themselves that this is so.

2.0.2 The general theory

The experimental fact which eventually led to the special theory was, as we have seen, the constancy of the velocity of light. The general theory, and the account it provides of gravitation, also spring from a crucial fact of observation, namely the equality of inertial and gravitational masses. In (2.4), the mass m_1 appears in two different guises. On the left-hand side, m_1 denotes the *inertial mass*, which governs the response of the body to a given force. On the right-hand side, it denotes the *gravitational mass*, which determines the strength of the gravitational force. The gravitational mass is analogous to the electric charge in Coulomb's law and, since the electrical charge on a body is not necessarily proportional to its mass, there is no obvious reason why the gravitational 'charge' should be determined by the mass either. The equality of gravitational and inertial masses is, of course, responsible for the fact that the acceleration of a body in the Earth's gravitational field is independent of its mass, and this has been familiar since the time of Galileo and Newton. It was checked in 1889 to an accuracy of about one part in 10^9 by Eötvös, whose method has been further refined more recently by R H Dicke and his collaborators.

It seemed to Einstein that this precise equality demanded some explanation, and he was struck by the fact that *inertial* forces such as centrifugal and Coriolis forces are proportional to the inertial mass of the body on which they act. These inertial forces are often regarded as 'fictitious', in the sense that they arise from the use of accelerating (and therefore non-inertial) frames of reference. Consider, for example, a spaceship far from any gravitating bodies such as stars and planets. When its motors are switched off, a frame of reference S fixed in it is inertial provided, as we assume, that the ship is not spinning relative to distant stars. Relative to this frame, the equation of motion of an object on which no forces act is $m\mathrm{d}^2\boldsymbol{x}/\mathrm{d}t^2 = 0$. Suppose the motors are started at time $t = 0$, giving the ship a constant acceleration a in the x direction. S is now not an inertial frame. If S′ is the inertial frame which coincided with S for $t < 0$, then the equation of motion is still $m\mathrm{d}^2x'/\mathrm{d}t'^2 = 0$, at least until the object collides with the cabin walls. Using Galilean relativity for simplicity, we have $x' = x + at^2/2$ and $t' = t$, so relative to S the equation of motion is

$$m\frac{\mathrm{d}^2x}{\mathrm{d}t^2} = -ma. \tag{2.5}$$

The force on the right-hand side arises trivially from the coordinate transformation and is definitely proportional to the *inertial* mass.

Einstein's idea is that gravitational forces are of essentially the same type as that appearing in (2.5), which means that the inertial and gravitational masses are necessarily identical. Suppose that the object in question is in fact a physicist, whose ship-board laboratory is completely soundproof and windowless. His sensation of weight, as expressed by (2.5), is equally consistent either with the ship's being accelerated by its motors or with its having landed on a planet at whose surface the acceleration due to gravity is a. Conversely, when he was apparently weightless, he would be unable to tell whether his ship was actually in deep space or freely falling towards a nearby planet. This illustrates Einstein's *principle of equivalence*, according to which the effects of a gravitational field can locally be eliminated by using a freely falling frame of reference. This frame is inertial and, relative to it, the laws of physics take the same form as they would have relative to any inertial frame in a region of space far removed from any gravitating bodies.

The word 'locally' indicates that the freely falling inertial frame can usually extend only over a small region. Let us suppose that our spaceship is indeed falling freely towards a nearby planet. (Readers may rest assured that the pilot, unlike the physicist, is aware of this and will eventually act to avert the impending disaster.) If he has sufficiently accurate apparatus, the physicist can detect the presence of the planet in the following way. Knowing the standard landing procedure, he allows two small objects to float freely on either side of his laboratory, so that the line joining them is perpendicular to the direction in which he knows that the planet, if any, will lie. Each of these objects falls towards the centre of the planet, and therefore their paths slowly converge. As observed in the freely falling laboratory, they do not accelerate in the direction of the planet, but they do accelerate towards each other, even though their mutual gravitational attraction is negligible. (The tendency of the walls of the laboratory to converge in the same manner is, of course, opposed by interatomic forces within them.) Strictly then, the effects of gravity are eliminated in the freely falling laboratory only to the extent that two straight lines passing through it, which meet at the centre of the planet, can be considered parallel. If the laboratory is small compared with its distance from the centre of the planet, then this will be true to a good approximation, but the principle of equivalence applies with complete exactness only to an infinitesimal region.

The principle of equivalence as stated above is not as innocuous as it might appear. We illustrated it by considering the behaviour of freely falling objects, and found that it followed in a more or less trivial manner from the equality of gravitational and inertial masses. A version restricted to such situations is sometimes called the *weak* principle of equivalence. The *strong* principle, applying to all the laws of physics, has much more profound implications. It led Einstein to the view that gravity is not a force of the usual kind. Rather, the effect of a massive body is to modify the geometry of space and time. Particles which are not acted on by any ordinary forces are not accelerated: they merely appear to be accelerated if we make the false assumption that the geometry is that of Galilean or Minkowski spacetime and interpret our observations accordingly.

Consider again the expression for proper time intervals given in (2.3). It is valid when (x, y, z, t) refer to Cartesian coordinates in an inertial frame. In the neighbourhood of a gravitating body, a freely falling inertial frame can be defined only in some small region, so we write it as

$$c^2(\mathrm{d}\tau)^2 = c^2(\mathrm{d}t)^2 - (\mathrm{d}\boldsymbol{x})^2 \tag{2.6}$$

where $\mathrm{d}t$ and $\mathrm{d}\boldsymbol{x}$ are infinitesimal coordinate differences. Now let us make a transformation to an arbitrary system of coordinates (x^0, x^1, x^2, x^3), each new coordinate being expressible as some function of x, y, z and t. Using the chain rule, we find that (2.6) becomes

$$c^2(\mathrm{d}\tau)^2 = \sum_{\mu,\nu=0}^{3} g_{\mu\nu}(x)\,\mathrm{d}x^\mu\,\mathrm{d}x^\nu \tag{2.7}$$

where the functions $g_{\mu\nu}(x)$ are given in terms of derivatives of the transformation functions. They are components of what is called the *metric tensor*. In the usual version of general relativity, it is the metric tensor which embodies all the geometrical structure of space and time.

Suppose we are given a set of functions $g_{\mu\nu}(x)$ which describe this structure in terms of some system of coordinates, $\{x^\mu\}$. According to the principle of equivalence, it is possible at any point (say X, with coordinates $\{X^\mu\}$) to construct a freely falling inertial frame, valid in a small neighbourhood surrounding X, relative to which there are no gravitational effects and all other physical processes occur as in special relativity. This means that it is possible to find a set of coordinates (ct, x, y, z) such that the proper time interval (2.7) reverts to the form of (2.6). Using a matrix representation of the metric tensor, we can write

$$g_{\mu\nu}(\mathrm{X}) = \eta_{\mu\nu} \equiv \begin{pmatrix} 1 & 0 & 0 & 0 \\ 0 & -1 & 0 & 0 \\ 0 & 0 & -1 & 0 \\ 0 & 0 & 0 & -1 \end{pmatrix} \tag{2.8}$$

where $\eta_{\mu\nu}$ is the special metric tensor corresponding to (2.6).

If the geometry is that of Minkowski spacetime, then it will be possible to choose (ct, x, y, z) in such a way that $g_{\mu\nu} = \eta_{\mu\nu}$ everywhere. Otherwise, the best we can usually do is to make $g_{\mu\nu} = \eta_{\mu\nu}$ at a single point, though that point can be anywhere, or at every point along a line, such as the path followed by an observer. Even when we do not have a Minkowski spacetime, it may be possible to set up an approximately inertial and approximately Cartesian coordinate system such that $g_{\mu\nu}$ differs only a little from $\eta_{\mu\nu}$ throughout a large region. In such a case, we can do much of our physics successfully by assuming that spacetime is exactly Minkowskian. If we do so, then, according to general relativity, we will interpret the slight deviations from the true Minkowski metric as gravitational forces.

This concludes our introductory survey of the theories of relativity. We have concentrated on the ways in which our common-sense ideas of spacetime geometry must be modified in order to accommodate two key experimental observations: the constancy of the velocity of light and the equality of gravitational and inertial masses. It is clear that the modified geometry leads to modifications in the laws which govern the behaviour of physical systems, but we have not discussed these laws in concrete terms. That we shall be better equipped to do after we have developed some mathematical tools in the remainder of this chapter. At that stage, we shall be able to see much more explicitly how gravity arises from geometry.

2.1 Spacetime as a Differentiable Manifold

Our aim is to construct a mathematical model of space and time which involves as few assumptions as possible, and to be explicitly aware of the assumptions we do make. In particular, we have seen that the theories of relativity call into question the meanings we attach to distances and time intervals, and we need to be clear about these. The mathematical structure which has proved to be a suitable starting point, at least for a non-quantum-mechanical model of space and time, is called a *differentiable manifold*. It is a collection of *points*, each of which will eventually correspond to a unique position in space and time, and the whole collection comprises the entire history of the model universe. It has two key features which represent familiar facts about our experience of space and time. The first is that any point can be uniquely specified by a set of four real numbers, so spacetime is four-dimensional. For the moment, the exact number of dimensions is not important. Later on, indeed, we shall encounter some recent theories which suggest that there may be more than four, the extra ones being invisible to us. Even in more conventional theories, we shall find that it is helpful to consider other numbers of dimensions as a purely mathematical device. The second feature is a kind of 'smoothness', meaning roughly that, given any two distinct points, there are more points in between them. This feature allows us to describe physical quantities such as particle trajectories or electromagnetic fields in terms of differentiable functions and hence to do theoretical physics of the usual kind. We do not know for certain that space and time are quite as smooth as this, but at least there is no evidence for any granularity down to the shortest distances we are able to probe experimentally.

Our first task is to express these properties in a more precise mathematical form. It is of fundamental importance that this can be done without recourse to any notion of length. The properties we require are *topological* ones, and we begin by introducing some elementary ideas of topology. Roughly speaking, we want to be able to say that some pairs of points are 'closer together' than others, without having any quantitative measure of distance. As an example, consider a sheet of rubber, marked off into different regions as in figure 2.2. There is no definite distance between two points in the sheet, because it can be deformed at will. No matter how it is deformed, however, any given region is always surrounded by the same neighbouring regions. Given a point in d and another in f, we can never draw a line between them which does not pass through at least one of regions b, e and h. The same holds, moreover, of more finely subdivided regions, as shown for subdivisions of a, each of which could be further subdivided, and so on. In this sense, points on the sheet are smoothly connected together. The smoothness would be lost if the rubber were vaporized, the individual molecules being considered as the collection of points. Mathematically, the kind of smoothness we want is a property of the real line (i.e. the set of all real numbers, denoted by R). So, as part of the definition of the manifold, we demand that it should be possible to set up correspondences (called 'maps') between points of the manifold and sets of real numbers. We shall next look at the topological properties of the real numbers, and then see how we can ensure that the manifold shares them.

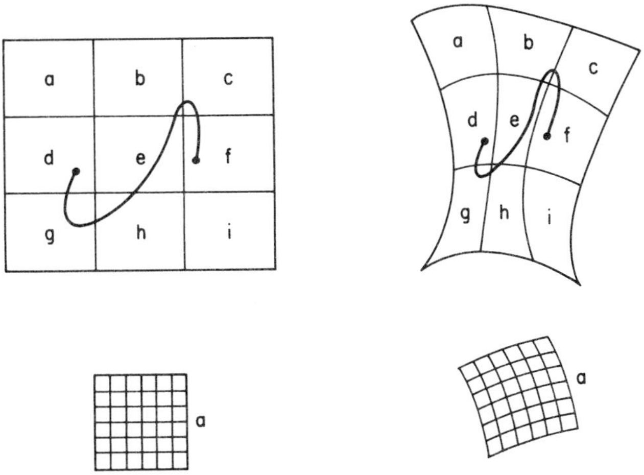

Figure 2.2 A deformable sheet of rubber, divided into several regions. Although there is no definite distance between the points indicated by •, there are always other points between them, because any curve joining them must pass through at least one of regions b, e and h.

Topology of the real line R and of R^d.

The topological properties we are interested in are expressed in terms of 'open sets', which are defined in the following way. An *open interval* (a, b) is the set of all points (real numbers) x such that $a < x < b$:

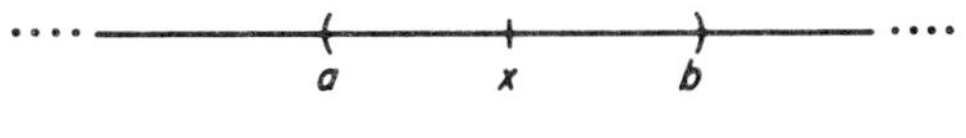

The end points $x = a$ and $x = b$ are excluded. Consequently, *any* point x in (a, b) can be surrounded by another open interval $(x - \varepsilon, x + \varepsilon)$, all of whose points are also in (a, b). For example, however close x is to a, it cannot be equal to a. There are always points between a and x, and if x is closer to a than to b, we can take $\varepsilon = (x - a)/2$. An *open set* of R is defined as any union of 1, 2, 3,. . . open intervals:

etc. (The *union*, $A \cup B \cup C \ldots$ of a number of sets is defined as the set of all points which belong to at least one of A, B, C,. . .. The *intersection*, $A \cap B \cap C \ldots$ is the set of all points which belong to all the sets A, B, C,. . ..) In addition, the empty set, which contains no points, is defined to be an open set.

The space R^2 is the set of all pairs of real numbers (x^1, x^2), which may be envisaged as an infinite plane. The definition of open sets is easily extended to R^2, as illustrated in figure 2.3. If x^1 and x^2 each lie in an open interval, then the pair lies in an open rectangle, and any union of open rectangles is an open set. Since the rectangles can be made arbitrarily small, we can say that any region bounded by a closed curve, but excluding points actually on the curve, is also an open set and so is any union of such regions. Obviously, the same ideas can be further extended to R^d, which is the set of d-tuples of real numbers $(x^1, x^2, \ldots, x^d)$.

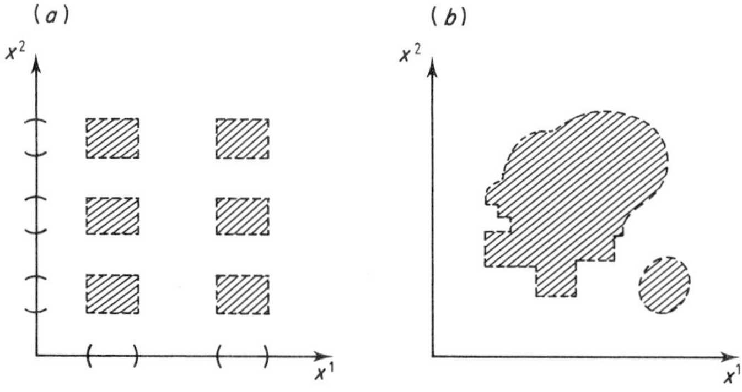

Figure 2.3 (*a*) An open set in R^2. It is a union of open rectangles constructed from unions of open intervals in the two copies of R which form the x^1 and x^2 axes. (*b*) Another open set in R^2, which can be constructed as a union of open rectangles.

An important use of open sets is to define continuous functions. Consider, for instance, a function f which takes real numbers x as arguments and has real number values $y = f(x)$. An example is shown in figure 2.4. The *inverse image* of a set of points on the y axis is the set of all points on the x axis for which $f(x)$ belongs to the original set. Then we say that f is continuous if the inverse image of any open set of the y axis is an open set of the x axis. The example shown fails to be continuous because the inverse image of any open interval containing $f(x_0)$ contains an interval of the type $(x_1, x_0]$ which includes the end point x_0 and is therefore not open. (Readers who are not at home with this style of argument should spend a short while considering the implications of these definitions: why, for example, is it necessary to include not only open intervals but also their unions and the empty set as open sets?)

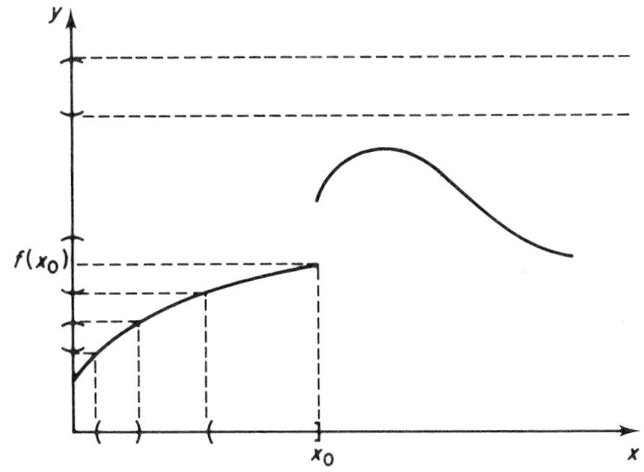

Figure 2.4 The graph $y = f(x)$ of a function which is discontinuous at x_0. Any open interval of y which includes $f(x_0)$ has an inverse image on the x axis which is not open. The inverse image of an interval in y which contains no values of $f(x)$ is the empty set.

The open sets of R^d have two fairly obvious properties: (i) any union of open sets is itself an open set; (ii) any intersection of a finite number of open sets is itself an open set. Given any space (by which we mean a set of points), suppose that a collection of subsets of its points is specified, such that any union or finite intersection of them also belongs to the collection and that every point of the space belongs to at least one of the subsets. Then the subsets in the collection may, by analogy, be called *open sets*. The collection of open sets is called a *topology* and the space, together with its topology, is called a *topological space*. It is, of course, possible to endow a given space with many different topologies. For example, the collection of all subsets of the space clearly satisfies all the above conditions. By endowing the real line with such a topology, we would obtain a new definition of continuity—it would not be a useful definition, however, as any function at all would turn out to be continuous. The particular topology of R^d described above is called the *natural topology* and is the one we shall always use.

It is important to realize that a topology is quite independent of any notion of distance. For instance, a sheet of graph paper may be regarded as a part of R^2. If it is used to draw figures in Euclidean geometry, then the distance D between two points is defined by the Pythagoras rule as $D = [(\Delta x)^2 + (\Delta y)^2]^{1/2}$. But it might equally well be used to plot the mean atmospheric concentration of carbon monoxide in central London (represented by y) as a function of time (represented by x), in which case D would have no sensible meaning.

A topology imposes two kinds of structure on the space. The *local topology*—the way in which open sets fit inside one another over small regions—determines the way in which notions like continuity apply to the space. The *global topology*—the way in which open sets can be made to cover the whole space—determines its overall structure. Thus the plane, sphere and torus have the same local structure but different global structures. Physically, we have no definite information about the global topology of spacetime, but its local structure seems to be very similar to that of R^4.

Differentiable spacetime manifold

In order that our model of space and time should be able to support continuous and differentiable functions of the sort that we rely on to do physics, we want it to have the same local topology as R^4. First of all, then, it must be a topological space, that is it must have a collection of open sets, in terms of which continuous functions can be defined. Second, the structure of these open sets must be similar, within small regions, to the natural topology of R^4. To this end, we demand that every point of the space belongs to at least one open set, all of whose points can be put into a one-to-one correspondence with the points of some open set of R^4. More technically, the correspondence is a one-to-one mapping of the open set of the space *onto* the open set of R^4, which is to say that every point of the open set in the space has a unique image point in the open set of R^4 and every point in the open set of R^4 has a unique inverse image point in the open set of the space. We further demand that this mapping be continuous, according to our previous definition. When these conditions are met, the space is called a *manifold*. The existence of continuous mappings between the manifold and R^4 implies that a function f defined on the manifold (i.e. one which has a value $f(P)$ for each point P of the manifold) can be re-expressed as a function g defined on R^4, so that $f(P) = g(x^0, \ldots, x^3)$, where $(x^0, \ldots, x^3)$ is the point of R^4 corresponding to P. In this way, continuous functions defined on the manifold inherit the characteristics of those defined on R^4.

This definition amounts to saying that the manifold can be covered by patches, in each of which a four-dimensional coordinate system can be set up, as illustrated in figure 2.5 for the more easily drawn case of the two-dimensional manifold. Normally, of course, many different coordinate systems can be set up on any part of the manifold. The definition also ensures that, within the range of coordinate values corresponding to a given patch, there exists a point of the manifold for each set of coordinate values—so there are no points 'missing' from the manifold.

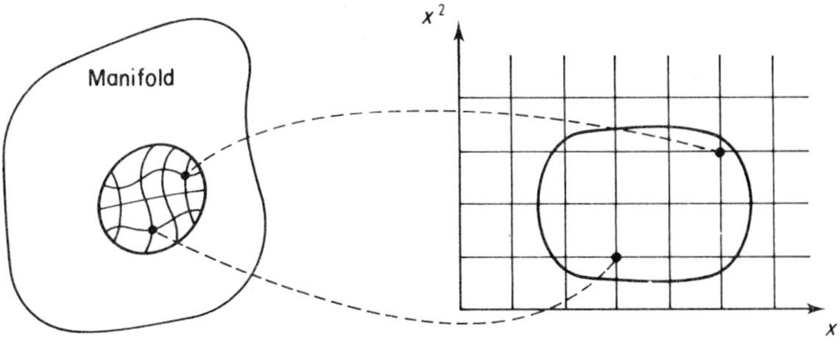

Figure 2.5 A coordinate patch on a two-dimensional manifold. Each point in the patch is mapped to a unique image point in a region of R^2 and vice versa.

Within a coordinate patch, a quantity such as an electric potential which has a value at each point of the manifold can be expressed as an ordinary function of the coordinates of the point. Often, we shall expect such functions to be *differentiable* (i.e. to posses unique partial derivatives with respect to each coordinate at each point of the patch).

Suppose we have two patches, each with its own coordinate system, which partly or wholly overlap, as in figure 2.6. Each point in the overlap region has two sets of coordinates, say $(x^0, \ldots, x^3)$ and $(y^0, \ldots, y^3)$, and the y coordinates can be expressed as functions of the x: $y^0 = y^0(x^0, \ldots, x^3)$, etc. If a function defined on the manifold is differentiable when expressed in terms of the x, then our usual idea of spacetime suggests that it ought also to be differentiable when expressed in terms of the y. This will indeed be true if the transformation functions are differentiable. Whether they are or not depends on the topology of the manifold. We shall assume that all transformation functions are differentiable, in which case we have a *differentiable manifold*. In order for a function to remain differentiable at least n times after a change of coordinates, at least the first n derivatives of all

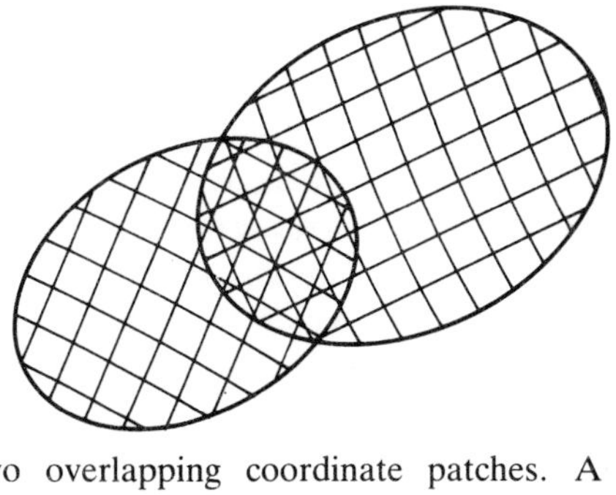

Figure 2.6 Two overlapping coordinate patches. A point in the overlap region can be identified using either set of coordinates.

transformation functions must exist. If they do, then we have what is called a *C^n manifold*. Intuitively, we might think it possible to define functions of space and time which can be differentiated any number of times, for which we would need $n = \infty$. We shall indeed take a C^∞ manifold as the basis of our model spacetime. Mathematically, though, this is a rather strong assumption, and for many physical purposes it would be sufficient to take, say, $n = 4$.

Summary and examples

Our starting point for a model of space and time is a C^∞ manifold. The essence of the technical definition described above is, first, that it is possible to set up a local coordinate system covering any sufficiently 'small' region and, second, that it is possible to define functions on the manifold which are continuous and differentiable in the usual sense. It is, of course, perfectly possible to define functions which are neither continuous nor differentiable. The point is that, if a function fails to be continuous or differentiable, this will be the fault of the function itself rather than of the manifold. The word 'small' appears in inverted commas because, as we have emphasized, there is as yet no definite notion of length: it simply means that it may well not be possible to cover the entire manifold with a single coordinate system. The coordinate systems themselves are not part of the structure of the manifold. They serve merely as an aid to thought, providing a practical means of specifying properties of sets of points belonging to the manifold.

The following examples illustrate, in terms of two-dimensional manifolds, some of the important ideas. Figure 2.7(a) shows a manifold, M, which is part of the surface of the paper on which it is printed. For the sake of argument, I am asking readers to suppose that this surface is perfectly smooth, rather than composed of tiny fibres. For the definitions to work, we must take the manifold to be the interior of the rectangular region, excluding points *on* the boundary. The interior of the roughly circular region is a coordinate patch. Inside it are drawn some of the grid lines by means of which we assign coordinates x^1 and x^2 to each point. Figure 2.7(b) is a pictorial representation of part of the space R^2 of pairs of coordinates. The interior of the shaded region represents the coordinates actually used. To every point of this region there corresponds a unique point of the coordinate patch in M, and vice versa. Figure 2.8 shows a similar arrangement, using a different coordinate system. Here, again, the *interior* of the shaded region of R^2 represents the open set of points which correspond uniquely to points of the coordinate patch. As before, the boundary of the coordinate patch and the corresponding line $x^1 = 4$ in R^2 are excluded . Also excluded, however, are the boundary lines $x^1 = 0$, $x^2 = 0$ and $x^2 = 2\pi$ in R^2, which means that points on the line labelled by $x^2 = 0$ in M do not, in fact, belong to the coordinate patch. Since the coordinate system is obviously usable, even when these points are included, their exclusion may seem like an annoying piece of bureaucracy: however, it is essential to apply the rules correctly if the definitions of continuity and differentiability are to work smoothly. For example, the function $g(x^1, x^2) = x^2$ is continuous throughout R^2, but the corresponding function on M is discontinuous at $x^2 = 0$.

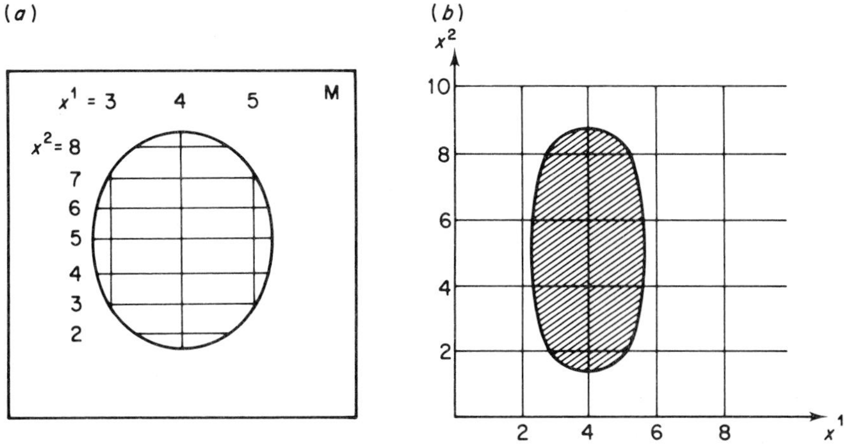

Figure 2.7 (a) A manifold M, part of the surface of this page, with a coordinate patch. (b) Part of R^2, showing the coordinate values used in (a).

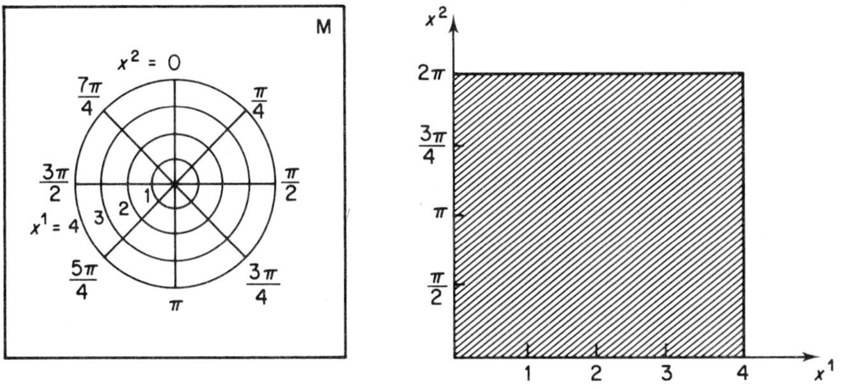

Figure 2.8 Same as figure 2.7, but using different coordinates.

It should be clear that, whereas a single coordinate patch like that in figure 2.7 can be extended to cover the whole of M, at least two patches of the kind in figure 2.8 would be needed. Readers should also be able to show that, if M were taken to be the two-dimensional surface of a sphere, no single patch of any kind could cover all of it. These examples also illustrate the fact that, although the coordinates which label the points of M have definite numerical values, these values do not, in themselves, supply any notion of a distance between two points. The distance along some curve drawn in M *may* be defined by some suitable rule, such as (i) 'use a ruler' or (ii) 'measure the volume of ink used by a standard pen to trace the curve' or, given a particular coordinate system, (iii) 'use the mathematical expression $D =$ (function of coordinates)'. Any such rule imposes an additional structure—called a *metric*—which is not inherent in the manifold. In particular, there is no naturally occurring function for use in (iii). Any specific function, such as the Pythagoras expression, would have quite different effects when applied to different coordinate systems, and the definition of the manifold certainly does not single out a special coordinate system to which that function would apply. We do have a more or less unambiguous means of determining distances on a sheet of paper, and this is because the paper, in addition to the topological properties it possesses as a manifold, had physical properties which enable us to apply a definite measuring procedure. The same is true of space and time and, although we have made some initial assumptions about their topological structure, we have yet to find out what physical properties determine their metrical structure.

2.2 Tensors

From our discussion so far, it is apparent that coordinate systems can be dangerous, even though they are often indispensable for giving concrete descriptions of a physical system. We have seen that the topology of a manifold such as that of space and time may permit the use of a particular coordinate system only within a small patch. Suppose, for the sake of argument, that the surface of the Earth is a smooth sphere. We encounter no difficulty in drawing, say, the street plan of a city on a flat sheet of paper using Cartesian coordinates, but we should obviously be misled if we assumed that this map could be extended straightforwardly to cover the whole globe. By assuming that two-dimensional Euclidean geometry was valid on the surface of the Earth, we should be making a mistake, owing to the curvature of the spherical surface, but the mistake would not become apparent as long as we made measurements only within a region the size of a city. Likewise, physicists before Einstein assumed that a frame of reference fixed on the Earth would be inertial, except for effects of the known orbital motion of the Earth around the Sun and its rotation about its own axis, which could be corrected for if necessary. According to Einstein, however, this assumption is also mistaken. It fails to take account of the true geometry of space and time in much the same way that, by treating a city plan as a Euclidean plane, we fail to take account of the true geometry of the Earth. The mistake only becomes apparent, however, when we make precise observations of gravitational phenomena.

The difficulty here is that we often express the laws of physics in the form which, we believe, applies to inertial frames. If we do not know, *a priori*, what the true geometry of space and time is, then we do not know whether any given frame of reference is truly inertial. Therefore, we need to express our laws in a way which does not rely on our making any special assumption about the coordinate system. There are two ways of achieving this. The method adopted by Einstein himself is to write our equations in a form which applies to *any* coordinate system: the mathematical techniques for doing this constitute what is called *tensor analysis*. The other, more recent method is to write them in a manner which makes no reference to coordinate systems at all: this requires the techniques of *differential geometry*. For our purposes, these two approaches are entirely equivalent, but each has its own advantages and disadvantages in terms of conceptual and notational clarity. So far as I can, I will follow a middle course which seems to me to maximize the advantages. Both techniques deal with objects called *tensors*. Tensor analysis, like elementary vector analysis, treats them as being defined by sets of components, referred to particular coordinate systems. Differential geometry treats them as entities in their own right, which may be described in terms of components, but need not be. When components are used, the two techniques become identical, so there is no difficulty in changing from one description to the other.

Many, though not all, of the physical objects which inhabit the spacetime manifold will be described by tensors. A *tensor* at a point P of the manifold refers only to that point. A *tensor field* assigns some property to every point of the manifold, and most physical properties will be described by tensor fields. Tensors and tensor fields are classified by their *rank*, a pair of numbers $\binom{a}{b}$.

Rank $\binom{0}{0}$ tensors, also called *scalars*, are simply real numbers. A *scalar field* is a real-valued function, say $f(\mathrm{P})$, which assigns a real number to each point of the manifold. If our manifold were just the three-dimensional space encountered in Newtonian physics, then at a particular instant in time an electric potential $V(\mathrm{P})$ or the density of a fluid $\rho(\mathrm{P})$ would be examples of scalar fields. In relativistic physics, these and all other simple examples I can think of are not true scalars, because their definitions depend in one way or another on the use of specific coordinate systems, or on metrical properties of the space which our manifold does not yet possess. For the time being, however, no great harm will be done if readers bear these examples in mind. If we introduce coordinates x^μ, then we can express $f(\mathrm{P})$ as an algebraic function $f(x^\mu)$. (Readers will realize that I am, for convenience, using the same symbol f to denote two different, though related functions: we have $f(x^\mu) = f(\mathrm{P})$ when x^μ are the coordinates of the point P.) In a different coordinate system, where P has the coordinates $x^{\mu'}$, the same quantity will be described by a new algebraic function $f'(x^{\mu'})$ related to the old one by

$$f'(x^{\mu'}) = f(x^\mu) = f(\mathrm{P}). \tag{2.9}$$

In tensor analysis, this transformation law is taken to *define* what is meant by a scalar field.

Rank $\binom{1}{0}$ tensors are called *vectors* in differential geometry. They correspond to what are called *contravariant vectors* in tensor analysis. The prototypical vector is the tangent vector to a curve. In ordinary Euclidean geometry, the equation of a curve may be expressed parametrically by giving three functions $x(\lambda)$, $y(\lambda)$ and $z(\lambda)$, so that each point of the curve is labelled by a value of λ and the functions give its coordinates. If λ is chosen to be the distance along the curve from a given starting point, then the tangent vector to the curve at the point labelled by λ has components $(\mathrm{d}x/\mathrm{d}\lambda, \mathrm{d}y/\mathrm{d}\lambda, \mathrm{d}z/\mathrm{d}\lambda)$. In our manifold, we have not yet given any meaning to 'distance along the curve', and we want to avoid defining vectors in terms of their components relative to a specific coordinate system. Differential geometry provides the following indirect method of generalizing the notion of a vector to any manifold. Consider, in Euclidean space, a differentiable function $f(x, y, z)$. This function has, in particular, a value $f(\lambda)$ at each point of the curve, which we obtain by substituting for x, y and z the appropriate functions of λ. The rate of change of f with respect to λ is

$$\frac{\mathrm{d}f}{\mathrm{d}\lambda} = \frac{\mathrm{d}x}{\mathrm{d}\lambda}\frac{\partial f}{\partial x} + \frac{\mathrm{d}y}{\mathrm{d}\lambda}\frac{\partial f}{\partial y} + \frac{\mathrm{d}z}{\mathrm{d}\lambda}\frac{\partial f}{\partial z} \tag{2.10}$$

so, by choosing $f = x$, $f = y$ or $f = z$, we can recover from this expression each component of the tangent vector. All the information about the tangent vector is contained in the differential operator $\mathrm{d}/\mathrm{d}\lambda$, and in differential geometry this operator is defined to *be* the tangent vector.

A little care is required when applying this definition to our manifold. We can certainly draw a continuous curve on the manifold and label its points continuously by a parameter λ. What we cannot yet do is select a special parameter which measures distance along it. Clearly, by choosing different parametrizations of the curve, we shall arrive at different definitions of its tangent vectors. It is convenient to refer to the one-dimensional set of points in the manifold as a *path*. Then each path may be parametrized in many different ways, and we regard each parametrization of the path as a distinct curve. This has the advantage that each curve, with parameter λ, has a unique tangent vector $\mathrm{d}/\mathrm{d}\lambda$ at every point. Suppose we have two curves, corresponding to the same path, but with parameters λ and μ, which are related by $\mu = a\lambda + b$, a and b being constants. The difference is obviously a rather trivial one and the two parameters are said to be *affinely related*.

If we now introduce a coordinate system, we can resolve a vector into components, in much the same way as in Euclidean geometry. At this point, it is useful to introduce two abbreviations into our notation. First, we use the symbol ∂_μ to denote the partial derivative $\partial/\partial x^\mu$. Second, we will use the summation convention according to which, if an index such as μ appears in an expression twice, once in the upper position and once in the lower position, then a sum over the values $\mu = 0 \ldots 3$ is implied. (More generally, in a d-dimensional manifold, the sum is over the values $0 \ldots (d-1)$.) If the same index appears more than twice, or appears twice in the same position, then the expression is likely to be meaningless. We will use bold capital letters to denote vectors, such as $\boldsymbol{V} = \mathrm{d}/\mathrm{d}\lambda$. If, then, a curve is represented in a particular coordinate system by the functions $x^\mu(\lambda)$, we can write

$$\boldsymbol{V} \equiv \frac{\mathrm{d}}{\mathrm{d}\lambda} = \sum_{\mu=0}^{3} \frac{\mathrm{d}x^\mu}{\mathrm{d}\lambda} \frac{\partial}{\partial x^\mu} \equiv V^\mu \partial_\mu \equiv V^\mu \boldsymbol{X}_\mu \tag{2.11}$$

where the quantities $\boldsymbol{X}_\mu = \partial/\partial x^\mu$ are identified as basis vectors in the particular coordinate system and V^μ are the corresponding components of the vector. Note that components of a vector are labelled by upper indices and basis vectors by lower ones. In a new coordinate system, with coordinates $x^{\mu'}$ and basis vectors $\boldsymbol{X}_{\mu'} = \partial/\partial x^{\mu'}$, the chain rule shows that the same vector has components

$$V^{\mu'} = \frac{\partial x^{\mu'}}{\partial x^\mu} V^\mu. \tag{2.12}$$

In tensor analysis, a vector is defined by specifying its components in some chosen coordinate system and requiring its components in any other system to be those given by the transformation law (2.12). It will be convenient to denote the transformation matrix by

$$\Lambda^{\mu'}{}_{\mu} = \frac{\partial x^{\mu'}}{\partial x^\mu}. \tag{2.13}$$

Note that the prime on the index μ' is used to indicate that the coordinates x^μ and $x^{\mu'}$ belong to two different coordinate systems. Using the chain rule, we find that

$$\Lambda^{\mu}{}_{\nu'}\,\Lambda^{\nu'}{}_{\sigma} = \frac{\partial x^{\mu}}{\partial x^{\nu'}}\frac{\partial x^{\nu'}}{\partial x^{\sigma}} = \frac{\partial x^{\mu}}{\partial x^{\sigma}} = \delta^{\mu}{}_{\sigma} \tag{2.14}$$

so the matrix $\Lambda^{\mu}{}_{\nu'}$ is the inverse of the matrix $\Lambda^{\nu'}{}_{\mu}$.

Rank $\binom{0}{1}$ tensors are called *one-forms* in differential geometry or *covariant vectors* in tensor analysis. Consider the scalar product $\boldsymbol{u}\cdot\boldsymbol{v}$ of two Euclidean vectors. Normally, we regard this product as a rule which combines the two vectors $\boldsymbol{u}$ and $\boldsymbol{v}$ to produce a real number. As we shall see, this scalar product involves metrical properties of Euclidean space which our manifold does not yet possess. There is, however, a different point of view which can be transferred to the manifold. For a given vector $\boldsymbol{u}$, the symbol $\boldsymbol{u}\cdot$ can be regarded as defining a *function*, whose argument is a vector, say $\boldsymbol{v}$, and whose value is the real number $\boldsymbol{u}\cdot\boldsymbol{v}$. The function $\boldsymbol{u}\cdot$ is *linear*. That is to say, if we give it the argument $a\boldsymbol{v} + b\boldsymbol{w}$, where $\boldsymbol{v}$ and $\boldsymbol{w}$ are any two vectors and a and b are any two real numbers, then $\boldsymbol{u}\cdot(a\boldsymbol{v} + b\boldsymbol{w}) = a\boldsymbol{u}\cdot\boldsymbol{v} + b\boldsymbol{u}\cdot\boldsymbol{w}$. This is, in fact, the definition of a one-form. In our manifold, a one-form, say ω, is a real-valued linear function whose argument is a vector: $\omega(\boldsymbol{V}) =$ real number. Because the one-form is a linear function, its value must be a linear combination of the components of the vector, in any coordinate system:

$$\omega(\boldsymbol{V}) = \omega_{\mu}V^{\mu}. \tag{2.15}$$

The coefficients ω_{μ} are the components of the one-form relative to the particular coordinate system. A *one-form field* is defined in the same way as a linear function of vector fields, whose value is a real scalar field. In the definition of linearity, a and b may be any two scalar fields.

The expression (2.15) is, of course, similar to the rule for calculating the scalar product of two Euclidean vectors from their components. Nevertheless, it is clear from their definitions that vectors and one-forms are quite different things, and (2.15) does not allow us to form a scalar product of two vectors. Later on we shall find that, when the manifold is endowed with a metric, it is possible to specify a unique one-form corresponding to any vector, and vice versa. It will then be possible to form the scalar product of two vectors by using in (2.15) the components of one vector and those of the one-form corresponding to the second vector.

An example of a one-form field is the gradient of a scalar field f, whose components are $\partial_{\mu}f = \partial f/\partial x^{\mu}$. Notice the consistency of the convention for the placing of indices: the components of a one-form have indices which naturally appear in the lower position. Call this gradient one-form ω_f. If $\boldsymbol{V} = \mathrm{d}/\mathrm{d}\lambda$ is the tangent vector to a curve $x^{\mu}(\lambda)$, then the new scalar field $\omega_f(\boldsymbol{V})$ is the rate of change of f along the curve:

$$\omega_f(V) = \frac{\partial f}{\partial x^\mu}\frac{dx^\mu}{d\lambda} = \frac{df}{d\lambda}. \tag{2.16}$$

Since vectors and one-forms exist independently of any coordinate system, the function $\omega(V)$ given in (2.15) must be a true scalar field—it must have the same value in any coordinate system. This means that the matrix which transforms the components of a one-form between two systems must be the inverse of that which transforms the components of a vector:

$$\omega_{\mu'} = \omega_\mu \Lambda^\mu{}_{\mu'} = \omega_\mu \frac{\partial x^\mu}{\partial x^{\mu'}}. \tag{2.17}$$

Then, on transforming (2.15), we get

$$\omega(V) = \omega_{\mu'} V^{\mu'} = \omega_\mu \Lambda^\mu{}_{\mu'} \Lambda^{\mu'}{}_\nu V^{\nu'} = \omega_\mu V^\mu. \tag{2.18}$$

In tensor analysis, a covariant vector is defined by requiring that its components satisfy the transformation law (2.17). Clearly, this is the correct way of transforming a gradient.

Rank $\binom{a}{b}$ tensors and tensor fields can be defined in a coordinate-independent way, making use of the foregoing definitions of vectors and one-forms. At this point, however, it becomes rather easier to adopt the tensor analysis approach of defining higher-rank tensors in terms of their components. A tensor of *contravariant rank a* and *covariant rank b* has, in a d-dimensional manifold, d^{a+b} components, labelled by a upper indices and b lower ones. The tensor may be specified by giving each of its components relative to some chosen coordinate system. In any other system, the components are then given by a transformation law which generalizes the transformations of vectors and one-forms in an obvious way:

$$T^{\alpha'\beta'\ldots}{}_{\mu'\nu'\ldots} = \Lambda^{\alpha'}{}_\alpha \Lambda^{\beta'}{}_{\beta\ldots} \Lambda^\mu{}_{\mu'} \Lambda^\nu{}_{\nu'\ldots} T^{\alpha\beta\ldots}{}_{\mu\nu\ldots}. \tag{2.19}$$

If ω is a one-form and V a vector, then the tensor whose components are $\omega_\mu V^\nu$ is a rank $\binom{1}{1}$ tensor. As we saw in (2.15), by setting $\mu = \nu$ and carrying out the implied sum, we obtain a scalar quantity, a tensor of rank $\binom{0}{0}$. This process is called *contraction*. Given any tensor of rank $\binom{1}{1}$ or higher, we may contract an upper index with a lower one to obtain a new tensor of one lower contravariant rank and one lower covariant rank then the original.

2.3 Extra Geometrical Structures

Two geometrical structures are needed to endow our manifold with the familiar properties of space and time: (i) the notion of *parallelism* is represented mathematically by an *affine connection*; (ii) the notions of *length* and *angle* are represented by a *metric*. In principle, these two structures are quite independent. In Euclidean geometry, of course, it is perfectly possible to define what we mean by parallel lines in terms of distances and angles, and this is also true of the structures which are most commonly used in general relativistic geometry. Thus there is, as we shall see, a special kind of affine connection which can be deduced from a metric. It is called a *metric connection*. We shall eventually assume that the actual geometry of space and time is indeed described by a metric connection. From a theoretical point of view, however, it is instructive to understand the distinction between those geometrical ideas which rely only on an affine connection and those which require a metric. Moreover, there are manifolds other then spacetime which play important roles in physics (in particular, those connected with the gauge theories of particle physics) which possess connections but do not necessarily possess metrics. To emphasize this point, therefore, I shall deal first with the affine connection, then with the metric, and finally with the metric connection.

2.3.1 The affine connection

There are four important geometrical tools provided by an affine connection: the notion of *parallelism*, the notion of *curvature*, the *covariant derivative* and the *geodesic*. Let us first understand what it is good for.

(a) Newton's first law of motion claims that 'a body moves at constant speed in a straight line unless it is acted on by a force'. In general relativity, we will replace this with the assertion that 'a test particle follows a geodesic curve unless it is acted on by a non-gravitational force'. As we saw earlier, gravitational forces are going to be interpreted in terms of spacetime geometry, which is itself modified by the presence of gravitating bodies. By a 'test particle', we mean one which responds to this geometry, but does not modify it significantly. A *geodesic* is a generalization of the straight line of Euclidean geometry. It is defined, roughly, as a curve whose tangent vectors at successive points are parallel, as illustrated in figure 2.9. Given a definition of 'parallel', as provided by the connection, this is, perhaps, intuitively recognizable as the natural state of motion for a particle which is not disturbed by external influences.

(b) The equations of physics, which we wish to express entirely in terms of tensors, frequently involve the derivatives of vector or tensor fields. Now the derivatives of a scalar field, $\partial_\mu f$, are, as we have seen, the components of a one-form. However, the derivatives of the components of a vector field, $\partial_\mu V^\nu$, are not the components of a tensor field, even though they are labelled by a contravariant and a covariant index.

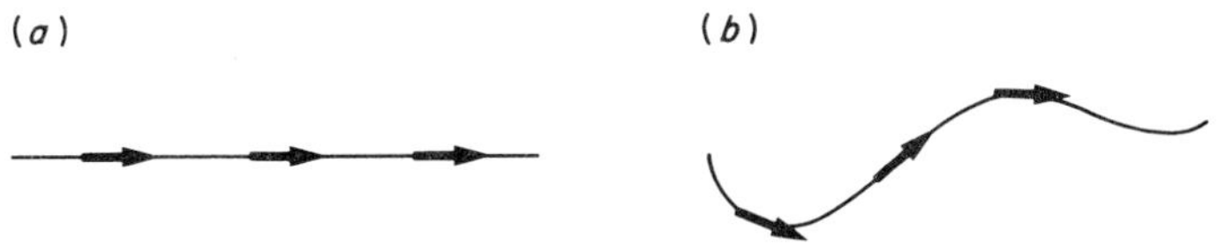

Figure 2.9 (*a*) A geodesic curve: successive tangent vectors are parallel to each other. (*b*) A non-geodesic curve: successive tangent vectors are not parallel.

On transforming these derivatives to a new coordinate system, we find

$$\begin{aligned}\partial_{\mu'}V^{\nu'} &= \Lambda^{\mu}{}_{\mu'}\partial_{\mu}(\Lambda^{\nu'}{}_{\nu}V^{\nu})\\ &= \Lambda^{\mu}{}_{\mu'}\Lambda^{\nu'}{}_{\nu}\partial_{\mu}V^{\nu} + \Lambda^{\mu}{}_{\mu'}(\partial_{\mu}\Lambda^{\nu'}{}_{\nu})V^{\nu}.\end{aligned} \tag{2.20}$$

Because of the last term, this does not agree with the transformation law for a second-rank tensor. The affine connection will enable us to define what is called a *covariant derivative*, ∇_μ, whose action on a vector field is of the form $\nabla_\mu V^\nu = \partial_\mu V^\nu$ + (connection term). The transformation of the extra term involving the affine connection will serve to cancel the unwanted part in (2.20), so that $\nabla_\mu V^\nu$ will be a tensor.

(c) The fact that the functions $\partial_\mu V^\nu$ do not transform as the components of a tensor indicates that they have no coordinate-independent meaning. To see what goes wrong, consider the derivative of a component of a vector field along a curve, as illustrated in figure 2.10(*a*), where P and Q are points on the curve with parameters λ and $\lambda + \delta\lambda$ respectively. The derivative at P is

$$\frac{\mathrm{d}V^\mu}{\mathrm{d}\lambda} = \frac{\mathrm{d}x^\nu}{\mathrm{d}\lambda}\frac{\partial V^\mu}{\partial x^\nu} = \lim_{\delta\lambda\to 0}\frac{V^\mu(Q) - V^\mu(P)}{\delta\lambda}. \tag{2.21}$$

For a scalar field, which has unique values at P and Q, such a derivative makes good sense. However, the values at P and Q of the components of a vector field depend on the coordinate system to which they are referred. It is easy to make a change of coordinates such that, for example, V^μ(Q) is changed while V^μ(P) is not, and so the difference of

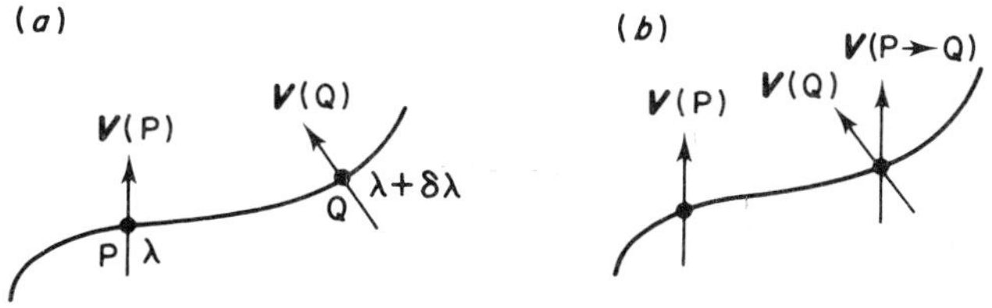

Figure 2.10 $\boldsymbol{V}$(P) and $\boldsymbol{V}$(Q) are the vectors at P and Q belonging to the vector field $\boldsymbol{V}$. $\boldsymbol{V}$(P $\to$ Q) is the vector at Q which results from parallelly transporting $\boldsymbol{V}$(P) along the curve.

these two quantities has no coordinate-independent meaning. If we try to find the derivative of the vector field itself, we will encounter the expression $\boldsymbol{V}(\mathrm{Q}) - \boldsymbol{V}(\mathrm{P})$. Now, $\boldsymbol{V}(\mathrm{P})$ is the tangent vector to some curve passing through P (though not necessarily to the curve shown in figure 2.10(*a*)) and $\boldsymbol{V}(\mathrm{Q})$ is the tangent vector to a curve passing through Q. The difference of two vectors at P is another vector at P: each vector is tangent to some curve passing through P. However, $\boldsymbol{V}(\mathrm{Q}) - \boldsymbol{V}(\mathrm{P})$ is not, in general, the tangent vector to a curve at a specific point. It is not, therefore, a vector and has, indeed, no obvious significance at all.

To define a meaningful derivative of a vector field, we need to compare two vectors at the same point, say Q. Therefore, we construct a new vector $\boldsymbol{V}(\mathrm{P} \to \mathrm{Q})$ which exists at Q but represents $\boldsymbol{V}(\mathrm{P})$. Then a new vector, $\mathrm{D}\boldsymbol{V}/\mathrm{d}\lambda$, which will be regarded as the derivative of $\boldsymbol{V}$ along the curve, may be defined as

$$\left.\frac{\mathrm{D}\boldsymbol{V}}{\mathrm{d}\lambda}\right|_{\mathrm{P}} = \lim_{\delta\lambda \to 0} \frac{\boldsymbol{V}(\mathrm{Q}) - \boldsymbol{V}(\mathrm{P} \to \mathrm{Q})}{\delta\lambda}. \tag{2.22}$$

In the limit, of course, Q coincides with P and this is where the new vector exists. There is no natural way in which a vector at Q corresponds to a vector at P, so we must provide a rule to define $\boldsymbol{V}(\mathrm{P} \to \mathrm{Q})$ in terms of $\boldsymbol{V}(\mathrm{P})$. This rule is the affine connection. In figure 2.10(*b*), $\boldsymbol{V}(\mathrm{P} \to \mathrm{Q})$ is shown as a vector at Q which is parallel to $\boldsymbol{V}(\mathrm{P})$. The figure looks this way because of the Euclidean properties of the paper on which it is printed. Mathematically, the affine connection *defines* what it means for a vector at Q to be parallel to one at P: it is said to define *parallel transport* of a vector along the curve. From a mathematical point of view, we are free to specify the affine connection in any way we choose. Physically, on the other hand, we will need to find out what the affine connection is, with which nature has actually provided us, and we will address this problem in due course. It might be thought that a vector which represents $\boldsymbol{V}(\mathrm{P})$ should not only be parallel to it but also have the same length. In Euclidean geometry, the magnitude of $\boldsymbol{v}$ is $(\boldsymbol{v}\cdot\boldsymbol{v})^{1/2}$ and, as we have seen, the scalar product needs a metric for its definition. The metric connection, mentioned above, does indeed define parallel transport in a manner which preserves the magnitude of the transported vector.

The concrete definition of parallel transport is most clearly written down by choosing a coordinate system. If P and Q lie on a curve $x^\mu(\lambda)$ and are separated by an infinitesimal parameter distance $\delta\lambda$, then the components of $\boldsymbol{V}(\mathrm{P} \to \mathrm{Q})$ are defined by

$$V^\mu(\mathrm{P} \to \mathrm{Q}) = V^\mu(\mathrm{P}) - \delta\lambda\Gamma^\mu{}_{\nu\sigma}(\mathrm{P})V^\nu(\mathrm{P})\frac{\mathrm{d}x^\sigma}{\mathrm{d}\lambda} \tag{2.23}$$

and the functions $\Gamma^\mu{}_{\nu\sigma}$ are called the affine connection coefficients.

These coefficients exist at each point on the manifold and are not associated with any particular curve. However, the rule (2.23) for parallel transport involves, in addition to the vector V itself, both the connection coefficients and the tangent vector $dx^\sigma/d\lambda$, so parallel transport is defined only along a curve. To transport V along a curve by a finite parameter distance, we have to integrate (2.23). If we wish to transport a vector from an initial point P to a final point Q, we must choose a curve, passing through both P and Q, along which to transport it. There will usually be many such curves and it is vital to realize that the vector which finally arrives at Q depends on the route taken. This fact lies at the root of the idea of the *curvature* of a manifold, as we shall see shortly.

The idea of parallel transport is illustrated in figure 2.11, which shows the surface of a Euclidean sphere. For the purposes of this example, we assume the usual metric properties of Euclidean space, so that distances and angles have their usual meanings. The manifold we consider is the two-dimensional surface of the sphere, so every vector is tangential to this surface. P and Q are points on the equator, separated by a quarter of its circumference, and N is the north pole. The equator and the curves PN and QN are parts of great circles on the sphere, and are 'straight lines' as far a geometry on the sphere is concerned: one would follow such a path by walking straight ahead on the surface of the Earth. Consider a vector $V(\mathrm{P})$ which points due north—it is a tangent vector at P to the curve PN. We will transport this vector to Q, first along the equator and second via the north pole. The role for parallel transport along a straight line is particularly simple: the angle between the vector and the line remains constant. For transport along the equator, the vector clearly points due north at each step and so $V(\mathrm{P} \to \mathrm{Q})$ also points north along QN. Along PN, the vector also points north, so on arrival at the pole it is perpendicular to QN. On its way south, it stays perpendicular to QN. Thus, the transported vector $V(\mathrm{P} \to \mathrm{Q})$ as defined by the polar route points along the equator.

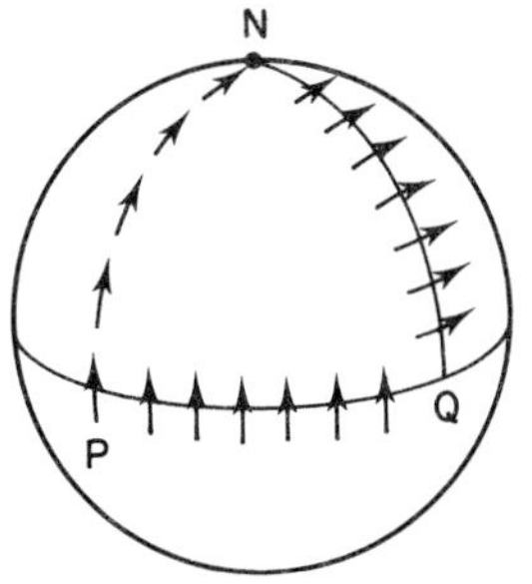

Figure 2.11 Parallel transport of a vector from P to Q on a spherical surface by two routes.

At this point, readers should consider parallel transport along the side of a plane equilateral triangle PNQ. It is easy to see that $V(\mathrm{P} \to \mathrm{Q})$ is independent of the route taken. Clearly, the difference between the two cases is that the spherical surface is curved while the plane surface is flat. The rule for parallel transport, embodied mathematically in the affine connection coefficients, evidently provides a measure of the curvature of a manifold, and we will later formulate this precisely. It should be emphasized that a manifold possesses a curvature *only when* it has an affine connection. If it has no connection, it is neither flat nor curved: the question just does not arise. Finally, returning to figure 2.11, suppose that we had chosen Q to lie close to P and considered only paths contained in a small neighbourhood of the two points. The surface would have been almost indistinguishable from a flat one and the transported vector would have been almost independent of the path. This is consistent with the mathematical expression (2.23). If P has coordinates x^μ and Q is infinitesimally close to P, with coordinates $x^\mu + \mathrm{d}x^\mu$, then we may substitute $\mathrm{d}x^\mu$ for $\delta\lambda \mathrm{d}x^\mu/\mathrm{d}\lambda$, and all reference to the path between P and Q disappears. The affine connection of two-dimensional Euclidean geometry is explored in exercise 2.10.

One of our motivations for introducing the affine connection was to be able to define a meaningful derivative of a vector field. The covariant derivative along a curve was to be defined, using the idea of parallel transport, by (2.22). As we have just seen, it is not actually necessary to specify a curve when P and Q are infinitesimally close. In terms of components, then, let us write $\mathrm{D}V^\mu/\mathrm{d}\lambda = (\mathrm{d}x^\sigma/\mathrm{d}\lambda)\, \nabla_\sigma V^\mu$ and calculate the covariant derivative $\nabla_\sigma V^\mu$ using (2.22) and (2.23). We find

$$\nabla_\sigma V^\mu = \partial_\sigma V^\mu + \Gamma^\mu{}_{\nu\sigma} V^\nu. \tag{2.24}$$

Notice that the three indices of the connection coefficient have different functions. There are, indeed, important situations in which the connection is *symmetric* in its two lower indices: $\Gamma^\mu{}_{\nu\sigma} = \Gamma^\mu{}_{\sigma\nu}$. In general, however, it is the last index which corresponds to that of ∇_σ. Since $\mathrm{D}V^\mu/\mathrm{d}\lambda$ and $\mathrm{d}x^\sigma/\mathrm{d}\lambda$ are both vectors, it follows from their transformation laws that the functions $\nabla_\sigma V^\mu$ are the components of a second-rank tensor, with the transformation law

$$\nabla_{\sigma'} V^{\mu'} = \Lambda^\sigma{}_{\sigma'} \Lambda^{\mu'}{}_\mu \nabla_\sigma V^\mu. \tag{2.25}$$

From this we can deduce the transformation law for the connection coefficients themselves, which may be written as

$$\Gamma^{\mu'}{}_{\nu'\sigma'} = (\Lambda^{\mu'}{}_\mu \Lambda^\nu{}_{\nu'} \Lambda^\sigma{}_{\sigma'})\, \Gamma^\mu{}_{\nu\sigma} + \Lambda^{\mu'}{}_\nu \left(\frac{\partial \Lambda^\nu{}_{\nu'}}{\partial x^{\sigma'}}\right). \tag{2.26}$$

Readers are urged to verify this in detail, bearing in mind that $\partial_{\sigma'}(\Lambda^{\mu'}{}_\nu \Lambda^\nu{}_{\nu'}) = \partial_{\sigma'}(\delta^{\mu'}{}_{\nu'}) = 0$.

Evidently, the affine connection is not itself a tensor. However, the covariant derivative which contains it acts on any tensor to produce another tensor of one higher rank. So far, we have defined only the covariant derivative of a vector, which was given in (2.24). The covariant derivative of a scalar field is just the partial derivative, $\nabla_\mu f = \partial_\mu f$, since this is already a tensor field. In order for the covariant derivative of a one-form field to be a second-rank tensor field, we must have

$$\nabla_\sigma \omega_\mu = \partial_\sigma \omega_\mu - \Gamma^\nu{}_{\mu\sigma}\omega_\nu. \tag{2.27}$$

Notice that the roles of the upper and first lower indices have been reversed, compared with (2.24), and that the sign of the connection term has changed. It is straightforward to check that these changes are vital if this derivative is to transform as a rank $\binom{0}{2}$ tensor. The covariant derivative of a tensor of arbitary rank is

$$\nabla_\sigma T^{\alpha\beta\ldots}{}_{\mu\nu\ldots} = \partial_\sigma T^{\alpha\beta\ldots}{}_{\mu\nu\ldots} + (\text{connection terms}). \tag{2.28}$$

There is one connection term for each index of the original tensor. For each upper index, it is a term like that in (2.24), and for each lower index it is like that in (2.27). Exercise 2.11 invites readers to consider in more detail how these definitions are arrived at.

There is a convenient notation which represents partial derivatives of tensor fields by a comma and covariant derivatives by a semicolon. That is:

$$\partial_\sigma T^\alpha{}_{\mu\nu} \equiv T^\alpha{}_{\mu\nu,\sigma} \quad \text{and} \quad \nabla_\sigma T^\alpha{}_{\mu\nu} \equiv T^\alpha{}_{\mu\nu;\sigma}. \tag{2.29}$$

2.3.2 Geodesics

As mentioned earlier, a geodesic is, in a sense, a generalization of the straight line of Euclidean geometry. Of course, we can reproduce only those properties of straight lines which make sense in our manifold with its affine connection. For example, the idea that a straight line is the shortest path between two points will only make sense when we have a metric to measure distances. The idea of a geodesic is that, if we are to walk along a straight line, each step we take must be parallel to the last. Consider, then the special case of the parallel transport equation (2.23) in which the vector transported from P to Q is the curve's own tangent vector at P: $V^\mu = \mathrm{d}x^\mu/\mathrm{d}\lambda$. If the curve is a geodesic, the transported vector $V(\mathrm{P} \to \mathrm{Q})$ will be proportional to $V(\mathrm{Q})$. Since the vectors have no definite length, the constant of proportionality may well depend on λ, but if P and Q are separated by an infinitesimal parameter distance, it will be only infinitesimally different from 1. So we may write

$$\left.\frac{\mathrm{d}x^\mu}{\mathrm{d}\lambda}\right|_{\mathrm{P}\to\mathrm{Q}} = [1 - \delta\lambda f(\lambda)] \left.\frac{\mathrm{d}x^\mu}{\mathrm{d}\lambda}\right|_{\mathrm{Q}} \tag{2.30}$$

where $f(\lambda)$ is an unknown function. Using this in (2.23) and taking the limit $\delta\lambda \to 0$, we obtain the equation

$$\frac{d^2x^\mu}{d\lambda^2} + \Gamma^\mu{}_{\nu\sigma}\frac{dx^\nu}{d\lambda}\frac{dx^\sigma}{d\lambda} = f(\lambda)\frac{dx^\mu}{d\lambda}. \tag{2.31}$$

A curve $x^\mu(\lambda)$ is a geodesic if and only if it satisfies an equation of this form, where $f(\lambda)$ can be any function.

Remember now that a given path through the manifold can be parametrized in many different ways, each one being regarded as a different curve. It is easy to see that if the curve given by one parametrization is a geodesic, then so is any other curve which results from another parametrization of the same path. We need only express the new parameter, say μ, as a function of λ and use the chain rule in (2.31):

$$\frac{d^2x^\mu}{d\mu^2} + \Gamma^\mu{}_{\nu\sigma}\frac{dx^\nu}{d\mu}\frac{dx^\sigma}{d\mu} = \left(\frac{d\mu}{d\lambda}\right)^{-2}\left(f(\lambda)\frac{d\mu}{d\lambda} - \frac{d^2\mu}{d\lambda^2}\right)\frac{dx^\mu}{d\mu}. \tag{2.32}$$

This has the same form as (2.31) but involves a different function of μ on the right-hand side. In particular, it is always possible to find a parameter for which the right-hand side of (2.32) vanishes. Such a parameter is called an *affine parameter* for the path. It is left as a simple exercise for the reader to show that if λ is an affine parameter, then any parameter μ which is affinely related to it (i.e. it is a linear function $\mu = a\lambda + b$) is also an affine parameter.

2.3.3 The Riemann curvature tensor

We saw in connection with figure 2.11 that parallel transport of a vector between two points along different curves can be used to detect curvature of the manifold. This is because both parallel transport and curvature are properties of the affine connection. The definition of curvature is made precise by the Riemann curvature tensor. Consider two points P and Q with coordinates x^μ and $x^\mu + \delta x^\mu$ respectively, such that $\delta x^\mu = 0$, except for $\mu = 1$ or 2. A region of the (x^1, x^2) surface near these points is shown in figure 2.12. By transporting a vector $V(\mathrm{P})$ to Q via R or S, we obtain at Q the two vectors $V(\mathrm{P} \to \mathrm{R} \to \mathrm{Q})$ and $V(\mathrm{P} \to \mathrm{S} \to \mathrm{Q})$. To first order in δx^μ these two vectors are the same, as we have seen. If we expand them to second order, however, they are different, and we obtain an expression of the form

$$V^\mu(\mathrm{P} \to \mathrm{S} \to \mathrm{Q}) - V^\mu(\mathrm{P} \to \mathrm{R} \to \mathrm{Q}) = R^\mu_{\nu 12}V^\nu\delta x^1\delta x^2 + \ldots \tag{2.33}$$

where the quantities $R^\mu{}_{\nu 12}$ depend on the connection coefficients and their derivatives. Readers are invited to verify that they are components of the Riemann tensor we are about to define.

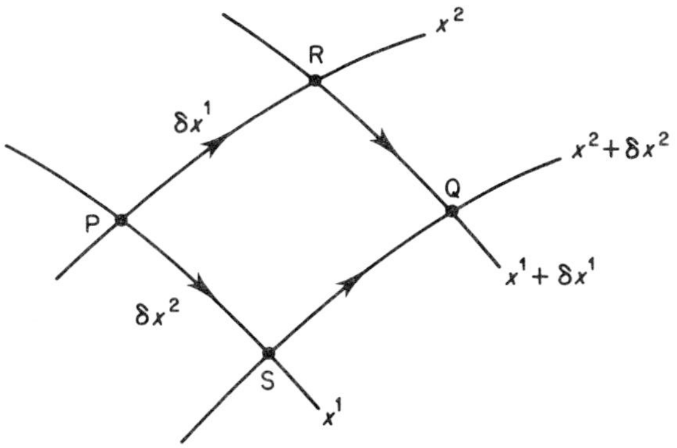

Figure 2.12 Two paths, PRQ and PSQ, for parallelly transporting a vector from P to Q.

It should be clear that the process of transporting the vector from P to Q along the two paths is related to that of taking two derivatives, with respect to x^1 and x^2 in either order. If we act on a vector field with the two covariant derivatives ∇_σ and ∇_τ in succession, the result depends on the order of the two operations, they do not commute. To work out the commutator, we use the definition (2.28), bearing in mind that $\nabla_\sigma V^\mu$ is itself a rank $\binom{1}{1}$ tensor. The result is

$$[\nabla_\sigma, \nabla_\tau]V^\mu \equiv \nabla_\sigma \nabla_\tau V^\mu - \nabla_\tau \nabla_\sigma V^\mu = R^\mu{}_{\nu\sigma\tau}V^\nu + (\Gamma^\lambda{}_{\sigma\tau} - \Gamma^\lambda{}_{\tau\sigma})\nabla_\lambda V^\mu \tag{2.34}$$

where

$$R^\mu{}_{\nu\sigma\tau} = \Gamma^\mu{}_{\nu\tau,\sigma} - \Gamma^\mu{}_{\nu\sigma,\tau} + \Gamma^\mu{}_{\lambda\sigma}\Gamma^\lambda{}_{\nu\tau} - \Gamma^\mu{}_{\lambda\tau}\Gamma^\lambda{}_{\nu\sigma}. \tag{2.35}$$

This formidable expression defines the Riemann tensor. As a fourth-rank tensor, it has $4^4 = 256$ components! Actually, owing to various symmetry properties, of which the most obvious is antisymmetry in the indices σ and τ, it can be shown that only 80 of these are independent. When $\Gamma^\mu{}_{\nu\sigma}$ is a *metric* connection of the kind described in §2.3.5, there is a further symmetry which reduces the number of independent components to 20. Even so, the Riemann tensor is clearly an inconvenient object to deal with. Readers should not panic yet, though. Many of the most important applications of general relativity (including all those to be discussed in this book) do not require the complete Riemann tensor. In practice, we shall need only a simpler tensor derived from it. This is the *Ricci tensor*, defined by contracting two indices of the Riemann tensor:

$$R_{\mu\nu} \equiv R^\lambda{}_{\mu\lambda\nu} = \Gamma^\lambda{}_{\mu\nu,\lambda} - \Gamma^\lambda{}_{\mu\lambda,\nu} + \Gamma^\lambda{}_{\sigma\lambda}\Gamma^\sigma{}_{\mu\nu} - \Gamma^\lambda{}_{\sigma\nu}\Gamma^\sigma{}_{\mu\lambda}. \tag{2.36}$$

Although the definition still looks complicated, the components of this tensor can often be calculated with just a little patience, and it is relatively simple to use thereafter.

The second term on the right-hand side of (2.34) involves the antisymmetric part of the affine connection, $\Gamma^{\nu}{}_{\sigma\tau} - \Gamma^{\nu}{}_{\tau\sigma}$, which is called the *torsion* tensor. In most versions of general relativity, it is assumed that spacetime has no torsion. We will always assume this too, since it makes things much simpler. I do not know, however, of any direct method of testing this assumption experimentally.

Some simple illustrations of the idea of curvature are given in the exercises. These make more obvious sense when we have a metric at our disposal, and we will turn to that topic forthwith.

2.3.4 The metric

Yes, we are finally going to give our manifold a metrical structure which will make the notion of length meaningful. To define the infinitesimal distance $\mathrm{d}s$ between two points with coordinates x^{μ} and $x^{\mu} + \mathrm{d}x^{\mu}$, we use a generalization of the Pythagoras rule:

$$\mathrm{d}s^2 = g_{\mu\nu}(x)\,\mathrm{d}x^{\mu}\,\mathrm{d}x^{\nu}. \tag{2.37}$$

Naturally, we want this distance to be a scalar quantity, independent of our choice of coordinate system, and it is easy to see that the coefficients $g_{\mu\nu}$ must therefore be the components of a rank $\binom{0}{2}$ tensor. It is called the *metric tensor* and, since an antisymmetric part would obviously make no contribution to $\mathrm{d}s$, it is taken to be symmetric in μ and ν. Any finite distance between two points can be uniquely defined only as the length of a specified curve joining them. For the distance between P and Q on a curve $x^{\mu}(\lambda)$, we have the integral

$$s_{\mathrm{PQ}} = \int_{\mathrm{P}}^{\mathrm{Q}} \frac{\mathrm{d}s}{\mathrm{d}\lambda}\,\mathrm{d}\lambda = \int_{\mathrm{P}}^{\mathrm{Q}} \left(g_{\mu\nu}(x(\lambda))\,\frac{\mathrm{d}x^{\mu}}{\mathrm{d}\lambda}\,\frac{\mathrm{d}x^{\nu}}{\mathrm{d}\lambda} \right)^{1/2} \mathrm{d}\lambda. \tag{2.38}$$

In the space of three-dimensional Euclidean geometry, the components of the metric tensor relative to a Cartesian coordinate system are of course

$$g_{\mu\nu} = \begin{pmatrix} 1 & 0 & 0 \\ 0 & 1 & 0 \\ 0 & 0 & 1 \end{pmatrix}. \tag{2.39}$$

The metric tensor has several other geometrical uses. First, it serves to define the magnitude $|\boldsymbol{V}|$ of a vector or vector field $\boldsymbol{V}$:

$$|\boldsymbol{V}(x)|^2 = g_{\mu\nu}(x)\,V^{\mu}(x)\,V^{\nu}(x). \tag{2.40}$$

Second, it provides a definition of the angle between two vectors:

$$g_{\mu\nu}U^{\mu}V^{\nu} = |U||V|\cos\theta. \tag{2.41}$$

In Euclidean space, this is clearly equivalent to the usual 'dot product' of two vectors. In non-Euclidean spaces, the magnitudes and angles defined in this way may have complex values. Because $g_{\mu\nu}$ is a tensor, both magnitudes and angles are scalar quantities, with coordinate-independent meanings.

The third use of the metric tensor is apparent from the first two. When introducing one-forms, we pointed out that the symbol $\boldsymbol{u}.$ which appears in the Euclidean dot product can be regarded as a linear function taking a vector as its argument, and is in fact a one-form. From (2.40) and (2.41), we see that $g_{\mu\nu}$ plays the role of the dot, and the numbers $g_{\mu\nu}U^{\mu}$ are the components of a unique one-form corresponding to the vector $\boldsymbol{U}$. These components are denoted by U_{ν}:

$$U_{\nu} = g_{\mu\nu}U^{\mu} \tag{2.42}$$

and the metric tensor is said to *lower the index* of the vector. In the same way, the metric tensor associates a unique vector with each one-form ω: it is the vector whose corresponding one-form is ω. Actually, this assumes that the matrix of components of the metric tensor is non-singular, that is it has an inverse matrix $g^{\mu\nu}$ with

$$g_{\mu\sigma}g^{\sigma\nu} = \delta^{\nu}{}_{\mu}. \tag{2.43}$$

The geometrical properties of the metric would be rather peculiar if this were not so, and the existence of the inverse matrix is sometimes included as part of the definition of a metric. As long as the inverse matrix does exist, we have

$$\omega^{\mu} = g^{\mu\nu}\omega_{\nu} \tag{2.44}$$

for the components of the vector corresponding to ω. Here, the inverse matrix *raises* the index of the one-form. Any index of any tensor can be raised or lowered in this way. Since the metric tensor is symmetric, it obviously does not matter which of its indices is contracted.

Now that we have a metric tensor at our disposal, it is clearly possible in practice to regard vectors and one-forms as different versions of the same thing—hence the terms contravariant and covariant vector. In Euclidean geometry, we do not notice the difference, as long as we use Cartesian coordinates, because the metric tensor is just the unit matrix. In non-Cartesian coordinates, the metric tensor is not the unit matrix, and some consequences of this are explored in the exercises. Does this mean that there is, after all, no real distinction between vectors and one-forms? This depends on our attitude towards the metric. In the relativistic theory of gravity, the metric embodies information about gravitational fields, and different metrics may represent different, but equally possible, physical situations. The relation between the contravariant and covariant versions of a given physical quantity depends on the metric, and it is legitimate to ask which version is intrinsic to the quantity itself and which is a compound of information about both the quantity itself and the metric. To decide this we must ask whether, without using a metric, the quantity would naturally have been defined as a vector or as a one-form. Since metrical notions are taken for granted in much of our physical thinking, the answer to this may not always be obvious. If, as in Euclidean geometry, the metric is taken to be fixed and unalterable, then such questions need not arise.

2.3.5 The metric connection

Now that the magnitude of a vector and the angle between two vectors have acquired definite meanings, it is natural to demand that the rule for parallel transport should be consistent with them. Thus, if two vectors are transported along a curve, each one remaining parallel to itself, then the magnitude of each vector and the angle between them should remain constant. This requirement leads to a relation between the metric and the affine connection which we will now derive. Consider a curve $x^\mu(\lambda)$ passing through the point P and two vectors $\boldsymbol{V}$ and $\boldsymbol{W}$ at P. We can define a vector field $\boldsymbol{V}(x)$ such that its value at any point Q on the curve is equal to the transported vector $\boldsymbol{V}(\mathrm{P} \to \mathrm{Q})$, and a similar vector field $\boldsymbol{W}(x)$. If $\boldsymbol{U}$ is the tangent vector to the curve, then $U^\sigma \nabla_\sigma V^\mu$ is the covariant derivative of V^μ along the curve. It is given by the expression (2.22) and is clearly zero, as is the corresponding derivative of $\boldsymbol{W}$. The consistency condition we want to impose is that the scalar product $g_{\mu\nu} V^\mu W^\nu$ has the same value everywhere along the curve. Recalling that the covariant derivative of a scalar is equal to the ordinary derivative, we may express this condition as

$$U^\sigma \nabla_\sigma (g_{\mu\nu} V^\mu W^\nu) = 0. \qquad (2.45)$$

Now, the covariant derivative of a product of tensors obeys the same Leibnitz rule as an ordinary derivative:

$$\nabla_\sigma (g_{\mu\nu} V^\mu W^\nu) = (\nabla_\sigma g_{\mu\nu}) V^\mu W^\nu + g_{\mu\nu} (\nabla_\sigma V^\mu) W^\nu + g_{\mu\nu} V^\mu (\nabla_\sigma W^\nu). \qquad (2.46)$$

Readers may verify this explicitly or turn to exercise 2.11 for some further enlightenment. If we use this in (2.45), the last two terms vanish and our condition becomes $U^\sigma (\nabla_\sigma g_{\mu\nu}) V^\mu W^\nu = 0$. This must hold for any three vectors $\boldsymbol{U}$, $\boldsymbol{V}$ and $\boldsymbol{W}$, and therefore the covariant derivative of $g_{\mu\nu}$ must be zero:

$$\nabla_\sigma g_{\mu\nu} = g_{\mu\nu,\sigma} - \Gamma^\lambda{}_{\mu\sigma} g_{\lambda\nu} - \Gamma^\lambda{}_{\nu\sigma} g_{\mu\lambda} = 0. \qquad (2.47)$$

By combining this equation with two others obtained by renaming the indices, we can get the relation

$$g_{\sigma\mu,\nu} + g_{\sigma\nu,\mu} - g_{\mu\nu,\sigma}$$
$$= (\Gamma^{\lambda}{}_{\sigma\nu} - \Gamma^{\lambda}{}_{\nu\sigma})g_{\lambda\mu} + (\Gamma^{\lambda}{}_{\sigma\mu} - \Gamma^{\lambda}{}_{\mu\sigma})\, g_{\lambda\nu} + (\Gamma^{\lambda}{}_{\mu\nu} + \Gamma^{\lambda}{}_{\nu\mu})g_{\lambda\sigma}. \quad (2.48)$$

Assuming, as we discussed above, that the connection is symmetric in its lower indices, the first two terms on the right-hand side vanish. Then, on multiplying by $g^{\sigma\tau}$, we find that the symmetric connection coefficient is completely determined by the metric:

$$\Gamma^{\lambda}{}_{\mu\nu} = \tfrac{1}{2}g^{\lambda\sigma}(g_{\sigma\mu,\nu} + g_{\sigma\nu,\mu} - g_{\mu\nu,\sigma}). \quad (2.49)$$

When Γ is used to denote this expression, it is often called a *Christoffel symbol*. This metric connection expresses the definition of parallelism which is implied by the metric. In principle, there is no reason why a manifold should not possess one or more additional affine connections which would be quite independent of the metric. Indeed, it might also possess several different metrics. In such a case, there would exist several different kinds of 'distance' and several different meanings of 'parallel'. It appears, however, that a single metric and its associated connection given by (2.49) are sufficient to describe the properties of space and time as we know them.

The *Ricci curvature scalar* R is defined by

$$R = g^{\mu\nu}R_{\mu\nu}. \quad (2.50)$$

It gives a measure of the local 'radius of curvature' of a manifold, as is illustrated in exercise 2.15.

2.4 What is the Structure of our Spacetime?

We have now invested considerable effort in understanding the mathematical nature of the affine and metrical structures which give precise meanings to our intuitive geometrical ideas. The question naturally arises, what are the particular structures that occur in our real, physical space and time? Let us first consider what kind of an answer is needed.

Before Einstein's theories of relativity, it had seemed obvious that the geometry of space was that described by Euclid. (The logical possibility of non-Euclidean geometry had, however, been investigated rather earlier by Gauss, Bolyai, Lobachevski, Riemann and others. The history of this subject is nicely summarized by Weinberg (1972).) The Galilean spacetime which incorporates Euclidean geometry does not have exactly the kind of metrical structure we have been considering. It is a combination (in mathematical jargon, a *direct product*) of two manifolds

T (time) and S (space), each of which has its own metric. This structure, illustrated in figure 2.13, is called a *fibre bundle*. It has a *base manifold*, T, to each point of which is attached a fibre. Each fibre is, of course, a copy of the three-dimensional Euclidean space S. A curve such as PQR passing through the spacetime has no well defined length, although its projection onto one of the fibres does have a definite length l and its projection onto T spans a definite time interval t.

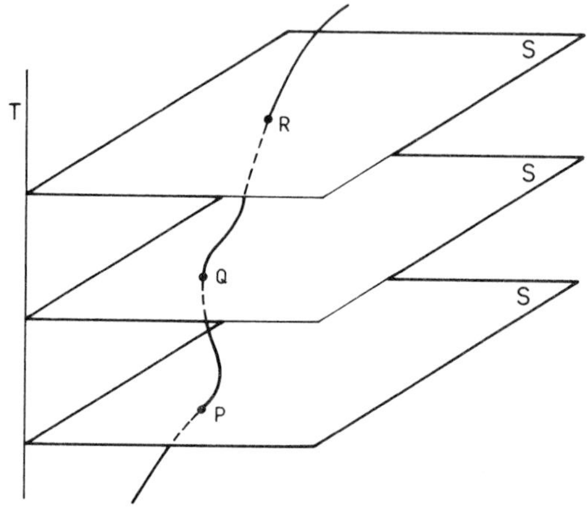

Figure 2.13 Fibre bundle structure of Galilean spacetime and the trajectory of a particle moving through it. Each fibre is a copy of three-dimensional Euclidean space S, which possesses a metric for measuring distances. The base manifold T has its own metric for measuring time intervals. There is no unique way of measuring the 'length' of the particle's trajectory.

The big difference between Galilean spacetime and the spacetimes of Einstein's theories is that the latter are *metric spaces*. That is, the spacetime is a manifold in which a single metric tensor field defines, as we saw in our initial survey, the arc length of any curve. This 'length' is a combination of temporal and spatial intervals, but there is no unique way in which the two can be separated. There is, of course, a profound difference between distance and time as we experience them, and we shall discuss in later chapters how this difference fits in with the mathematics.

An important similarity between Galilean spacetime and the Minkowski spacetime of special relativity is that their metrical properties are assumed to be known *a priori*, as specified either by (2.39) or by (2.8). Readers may be puzzled to see that the spatial components in (2.8) have changed sign relative to (2.39). This is purely a matter of convention:

the squared proper time intervals in (2.3) or (2.6) are taken to be positive if the separation of two events in time is greater that $1/c$ times their spatial separation, and negative otherwise. (Since proper time intervals are scalar quantities, having the same values in all frames of reference, this distinction is also independent of the frame in which the time and distance measurements are made.) If we chose to think in terms of proper distance rather than proper time, the opposite convention would be more natural, and every component in (2.8) would have the opposite sign. In fact, both conventions are used in the literature, although the one we are using is more popular among high-energy physicists than among relativity theorists.

The crux of the general relativistic theory of gravity is that neither of these simple assumptions about the metric tensor is in fact correct. Indeed, the most important conceptual step we have taken in this chapter is to recognize that the metric tensor is not an intrinsic part of the spacetime manifold, but rather an object which lives in the manifold. It is the same sort of thing as an electric or magnetic field. Electric and magnetic fields vary with position and time in accordance with definite physical laws which relate them to distributions of charged particles and currents. In the same way, the metric tensor field varies in accordance with its own laws of motion and depends on the distribution of matter. So far, we have no idea what the laws of motion for the metric tensor field are. Electromagnetic fields are easy to produce and control under laboratory conditions, and the laws which govern them were, for the most part inferred from comprehensive experimental investigations. In contrast, the gravitational forces which are the observable manifestation of the metric tensor field are immeasurably small, unless they are produced by bodies of planetary size, and there is little hope of deducing the laws which govern them from a series of controlled experiments. What Einstein did was to guess at what these laws might be, assuming that they would be reasonably similar to other known laws of physics. After one or two false guesses, he arrived at a set of equations, the *field equations* of general relativity, which are consistent with the most precise astronomical observations it has so far been possible to make.

With the benefit of hindsight, it is possible to see that these equations and all the other laws of classical (non-quantum-mechanical) physics can be deduced in exactly the same way from a single basic principle, called an *action principle*. This seems to me to be most satisfactory. I should be vastly more satisfied if I could explain why an action principle rather than something else is what actually works, but I cannot imagine how that would be done. At this point then, I propose to interrupt our study of geometry to examine how classical physics works in Galilean and Minkowski spacetimes. This is an important topic in its own right, because classical physics and the simple spacetimes often provide excellent approximations to the real world. In the course of understanding them, however, we shall also meet the action principle, whereupon we shall be equipped to embark upon general relativity and the theory of gravity.

Exercises

2.1. Consider two coordinate systems S and S′ whose spatial Cartesian axes lie in the same three directions. The origin of S′ moves with constant velocity $\boldsymbol{v}$ relative to S, and the origins of S and S′ coincide at $t = t' = 0$. Assume that the relation between the two sets of coordinates is linear and that space is isotropic. The most general form of the transformation law can then be written as

$$\boldsymbol{x}' = \alpha[(1 - \lambda v^2)\boldsymbol{x} + (\lambda\boldsymbol{v}\cdot\boldsymbol{x} - \beta t)\boldsymbol{v}] \qquad t' = \gamma[t - (\delta/c^2)\boldsymbol{v}\cdot\boldsymbol{x}]$$

where α, β, γ, δ and λ are functions of v^2. For the case that $\boldsymbol{v}$ is in the positive x direction, write out the transformations for the four coordinates. Write down the trajectory of the S′ origin as seen in S and that of the S origin as seen in S′ and show that $\beta = 1$ and $\alpha = \gamma$. Write down the trajectories as seen in S and S′ of a light ray emitted from the origin at $t = t' = 0$ which travels in the positive x direction, assuming that it is observed to travel with speed c in each case. Show that $\delta = 1$. The transformation from S′ to S should be the same as the transformation from S to S′, except for the replacement of v by $-v$. Use this to find γ. By considering the equation of the spherical wavefront of a light wave emitted from the origin at $t = t' = 0$, complete the derivation of the Lorentz transformations (2.2).

2.2. Two coordinate frames are related by the Lorentz transformations (2.2). A particle moves along the x axis with velocity u and acceleration a as measured in S. Show that its velocity and acceleration as measured in S′ are $u' = (u - v)/(1 - uv/c^2)$ and $a' = (1 - v^2/c^2)^{3/2}a/(1 - uv/c^2)^3$.

2.3. A rigid rod of length L is at rest in S′, with one end at $x' = 0$ and the other at $x' = L$. Find the trajectories of the two ends of the rod as seen in S and show that the length of the rod as measured in S is L/γ, where $\gamma = (1 - v^2/c^2)^{-1/2}$. This is the *Fitzgerald contraction*. If the rod lies along the y axis of S′, what is its apparent length in S? A clock is at rest at the origin of S′. It ticks at $t' = 0$ and again at $t' = \tau$. Show that the interval between these ticks as measured in S is $\gamma\tau$. This is *time dilation*.

2.4. As seen in S, a signal is emitted from the origin at $t = 0$, travels along the x axis with speed u, and is received at time τ at $x = u\tau$. Show that, if $u > c^2/v$ then, as seen in S′, the signal is received before being sent. Show that if such a paradox is to be avoided, no signal can travel faster than light.

2.5. A wheel has a perfectly rigid circular rim connected by unbreakable joints to perfectly rigid spokes. When measured at rest, its radius is r and its circumference $2\pi r$. When the wheel is set spinning with angular velocity ω, what, according to exercise 2.3, is the apparent circumference of its rim and the apparent length of its spokes? What is the speed of sound in a solid material of density ρ whose Young's modulus is Y? Is the notion of a perfectly rigid material consistent with the conclusion of exercise 2.4?

2.6. Consider the following three curves in the Euclidean plane with Cartesian coordinates x and y: (i) $x = 2\sin\lambda$, $y = 2\cos\lambda$, $0 \leqslant \lambda < 2\pi$; (ii) $x = 2\cos(s/2)$, $y = 2\sin(s/2)$, $0 \leqslant s < 4\pi$; (iii) $x = 2\cos(\mathrm{e}^{\mu})$, $y = 2\sin(\mathrm{e}^{\mu})$, $-\infty < \mu \leqslant \ln(2\pi)$. Show that all three curves correspond to the same path, namely a circle of radius 2. Show that λ and s are affinely related. What is the special significance of s? Find the components of the tangent vectors to each curve. Compare the magnitudes and directions of the three tangent vectors at various points on the circle. What is special about the tangent vectors to curve (ii)?

2.7. Consider a four-dimensional manifold and a particular system of coordinates x^{μ}. You are given four functions $a(x^{\mu})$, $b(x^{\mu})$, $c(x^{\mu})$ and $d(x^{\mu})$. Can you tell whether these are (i) four scalar fields, (ii) the components of a vector field, (iii) the components of a one-form field or (iv) none of these? If not, what further information would enable you to do so?

2.8. In the Euclidean plane, with Cartesian coordinates x and y, consider the vector field $\boldsymbol{V}$ whose components are $V^x = 2x$ and $V^y = y$ and the one-form ω_f which is the gradient of the function $f = x^2 + y^2/2$. Show that in any system of Cartesian coordinates $x' = x\cos\alpha + y\sin\alpha$, $y' = y\cos\alpha - x\sin\alpha$, where α is a fixed angle, the components of ω_f are identical to those of $\boldsymbol{V}$. In polar coordinates (r, θ), such that $x = r\cos\theta$ and $y = r\sin\theta$, show that $\boldsymbol{V}$ has components $(r(1 + \cos^2\theta)$, $-\sin\theta\cos\theta)$ while ω_f has components $(r(1 + \cos^2\theta)$, $-r^2\sin\theta\cos\theta)$. Note that the 'gradient vector' defined in elementary vector calculus to have polar components $(\partial f/\partial r, r^{-1}\partial f/\partial\theta)$ does not correspond to either $\boldsymbol{V}$ or ω_f.

2.9. Given a rank $\binom{a}{b}$ tensor, show that the result of contracting any upper index with any lower index is a rank $\binom{a-1}{b-1}$ tensor.

2.10. In the Euclidean plane, parallel transport is defined in the obvious way. If, in Cartesian coordinates, the components of $V(\mathrm{P})$ are (u, v), then the components of $V(\mathrm{P} \to \mathrm{Q})$ are also (u, v). Thus, the affine connection coefficients in Cartesian coordinates are all zero. Work out the matrices $\Lambda^{\mu'}{}_{\mu}$ for transforming between Cartesian and polar coordinates related by $x = r\cos\theta$ and $y = r\sin\theta$. Show that in polar coordinates, the only non-zero connection coefficients are $\Gamma^{r}{}_{\theta\theta} = -r$ and $\Gamma^{\theta}{}_{r\theta} = \Gamma^{\mu}{}_{\theta r} = 1/r$. Let P and Q be the points with Cartesian coordinates $(a, 0)$ and $(a\cos\alpha, a\sin\alpha)$ respectively, and let $V(\mathrm{P})$ have Cartesian components $(1, 0)$. Using polar coordinates and parallel transport around the circle of radius a centred at the origin, parametrized by the polar angle θ, show that $V(\mathrm{P} \to \mathrm{Q})$ has polar components $(\cos\alpha, -a^{-1}\sin\alpha)$. Show that $V(\mathrm{P} \to \mathrm{Q})$ has Cartesian components $(1, 0)$.

2.11. The covariant derivatives of tensors of arbitrary rank can be defined recursively by the following rules: (i) for any scalar field f, $\nabla_\sigma f = \partial_\sigma f$; (ii) the covariant derivative of a vector field is given by (2.24); (iii) the covariant derivative of a rank $\binom{a}{b}$ tensor field is a rank $\binom{a}{b+1}$ tensor field; (iv) for any two tensors A and B, the Leibnitz rule $\nabla_\sigma(AB) = (\nabla_\sigma A)B + A(\nabla_\sigma B)$ holds. By considering the fact that $\omega(V)$ is a scalar field, show that the covariant derivative of a one-form field is given by (2.27). Convince yourself that the recursive definition leads to (2.28) for an arbitrary tensor field.

2.12 In the Euclidean plane, consider the straight line $x = a$. Using $\lambda = y$ as a parameter, show, in both Cartesian and polar coordinates, that the geodesic equation (2.31) is satisfied and that λ is an affine parameter. Repeat the exercise using both affine and non-affine parameters of your own invention.

2.13. Write down the components of the metric tensor field of the Euclidean plane in the polar coordinates of exercise 2.8. Show, using both Cartesian and polar coordinates, that the vector V is obtained by raising the indices of ω_f and vice versa. Show that $|V|^2 = \omega_f(V)$. What is the magnitude of the 'gradient vector'? How does it involve the metric? Can a 'gradient vector' be defined in a non-Euclidean metric space or in a manifold which possesses no metric?

2.14. Show that the affine connection of exercise 2.10 is the metric connection.

2.15. In three-dimensional Euclidean space, define polar coordinates in the usual way by $x = r\sin\theta\cos\phi$, $y = r\sin\theta\sin\phi$ and $z = r\cos\theta$. The spherical surface $r = a$ is called a 2-sphere, and the angles θ and ϕ can be used as coordinates for this two-dimensional curved surface. Show that the line element on the sphere is $ds^2 = a^2(d\theta^2 + \sin^2\theta \, d\phi^2)$. Show that the only non-zero coefficients of the metric connection are $\Gamma^{\theta}{}_{\theta\phi} = -\sin\theta\cos\theta$ and $\Gamma^{\phi}{}_{\theta\phi} = \Gamma^{\phi}{}_{\phi\theta} = \cot\theta$. Show that the Ricci tensor is diagonal with elements $R_{\theta\theta} = 1$ and $R_{\phi\phi} = \sin^2\theta$ and that the Ricci scalar is $R = 2/a^2$.

3

Classical Physics in Galilean and Minkowski Spacetimes

This chapter is mostly about classical mechanics. By 'classical', I mean to indicate that we are not yet going to take any account of quantum mechanics. (In the literature, 'classical' is sometimes used to mean that no account is taken of special relativity either, and sometimes also to describe any venerable theory which has been superseded by a more 'modern' one.) I shall actually be assuming that readers already have a fair understanding of the elementary aspects of Newtonian mechanics: for example, we will not spend time developing techniques for calculating the trajectories of projectiles or planetary orbits, important though these topics undoubtedly are. The aim of this chapter is to set out the mathematics of classical mechanics in a way which makes clear the nature of the basic physical laws embodied in it and which, to a large extent, will enable us to see the principles of general relativity and of the quantum theory as natural generalizations of these laws. In a later chapter, this mathematical formalism will also help us towards setting up a statistical description of the macroscopic behaviour of large assemblages of particles.

There is, of course, nothing final or unalterable about the 'laws' of physics as they appear to physicists at any particular time. It is possible, however, to identify two mathematical ideas which lie at the heart of all theories which have so far had success in describing how the world is at a fundamental level. The first is a function called the *action* which, as we shall soon see, summarizes all the equations of motion for a given system. It is easy to invent equations of motion which cannot be summarized in this way. For example, equations which involve dissipative effects such as friction usually cannot be. These effects, however, can be understood as arising only on a macroscopic scale, and the fundamental equations which apply at the microscopic level always do seem to be derivable from an action. Why this should be so, I do not know.

The action is fundamental to both classical and quantum theories, although in somewhat different guises. It is a function of all the dynamical variables (in classical mechanics of particles, the positions and velocities of all the particles) which are needed to specify the state of a system. Once we know what this function is, we know what the laws of motion are, but the only way of finding this out is by guesswork. It seems that the possibilities amongst which we have to choose are quite considerably restricted by a variety of *symmetries* which are respected by nature. This is the second of the ideas mentioned above, and the role of symmetry in theoretical physics will be a recurring theme. Symmetry is, of course, an aesthetically pleasing feature of any theory, and this has come to weigh heavily with many physicists. At the same time, it is not really clear why nature should share our aesthetic tastes, if indeed she really does. But symmetry is more than a theoretician's fancy. As we shall soon discover, every symmetry leads to a conservation law, the best known examples being, perhaps, the conservation of energy, momentum and electric charge. These conservation laws are amenable to quite rigorous experimental checks and, conversely, the empirical discovery of conserved quantities may point to new symmetries which should be incorporated in mathematical models.

These, then, are the issues to which the present chapter is primarily addressed.

3.1 The Action Principle in Galilean Spacetime

The basic problem we set ourselves in classical mechanics is, given the state of a system at some initial time, to predict what its state will be at any later time. If we can do this correctly or if, at least, we are satisfied that only computational difficulties stand in the way of our doing it, then we feel that we understand how the system works. We will be concerned more or less exclusively with systems consisting of particles which are small enough to be considered as points. Large rigid bodies can be treated as being composed of such particles and introduce no new questions of principle.

Let us consider first what information we need in order to specify uniquely the instantaneous state of such a system. It is normally taken for granted that we have to know the positions and velocities of all the particles—whether these are given in Cartesian coordinates for each particle, in polar coordinates, in terms of relative positions and velocities for some of the particles, etc does not matter. But why is this? A snapshot of the system can be completely described by giving just the positions of the particles. Evidently, this is not enough, but if we go on to specify the velocities, then why not include accelerations and higher-order time derivatives as well? By saying that the state of the system is uniquely specified, we imply that, given the equations of motion, any future state of the system is uniquely determined. The equations of motion come simply from Newton's second law, which gives a set of second-order differential equations for the positions of the particles as functions of time. They have unique solutions if the initial positions and velocities are given. I emphasize this point because I am going to illustrate the role of symmetries by using them to *derive* Newton's second law, and I want to be clear about the assumptions which are needed to do this. The first assumption is that the state of the system is uniquely specified by giving the positions and velocities, and it is more or less equivalent to assuming that the equations of motion will be of second order in the time derivatives. I do not know of any justification for this beyond the fact that it works.

At this point, we must introduce the action principle. As a simple example, consider a single particle in a Galilean spacetime with one spatial dimension. If its mass is m and it has potential energy $V(x)$, then Newton's law gives

$$m\ddot{x} = -\mathrm{d}V/\mathrm{d}x. \tag{3.1}$$

This is equivalent to the statement that the quantity

$$S = \int_{t_1}^{t_2} [\tfrac{1}{2}m\dot{x}^2 - V(x)]\,\mathrm{d}t \tag{3.2}$$

called the *action*, is stationary with respect to variations in the path $x(t)$. That is to say, if $x(t)$ is the actual path of the particle, and we imagine changing it by a small but otherwise arbitrary amount $x(t) \to x(t) + \delta x(t)$, then the resulting first-order change in S is zero:

$$\delta S = \int_{t_1}^{t_2} [m\dot{x}\delta\dot{x} - (\mathrm{d}V/\mathrm{d}x)\delta x]\,\mathrm{d}t = 0. \tag{3.3}$$

If, in particular, we choose $\delta x(t)$ to vanish at t_1 and t_2, and take into account that $\delta\dot{x} = \mathrm{d}(\delta x)/\mathrm{d}t$, then we may integrate the first term by parts, giving

$$\int_{t_1}^{t_2} [m\ddot{x} + (\mathrm{d}V/\mathrm{d}x)]\delta x\,\mathrm{d}t = 0. \tag{3.4}$$

Since $\delta x(t)$ is an arbitrary function, the expression in square brackets must be zero, and in this way we recover the equation of motion (3.1). The integrand in (3.2) is called the *Lagrangian* and is seen to be (kinetic energy − potential energy).

In general, for a system of N particles in three-dimensional space, its instantaneous state is specified by a set of $3N$ quantities $\{q_i\}$, called *generalized coordinates*, which may be distances, angles, or any other quantities which serve to specify all the positions, together with the $3N$ *generalized velocities* $\{\dot{q}_i\}$. Then the Lagrangian may be a function of all $6N$ of these quantities and time, $L = L(\{q_i\}, \{\dot{q}_i\}, t)$. By repeating the above calculation, but allowing for independent variations in all the coordinates, the reader may easily verify that the resulting equations of motion are the $3N$ equations

$$\frac{\mathrm{d}}{\mathrm{d}t}\left(\frac{\partial L}{\partial \dot{q}_i}\right) = \frac{\partial L}{\partial q_i}. \tag{3.5}$$

These are called the *Euler–Lagrange* equations. The quantity $p_i = \partial L/\partial \dot{q}_i$ is called the *generalized momentum* conjugate to the coordinate q_i, and $\partial L/\partial q_i$ is the *generalized force*. The rate of change of a generalized momentum is thus a generalized force and, by choosing the Lagrangian function correctly, these equations can be made to reproduce those given by Newton's law.

Suppose, however, we do not assume Newton's law to be valid. Can we deduce what the Lagrangian is on *a priori* grounds? In fact, quite a lot can be deduced on the basis of *spacetime symmetries*, as we shall now see. Consider first the case of a single, isolated particle. Since it is free from any external influences, its equation of motion can depend only on the structure of spacetime itself: any symmetry of this structure must also be a symmetry of the equation of motion. In Galilean spacetime, there are three quite obvious symmetries, which place definite constraints on the form of the Lagrangian.

(i) *Invariance under time translations*. In terms of the geometrical ideas in the last chapter, Galilean time has its own metric, which gives a definite quantitative meaning to time intervals. We assume that the time coordinate t, as well as labelling instants of time, is a linear measure of time intervals. This means that, given any other parameter t' which labels instants of time (say, the readings of an imperfect clock), there is a temporal metric tensor with a single component $g(t')$ such that $\mathrm{d}t = g(t')\mathrm{d}t'$. In terms of t itself, $g = 1$, so there is nothing to distinguish one moment from any other. Thus, the equation of motion of an isolated particle must be the same at any instant, and therefore L cannot depend explicitly on time. Another way of saying this is that L is invariant under the coordinate transformation which shifts or 'translates' the origin of time measurement by an amount t_0: $L(\boldsymbol{x}, \dot{\boldsymbol{x}}, t + t_0) = L(\boldsymbol{x}, \dot{\boldsymbol{x}}, t)$ so the t argument can be omitted.

(ii) *Invariance under spatial translations*. In Cartesian coordinates, the Pythagoras rule for finding the distance between two points is unchanged by a translation of the origin $\boldsymbol{x} \to \boldsymbol{x} + \boldsymbol{x}_0$ or, in the terminology of the last chapter, the spatial metric tensor (2.39) is unchanged. By the same reasoning as above, we conclude that $L(\boldsymbol{x} + \boldsymbol{x}_0, \dot{\boldsymbol{x}}) = L(\boldsymbol{x}, \dot{\boldsymbol{x}})$ or that L must be a function of $\dot{\boldsymbol{x}}$ only.

(iii) *Invariance under rotations*. Similarly, the Pythagoras rule or the metric tensor is unchanged by a rotation to a new Cartesian coordinate system. Therefore, L must be invariant under rotations. This means that it cannot depend on individual components of $\dot{\boldsymbol{x}}$ but only on the magnitude $|\dot{\boldsymbol{x}}| = (\dot{\boldsymbol{x}}.\dot{\boldsymbol{x}})^{1/2}$ which is unchanged by rotations.

In order to tie down the Lagrangian completely, we have to assume a further symmetry which does not follow directly from the spacetime structure:

(iv) *Invariance under Galilean transformations*. This is the assumption that the equation of motion has the same form in two frames of reference which have a constant relative velocity $\boldsymbol{v}$. The exact significance of this assumption and its relation to the structure of Galilean spacetime are slightly complicated and are discussed in an appendix to this chapter. Clearly, though, it involves assuming the existence of privileged unaccelerated or inertial frames in which the equation of motion has a special form. We found above that L can depend only on $|\dot{\boldsymbol{x}}|$, and it is now convenient to express L as a function of the variable $X = |\dot{\boldsymbol{x}}|^2/2$. If we choose the generalized coordinates in (3.5) to be Cartesian, then the equation of motion can be written as

$$\frac{\mathrm{d}}{\mathrm{d}t}\left(\dot{\boldsymbol{x}}\frac{\mathrm{d}L}{\mathrm{d}X}\right) = \ddot{\boldsymbol{x}}\frac{\mathrm{d}L}{\mathrm{d}X} + \dot{\boldsymbol{x}}(\ddot{\boldsymbol{x}}.\dot{\boldsymbol{x}})\frac{\mathrm{d}^2L}{\mathrm{d}X^2} = 0. \tag{3.6}$$

If we make a Galilean transformation, replacing $\boldsymbol{x}$ by $\boldsymbol{x} - \boldsymbol{v}t$, $\dot{\boldsymbol{x}}$ and X are changed, but $\ddot{\boldsymbol{x}}$ is not. To ensure that the form of (3.6) remains unchanged, we must take L to be such that $\mathrm{d}L/\mathrm{d}X$ is simply a constant, which means that $\mathrm{d}^2L/\mathrm{d}X^2$ is zero. This constant is, of course, what we usually call the mass of the particle, and the Lagrangian has turned out to be just the kinetic energy, $L = m\dot{\boldsymbol{x}}^2/2$, as it ought to be.

The Lagrangian for a system of non-interacting particles will clearly be the sum of the kinetic energies of all the particles. If the particles interact with each other, it will contain further terms to account for the forces. It is not difficult to see that, if the above symmetries are still to hold, these terms can depend only on the magnitudes of separations $|\boldsymbol{x}_i - \boldsymbol{x}_j|$ and relative velocities $|\dot{\boldsymbol{x}}_i - \dot{\boldsymbol{x}}_j|$ of pairs of particles. Thus, the general form of the Lagrangian is

$$L = \sum_i \tfrac{1}{2}m_i\dot{\boldsymbol{x}}_i^2 - V(\{|\boldsymbol{x}_i - \boldsymbol{x}_j|\}, \{|\dot{\boldsymbol{x}}_i - \dot{\boldsymbol{x}}_j|\}) \tag{3.7}$$

but no more can be said *a priori* about the function V.

Our original example (3.2) is not of this form and, unless V is a trivial constant, $V(x + x_0)$ does not equal $V(x)$. If our symmetry arguments are correct, then a Lagrangian of this kind can arise only when the potential is produced by some external system whose own behaviour is not taken properly into account. This may well be an excellent approximation. For example, the motion of a small object (mass m, position $\boldsymbol{x}$) near the Earth (mass M, position $\boldsymbol{X}$) would, according to Newtonian gravity, be described by a Lagrangian of the form (3.7) with $V = -GmM/|\boldsymbol{x} - \boldsymbol{X}|$. For many purposes, we can simply take the Earth to be fixed, say at $\boldsymbol{X} = \boldsymbol{0}$, so that V becomes a function of $\boldsymbol{x}$ only. For the small object on its own, translational invariance does not hold because of the presence of the Earth, but for the combined system of object + Earth, translational invariance does hold, so long as we neglect any influence of the rest of the universe. Thus, we expect the symmetries to be valid for any *isolated* system.

3.2 Symmetries and Conservation Laws

We saw above that the symmetry of invariance under time translations implied that the Lagrangian could not depend explicitly on time. Therefore, all the time dependence of L is through the generalized coordinates and velocities, and we may write

$$\frac{\mathrm{d}L}{\mathrm{d}t} = \sum_i \left(\frac{\mathrm{d}q_i}{\mathrm{d}t}\frac{\partial L}{\partial q_i} + \frac{\mathrm{d}\dot{q}_i}{\mathrm{d}t}\frac{\partial L}{\partial \dot{q}_i}\right) = \sum_i \left(\dot{q}_i\frac{\partial L}{\partial q_i} + \ddot{q}_i\frac{\partial L}{\partial \dot{q}_i}\right). \tag{3.8}$$

When the functions $q_i(t)$ represent the actual trajectories of the particles, and therefore obey the equations of motion (3.5), this becomes

$$\frac{\mathrm{d}L}{\mathrm{d}t} = \frac{\mathrm{d}}{\mathrm{d}t}\left(\sum_i \dot{q}_i\frac{\partial L}{\partial \dot{q}_i}\right) \tag{3.9}$$

which shows that $\mathrm{d}E/\mathrm{d}t = 0$, where

$$E = \sum_i \dot{q}_i\frac{\partial L}{\partial \dot{q}_i} - L. \tag{3.10}$$

This quantity, therefore, is conserved: it is a 'constant of the motion'. When the Lagrangian is that in (3.2), we see that E is the total energy. In general, since the concept of energy is useful only because of the conservation law, we might as well regard (3.10) as *defining* the energy of the system. (There are awkward cases in which other definitions of energy give a result different from (3.10), but we shall not be meeting them.) Thus, *if the Lagrangian does not depend explicitly on time, or is invariant under time translations, then energy is conserved*. As discussed above, we would expect this to be true for any isolated system.

A variety of other conservation laws can be deduced from symmetry or invariance properties of the Lagrangian. Mathematically, this works in the following way. We replace the coordinates q_i by $q_i + \varepsilon f_i$ and the velocities by $\dot{q}_i + \varepsilon \mathrm{d}f_i/\mathrm{d}t$, where each f_i is a function of the original coordinates, velocities and time, and ε is a small constant parameter. The Lagrangian can be expanded as a Taylor series in ε:

$$L\left(q_i + \varepsilon f_i, \dot{q}_i + \varepsilon \frac{\mathrm{d}f_i}{\mathrm{d}t}, t\right)$$
$$= L(q_i, \dot{q}_i, t) + \varepsilon \sum_j \left(\frac{\mathrm{d}L}{\partial q_j} f_j + \frac{\partial L}{\partial \dot{q}_j}\frac{\mathrm{d}f_j}{\mathrm{d}t}\right) + \mathrm{O}(\varepsilon^2) \tag{3.11}$$

and if the first-order term is zero, we say that L is invariant under the infinitesimal transformation specified by the functions f. I will discuss the meaning of this shortly, but let us first derive its consequences. Using the equations of motion (3.5), and the fact that the coefficient of ε in (3.11) vanishes, we find that $\mathrm{d}F/\mathrm{d}t = 0$, where

$$F = \sum_i f_i \frac{\partial L}{\partial \dot{q}_i} = \sum_i f_i p_i \tag{3.12}$$

where p_i are the generalized momenta defined earlier. The quantity F is therefore conserved.

The simplest conservation law of this kind is the conservation of linear momentum, which follows from invariance under spatial translations. If we use Cartesian coordinates a Lagrangian of the form (3.7) is unchanged if we replace each $\boldsymbol{x}_i$ by $\boldsymbol{x}_i + \varepsilon \boldsymbol{a}$, where $\boldsymbol{a}$ is any constant vector but the same for each particle. The velocities are unaffected because $\boldsymbol{a}$ is constant, and $\boldsymbol{a}$ cancels out of all the differences of pairs of coordinates. Thus, not only the first-order term but also all the higher-order terms in (3.11) vanish. The conserved quantity F is $\boldsymbol{a}\cdot\boldsymbol{P}$, where $\boldsymbol{P}$ is the sum of the momenta of all the particles, or the total momentum of the system. Since $\boldsymbol{a}$ can be any vector, each component of the total momentum is conserved. So *if the Lagrangian is invariant under spatial translations, then total linear momentum is conserved*. In the same way, invariance under rotations leads to the conservation of angular momentum, details of which are explored in exercise 3.1.

The symmetry transformations we have been using can be interpreted in two ways. According to what is called the *active* point of view, by making the mathematical transformation $\boldsymbol{x} \to \boldsymbol{x} + \boldsymbol{a}$, we are comparing the behaviour of the system when it occupies one or other of two regions of space, separated by the vector $\boldsymbol{a}$. Because the geometrical properties of our Galilean spacetime are the same everywhere, we expect that the laws of physics will be too. So the behaviour of the system, and therefore the form of its Lagrangian, should be the same in each location, so long as the system is isolated from any external influence. According to the *passive* point of view, we are comparing descriptions of the system referred to two sets of coordinates whose origins are separated by the vector $\boldsymbol{a}$. Again, since geometry is the same everywhere, equations of motion should have the same form, regardless of where we choose to place the origin of coordinates. Similar remarks apply to time translations and rotations.

Of course, these considerations apply to displacements or rotations of any size, not just the infinitesimal ones we used to derive the conservation law. In fact, if the Lagrangian is unchanged at first order by an infinitesimal transformation, it will also be unchanged by a large transformation which can be built from a sequence of infinitesimal ones. In general, however, it is only the infinitesimal ones which have the right form for the derivation to work. For example, the rotation $(x, y) \rightarrow (x \cos \varepsilon + y \sin \varepsilon, y \cos \varepsilon - x \sin \varepsilon)$ can be written, when ε is infinitesimal, as $(x, y) \rightarrow (x + \varepsilon y, y - \varepsilon x)$, and only the infinitesimal version can be used in (3.11). However, a rotation through a finite angle can obviously be built up from many infinitesimal ones. If the first-order change in L vanishes, then $x\partial L/\partial y = y\partial L/\partial x$, from which it is easy to show that L must be a function only of $(x^2 + y^2)$ and therefore invariant under rotations through any angle.

3.3 The Hamiltonian

At the beginning of our discussion, we assumed that the state of a system would be uniquely specified by the coordinates and velocities of all its particles. For most theoretical purposes, however, the momenta play a more fundamental role than the velocities, and it is convenient to reformulate the theory in terms of them. To do this, we introduce a new function $H(\{p_i\}, \{q_i\})$ called the *Hamiltonian*. In terms of this function, a new set of equations of motion can be derived which are equivalent to the Euler–Lagrange equations, but which involve the momenta instead of the velocities.

The mathematical process of exchanging one set of variables for another is called a *Legendre transformation* and works as follows. We consider a set of small changes $\mathrm{d}q_i$ and $\mathrm{d}\dot{q}_i$ in the coordinates and velocities and write the corresponding change in the Lagrangian as

$$\mathrm{d}L = \sum_i \left(\frac{\partial L}{\partial q_i} \mathrm{d}q_i + p_i \mathrm{d}\dot{q}_i \right) \tag{3.13}$$

where we have used the definition $p_i = \partial L/\partial \dot{q}_i$. Next, we define the Hamiltonian as

$$H(\{p_i\}, \{q_i\}) = \sum_i p_i \dot{q}_i - L \tag{3.14}$$

which implies that, on the right-hand side, all the velocities have been expressed in terms of the coordinates and momenta. Apart from this last step, the Hamiltonian is, of course, just the same as the total energy function defined by (3.10). We can now use (3.13) to write down the change in the Hamiltonian that results from a small change in the state of the system:

$$\begin{aligned} \mathrm{d}H &= \sum_i (p_i \mathrm{d}\dot{q}_i + \dot{q}_i \mathrm{d}p_i) - \mathrm{d}L \\ &= \sum_i \left(\dot{q}_i \mathrm{d}p_i - \frac{\partial L}{\partial q_i} \mathrm{d}q_i \right). \end{aligned} \tag{3.15}$$

According to the Euler–Lagrange equations (3.5), $\partial L/\partial q_i$ is equal to $\mathrm{d}p_i/\mathrm{d}t$. So, by allowing independent variations in each of the coordinates and momenta in turn, we may deduce from (3.15) the equations of motion

$$\dot{q}_i = \frac{\partial H}{\partial p_i} \qquad \dot{p}_i = -\frac{\partial H}{\partial q_i}. \tag{3.16}$$

These are *Hamilton's equations*.

3.4 Poisson Brackets and Translation Operators

It may not be obvious that we have gained anything from these formal manipulations. In fact, when it comes to solving equations of motion for specific systems containing a few particles, it makes no practical difference whether we use the original equations of Newton, the Euler–Lagrange equations or Hamilton's equations: they all amount to the same thing, and exercise 3.2 invites readers to explore this equivalence in detail. However, the Lagrangian and Hamiltonian formulations of classical mechanics do reveal some mathematical features which are important for further developments. In modern theoretical physics, there are two situations in which an understanding of the mathematical structure of classical mechanics is particularly useful. The first is that, when we deal with large collections of particles, it rapidly becomes impractical to solve the equations of motion directly. We must resort to a statistical description of such systems, and the Hamiltonian formulation is, as we shall discover in chapter 10, an indispensable tool for setting up this description.

An appreciation of the formal structure of classical mechanics is also useful when making the transition to quantum mechanics, which appears to supersede classical mechanics as a means of accounting for the behaviour of physical systems on atomic or subatomic scales. It is very difficult to infer directly from our experience what the rules of quantum mechanics should be. However, it turns out that the formal mathematical structures of classical and quantum mechanics have quite a lot in common. From a theoretical point of view, it seems to me that the most satisfactory way of approaching quantum theory is by exploiting the mathematical analogy with classical mechanics, which we shall explore in chapter 5. In this section, we shall construct some of the mathematical tools which make this analogy clear.

We saw in §3.2 that when the equations of motion are invariant under time translations, the total energy of the system, which is obtained by substituting into the Hamiltonian the actual coordinates and momenta of the particles, is conserved. Now, Hamilton's equations (3.16) offer us a deeper understanding of the role played by this quantity in the evolution of the state of the system in time. Suppose we wish to know how some quantity A changes with time, and that A can be expressed in terms of the coordinates and momenta as $A(\{p_i\}, \{q_i\})$. Using Hamilton's equations, we can write

$$\frac{\mathrm{d}A}{\mathrm{d}t} = \sum_i \left(\frac{\partial A}{\partial q_i} \dot{q}_i + \frac{\partial A}{\partial p_i} \dot{p}_i \right) = \{A, H\}_{\mathrm{P}} \tag{3.17}$$

where, for any two quantities A and B, the *Poisson bracket* $\{A, B\}_{\mathrm{P}}$ is defined as

$$\{A, B\}_{\mathrm{P}} = \sum_i \left(\frac{\partial A}{\partial q_i} \frac{\partial B}{\partial p_i} - \frac{\partial A}{\partial p_i} \frac{\partial B}{\partial q_i} \right). \tag{3.18}$$

It is implied, of course, that we treat the p and the q as independent variables to evaluate the Poisson bracket and then substitute their actual values at time t to find the rate of change of A at that time.

Alternatively, we can define the differential operator $\mathscr{H}$ by

$$\mathscr{H} = \mathrm{i}\{\mathrm{H}, \ \}_{\mathrm{P}} = \mathrm{i}\sum_i \left(\frac{\partial H}{\partial q_i} \frac{\partial}{\partial p_i} - \frac{\partial H}{\partial p_i} \frac{\partial}{\partial q_i} \right) \tag{3.19}$$

which means that $\mathscr{H}A = \mathrm{i}\{H, A\}_{\mathrm{P}} = -\mathrm{i}\{A, H\}_{\mathrm{P}}$ for any function A. The factor of i has no significance in classical mechanics, and I have included it just in order to bring out the quantum-mechanical analogy. Let us now make explicit the procedure for evaluating (3.17). We denote by $A(t)$ the value of A at time t, obtained by substituting into $A(\{p_i\}, \{q_i\})$ the functions $p_i(t)$ and $q_i(t)$ which describe the actual state of the system. This substitution can be represented by using the Dirac delta function, which is described in appendix 1 for readers unfamiliar with its use. If we define

$$\rho(\{p_i\}, \{q_i\}, t) = \prod_i \delta(p_i - p_i(t))\delta(q_i - q_i(t)) \tag{3.20}$$

then $A(t)$ can be written as

$$A(t) = \int \prod_i \mathrm{d}p_i \mathrm{d}q_i \rho(\{p_j\}, \{q_j\}, t) A(\{p_j\}, \{q_j\}). \tag{3.21}$$

To find $\mathrm{d}A/\mathrm{d}t$ from this expression, we can proceed in two ways. One is simply to differentiate, which gives $\partial\rho/\partial t$ inside the integral, since $A(\{p_i\}, \{q_i\})$ does not depend on time. The other, according to (3.17), is to act on $A(\{p_i\}, \{q_i\})$ with $\mathrm{i}\mathscr{H}$. By integrating by parts, we see that this is equivalent to acting on ρ with $-\mathrm{i}\mathscr{H}$. The two results must be identical, so we find that ρ satisfies the equation

$$\mathrm{i}\frac{\partial \rho}{\partial t} = \mathscr{H}\rho \tag{3.22}$$

as readers may verify directly using (3.20), (3.19) and (3.17).

Readers who are familiar with elementary quantum mechanics will recognize (3.22) as having a similar form to Schrödinger's equation which is, of course, the point of the exercise. Equation (3.17) can be written as $\mathrm{i}\mathrm{d}A/\mathrm{d}t = -\mathscr{H}A$, but it should be clear that this is not to be interpreted in the same way as (3.22). In (3.17) we use $\mathscr{H}$ to differentiate with respect to the p and the q, treating them as dummy variables, and then substitute the appropriate functions of time. On the other hand, ρ is a function of the p and the q which appear as dummy integration variables in (3.21) and also of time, and (3.22) is to be taken at face value as a partial differential equation in all of these variables. Bearing these points in mind, we can express $A(t)$ as a Taylor series

$$\begin{aligned} A(t) &= \sum_{n=0}^{\infty} \frac{1}{n!} t^n A^{(n)}(0) \\ &= \sum_{n=0}^{\infty} \frac{1}{n!} (\mathrm{i}t\mathscr{H})^n A \\ &= \exp\{\mathrm{i}t\mathscr{H}\}A. \end{aligned} \tag{3.23}$$

Here, the nth time derivative of $A(t)$ evaluated at time $t = 0$ is denoted by $A^{(n)}(0)$, and the derivative can be replaced by $\mathrm{i}\mathscr{H}$ in the manner we have described. The exponential of the differential operator is a convenient shorthand for the power series. Obviously, we evaluate the last expression by substituting the p and the q corresponding to the state of the system at time $t = 0$ after acting with $\mathscr{H}$. The exponential operator is responsible for transforming $A(0)$ into $A(t)$ and in this context $\mathscr{H}$ is called the *generator of time translations*.

In Cartesian coordinates, we can transform any function $f(\{\boldsymbol{x}_i\})$ of the coordinates into $f(\{\boldsymbol{x}_i + \boldsymbol{a}\})$ by means of a similar Taylor series using the operator $\exp\{\mathrm{i}\boldsymbol{a}\cdot\boldsymbol{\mathcal{P}}\}$, where the *generator of spatial translations* is

$$\boldsymbol{\mathcal{P}} = -\mathrm{i}\sum_i \boldsymbol{\nabla}_i. \tag{3.24}$$

The sum here is over the N particles in the system rather than the $3N$

coordinates. It is easy to see that this generator may be written in a form similar to (3.19) as $\mathcal{P} = \mathrm{i}(\boldsymbol{P}, \;\}_{\mathrm{P}}$, where $\boldsymbol{P}$ is the total linear momentum, and we recall that $\boldsymbol{P}$ is the quantity whose conservation law follows from invariance under spatial translations. Again, knowledgeable readers will recognize (3.24) as being closely related to the momentum operator which acts on quantum-mechanical wavefunctions.

Equation (3.22) also serves as the starting point of classical statistical mechanics, if we regard ρ as expressing the probability that the momenta and coordinates have, at time t, the values $\{p_i\}$ and $\{q_i\}$. Then (3.21) is the usual expression for the mean value of A. In the case we have considered, the probability is zero unless momenta and coordinates correspond to the evolution of the system from a definite initial state, but more general probability distributions can be constructed as we shall see in chapter 10. In this context, (3.22) is called the *Liouville equation* and $\mathcal{H}$ the Liouville operator.

3.5 The Action Principle in Minkowski Spacetime

In earlier sections of this chapter, we have investigated the way in which the geometrical structure of Galilean spacetime constrains the possible kinds of behaviour of particles which live there. A source of difficulty was the fact that the geometrical roles of space and time are quite different. This leads to a certain amount of confusion about the exact significance of invariance under Galilean transformations and the meaning of inertial frames of reference. In particular, it does not appear to be possible to arrive at a purely geometrical definition of inertial frames which is independent of considerations about the way in which physical objects are actually observed to behave. In the Minkowski spacetime of special relativity, and in the more general spacetimes envisaged in general relativity and similar theories, space and time appear on much the same footing, and a more clear-cut discussion is possible. Conversely, to my mind, the relativistic view makes it rather more difficult to understand the obvious dissimilarity of space and time as they enter our conscious experience. I do not propose to enter into the philosophical perplexities of this question here, but interested readers may like to consult, for example, the books by Landsberg (1982), Lucas (1973), Ornstein (1969), Prigogine (1980), Smart (1964) and Whitrow (1975).

We learned in chapter 2 that the relativistic spacetimes are manifolds whose points may be labelled by a set of four coordinates $x^\mu (\mu = 0, 1, 2, 3)$. The separation of two points cannot be uniquely decomposed into spatial and temporal components. What we can do is to assign a *proper time interval* to a specific curve which joins them. The proper time interval $\mathrm{d}\tau$ for an infinitesimal segment of the curve is given by (2.7), in which the coefficients $g_{\mu\nu}$ are called components of the metric tensor. The metric tensor contains all our information about the geometrical structure. In general, its components vary from point to point and their values depend on the coordinate system we are using. The value of $d\tau$ is the same in all coordinate systems, and to find the components of the metric tensor in a new system we express the old coordinates in terms of the new ones and apply the chain rule for differentiation to (2.7). If the metric tensor is that of Minkowski spacetime then, by definition, it will be possible to find a Cartesian coordinate system (and, in fact, infinitely many of them) such that its components are given by the matrix (2.8). Relative to such a system, time is measured by x^0/c, where c is the speed of light, while the other three coordinates measure spatial distances.

We may now define an *inertial system of Cartesian coordinates* as one where the metric tensor has the special form (2.8). More generally, an inertial system is one which can be obtained from an inertial Cartesian system by keeping the time coordinate and redefining the spatial ones in a time-independent manner. For example, if we simply exchange (x^1, x^2, x^3) for polar coordinates (r, θ, ϕ) we still have an inertial system, but if we exchange them for a set of rotating axes, we get a non-inertial one. In the rest of this chapter, we will use only inertial Cartesian coordinates.

As with Galilean spacetime, we want to see how geometrical symmetries constrain the behaviour of physical systems. These symmetries consist of all the coordinate transformations which leave the form of the metric tensor unchanged: that is they convert one inertial Cartesian system into another. They are called *isometries*, meaning 'same metric'. Space and time translations can now be considered together. They are transformations of the type $x^{\mu'} = x^\mu + a^\mu$, where a^μ are a set of four constants. We see from (2.7) and (2.8) that this leaves $g_{\mu\nu}$ unchanged, since $dx^{\mu'} = dx^\mu$. The other isometries are *Lorentz transformations*. These include spatial rotations and 'boosts', which relate two systems with a constant relative velocity. They can be expressed in the form

$$x^{\mu'} = \Lambda^{\mu'}{}_{\mu} x^\mu \tag{3.25}$$

where, as in chapter 2, we are using a prime on the index μ to indicate the new coordinates. For example, a rotation about the x^1 axis through an angle θ corresponds to the transformation matrix

$$\Lambda^{\mu'}{}_{\mu} = \begin{pmatrix} 1 & 0 & 0 & 0 \\ 0 & 1 & 0 & 0 \\ 0 & 0 & \cos\theta & \sin\theta \\ 0 & 0 & -\sin\theta & \cos\theta \end{pmatrix} \tag{3.26}$$

while the boost written in (2.2) is represented by

$$\Lambda^{\mu'}{}_{\mu} = \begin{pmatrix} \cosh\alpha & -\sinh\alpha & 0 & 0 \\ -\sinh\alpha & \cosh\alpha & 0 & 0 \\ 0 & 0 & 1 & 0 \\ 0 & 0 & 0 & 1 \end{pmatrix} \tag{3.27}$$

with $\sinh\alpha = (1 - v^2/c^2)^{-1/2}v/c$ (and so $\cosh\alpha = (1 - v^2/c^2)^{-1/2}$). The set of all rotations and boosts is called the *proper Lorentz group*. The set of rotations, boosts and translations is called the *Poincaré group*, and it is the *isometry group* of Minkowski space.

Any Poincaré transformation—that is the net effect of any sequence of translations, rotations and boosts—can be expressed in the form $x^{\mu'} = \Lambda^{\mu'}{}_{\mu}x^{\mu} + a^{\mu'}$. Let $f(x)$ be a scalar function of the coordinates (one whose value depends on the spacetime point, but not on the choice of coordinate system). Under a Poincaré transformation, both infinitesimal coordinate differences and derivatives of scalar functions transform in a manner which depends only on Λ:

$$\mathrm{d}x^{\mu'} = \Lambda^{\mu'}{}_{\mu}\mathrm{d}x^{\mu} \tag{3.28}$$

$$\partial_{\mu'}f = \Lambda^{\mu}{}_{\mu'}\partial_{\mu}f. \tag{3.29}$$

(Recall the following from chapter 2: repeated indices, occurring once in the upper and once in the lower position are summed over; ∂_μ is an abbreviation for $\partial/\partial x^\mu$; the matrix $\Lambda^{\mu'}{}_{\mu}$ is the inverse of $\Lambda^{\mu}{}_{\mu'}$—see (2.14).) A set of four components V^μ which transform like $\mathrm{d}x^\mu$ is called a *contravariant 4-vector*; a set V_μ which transforms like $\partial_\mu f$ is a *covariant 4-vector*. More complicated entities, with transformation laws similar to (2.19), are *4-tensors*: for example, the metric tensor, with two lower indices, is said to have covariant rank 2. These 4-tensors are not necessarily true tensors as defined in chapter 2, because we are considering only Λ matrices with constant components. For example, $\partial_\mu V^\nu$ is a 4-tensor but not a true tensor. Readers may readily verify that any expression such as $\eta_{\mu\nu}U^\mu V^\nu$ composed of tensors, in which all indices appear in pairs and the implied summations have been carried out (the process called *contraction* in chapter 2), is invariant under Lorentz transformations: it is a Lorentz scalar.

The path of a particle through Minkowski spacetime may be described parametrically by a set of four functions $x^\mu(\tau)$, each point on the path being labelled by a value of the proper time τ. Since τ is a scalar, the set of functions $\mathrm{d}x^\mu/\mathrm{d}\tau$ are the components of a 4-vector, the tangent vector to the path. As in our discussion of Galilean spacetime, we expect the equations of motion for an isolated system to have the same form in any two coordinate systems in which the metric tensor is the same. Thus, the form of these equations should be unchanged by any Poincaré transformation: we say that they should be *covariant* under these transformations. To achieve this, we need an action which is Poincaré invariant.

That is, the action must be a Lorentz scalar and translationally invariant. Following the arguments of §3.1, we see that for a single particle it must be of the form

$$S = \int \mathrm{d}\tau L(\eta_{\mu\nu}\dot{x}^{\mu}\dot{x}^{\nu}) \tag{3.30}$$

where $\dot{x}^{\mu}$ denotes $\mathrm{d}x^{\mu}/\mathrm{d}\tau$. Using the notation $X = \eta_{\mu\nu}\dot{x}^{\mu}\dot{x}^{\nu}$, we find that the Euler–Lagrange equations are

$$\frac{\mathrm{d}^2x^{\mu}}{\mathrm{d}\tau^2}\frac{\mathrm{d}L(X)}{\mathrm{d}X} + \frac{\mathrm{d}x^{\mu}}{\mathrm{d}\tau}\frac{\mathrm{d}X}{\mathrm{d}\tau}\frac{\mathrm{d}^2L(X)}{\mathrm{d}X^2} = 0. \tag{3.31}$$

In Galilean spacetime, the function $L(X)$ could be determined by requiring invariance under Galilean transformations. Here, this symmetry is replaced by Lorentz invariance, which has already been taken into account. In fact, the form of $L(X)$ is quite irrelevant! According to (2.7), X has the constant value c^2 for any path. The only quantity with any physical significance is the value of $\mathrm{d}L/\mathrm{d}X$ at $X = c^2$ and, as long as this is non-zero, the equation of motion is simply $\mathrm{d}^2x^{\mu}/\mathrm{d}\tau^2 = 0$. We may as well make the simplest choice

$$L = \tfrac{1}{2}m\eta_{\mu\nu}\frac{\mathrm{d}x^{\mu}}{\mathrm{d}\tau}\frac{\mathrm{d}x^{\nu}}{\mathrm{d}\tau} \tag{3.32}$$

where, as before, m will be identified as the mass of the particle. (Many authors refer to m as the 'rest mass' to distinguish it from a velocity-dependent 'mass', which is in fact the energy divided by c^2. I do not recommend this practice and will not follow it in this book.)

The canonical momenta obtained from this Lagrangian, which are conserved as a consequence of translational invariance, are the four components of the *energy–momentum 4-vector*, or *4-momentum*

$$p^{\mu} = \eta^{\mu\nu}\frac{\partial L}{\partial \dot{x}^{\nu}} = m\frac{\mathrm{d}x^{\mu}}{\mathrm{d}\tau}. \tag{3.33}$$

For convenience, I have written the contravariant version of this 4-vector, 'raising' its index with $\eta^{\mu\nu}$. This is the inverse matrix to $\eta_{\mu\nu}$, which is, of course, exactly the same matrix, so long as we stick to Cartesian coordinates. Readers should satisfy themselves that this result is unaffected by the arbitrariness in the function $L(X)$. In a particular coordinate system, denote the three components of the velocity of the particle by $\boldsymbol{u} = \mathrm{d}\boldsymbol{x}/\mathrm{d}t$. From (2.6), we see that $\mathrm{d}\tau/\mathrm{d}t = (1 - u^2/c^2)^{1/2}$, so the 4-momentum may be written as

$$(p^0, \boldsymbol{p}) = \left(\frac{mc}{(1 - u^2/c^2)^{1/2}}, \frac{m\boldsymbol{u}}{(1 - u^2/c^2)^{1/2}}\right). \tag{3.34}$$

Since this is conserved, we may identify the zeroth, time-like, component as $1/c$ times the energy (to make its dimensions agree with the non-relativistic definition) and the other three as the linear momentum.

Because there is no unique time in Minkowski space, the integration variable τ in (3.30) is associated with the path of a specific particle. The action for a collection of non-interacting particles, labelled by i, following paths $x_i^\mu(\tau_i)$ is therefore

$$S = \sum_i \int d\tau_i \, \frac{1}{2} m_i \eta_{\mu\nu} \frac{dx_i^\mu}{d\tau_i} \frac{dx_i^\nu}{d\tau_i}. \tag{3.35}$$

It is useful to have expressions for the number density $n(x)$ (number per unit volume) and current density $\boldsymbol{j}(x)$ (number crossing unit area per unit time) of these particles. At the microscopic level, these are zero unless the point x lies exactly on the path of one of the particles. They may be written as

$$n(t, \boldsymbol{x}) = \sum_i \delta^3(\boldsymbol{x} - \boldsymbol{x}_i(t)) \tag{3.36}$$

$$\boldsymbol{j}(t, \boldsymbol{x}) = \sum_i \frac{d\boldsymbol{x}_i}{dt} \delta^3(\boldsymbol{x} - \boldsymbol{x}_i(t)). \tag{3.37}$$

So long as no particles are created or destroyed, they should satisfy the equation of continuity $\partial n/\partial t + \operatorname{div}\boldsymbol{j} = 0$. Readers are invited to verify this and to consider what happens if particles are created or destroyed. Using the fact that $dx^0/dt = c$, we can assemble the quantities (3.36) and (3.37) into a 4-vector

$$j^\mu(\boldsymbol{x}, t) = (cn(\boldsymbol{x}, t), j^1(\boldsymbol{x}, t), j^2(\boldsymbol{x}, t), j^3(\boldsymbol{x}, t)) = \sum_i \frac{dx_i^\mu}{dt} \delta^3(\boldsymbol{x} - \boldsymbol{x}_i(t)). \tag{3.38}$$

Although dx^μ is a 4-vector, neither dt nor the δ function is a scalar, so it is not obvious that this really is a 4-vector. It is left as an exercise for readers to show that the 4-vector current can be rewritten in the form

$$j^\mu(x) = c\sum_i \int d\tau_i \, \frac{dx_i^\mu}{d\tau_i} \delta^4(x - x_i(\tau_i)) \tag{3.39}$$

which manifestly is a 4-vector. In terms of j^μ, the equation of continuity reads

$$\partial_\mu j^\mu = 0. \tag{3.40}$$

A current which satisfies this equation is said to be a *conserved current*.

If A is some physical quantity carried by the particles, we can define a current whose zeroth component is the density of A (the amount of A per unit volume) and whose spatial components represent the rate at which A is transported by the flow of particles (the amount of A carried across unit area per unit time). It is

$$j_A^\mu(x) = c\sum_i \int \mathrm{d}\tau_i A_i \frac{\mathrm{d}x_i^\mu}{\mathrm{d}\tau_i}\,\delta^4(x - x_i(\tau_i)) \tag{3.41}$$

where A_i is the amount of A carried by the ith particle. Two important examples are the *electromagnetic current*, obtained by taking A to be electric charge, and the *stress–energy–momentum tensor*, which I will refer to as the *stress tensor* for brevity. This tensor is formed from the four currents obtained by taking A to be the components of the 4-momentum:

$$T^{\mu\nu}(x) = c\sum_i \int \mathrm{d}\tau_i m_i \frac{\mathrm{d}x_i^\mu}{\mathrm{d}\tau_i}\frac{\mathrm{d}x_i^\nu}{\mathrm{d}\tau_i}\,\delta^4(x - x_i(\tau_i)). \tag{3.42}$$

The stress tensor plays a vital role in the general relativistic theory of gravity. It is symmetric in the indices μ and ν and is conserved, since $\partial_\nu T^{\mu\nu} = 0$, as readers are invited to prove. This simply reflects the fact that energy and momentum are conserved quantities, so their densities and currents must obey equations of continuity. It should be borne in mind, however, that (3.42) is the stress tensor for a collection of non-interacting particles. If, for example, the particles interact via electromagnetic fields, then energy and momentum can be transferred to and from these fields and the stress tensor will be conserved only when a suitable electromagnetic contribution is included. The same goes for fields associated with other forces, including gravitational fields, but the nature of conservation laws in non-Minkowski spacetimes can be a little obscure.

A simple example of a stress tensor is afforded by what cosmologists call a *perfect fluid*. This is a fluid which has a rest frame, in which its density is spatially uniform and the average velocity of its particles is zero. For such a fluid, as discussed in exercise 3.4, the stress tensor is

$$T^{\mu\nu} = \begin{pmatrix} \rho & 0 & 0 & 0 \\ 0 & p & 0 & 0 \\ 0 & 0 & p & 0 \\ 0 & 0 & 0 & p \end{pmatrix} \tag{3.43}$$

where ρ is the density and p the pressure.

3.6 Classical Electrodynamics

The only fully fledged classical theory of interacting particles in Minkowski spacetime is electrodynamics, in which the forces are described by electric and magnetic fields $\boldsymbol{E}(\boldsymbol{x}, t)$ and $\boldsymbol{B}(\boldsymbol{x}, t)$ which obey *Maxwell's equations*. In a suitable system of units, these equations read

$$\nabla \cdot \boldsymbol{E} = \rho_e \tag{3.44}$$

$$\nabla \cdot \boldsymbol{B} = 0 \tag{3.45}$$

$$\nabla \times \boldsymbol{E} + \frac{1}{c}\frac{\partial \boldsymbol{B}}{\partial t} = 0 \tag{3.46}$$

$$\nabla \times \boldsymbol{B} - \frac{1}{c}\frac{\partial \boldsymbol{E}}{\partial t} = \frac{1}{c}\boldsymbol{j}_e \tag{3.47}$$

where ρ_e is the electric charge density and $\boldsymbol{j}_e$ is the electric current density. The first of these equations is Gauss' law which, for a static charge distribution, is a simple consequence of the Coulomb force law. The second asserts that there are no magnetic monopoles, which would be the magnetic analogues of electric charges. The *grand unified theories* of fundamental forces discussed in chapter 12 predict that such monopoles should exist but, at the time of writing, there is no firm experimental evidence of their existence. The third equation (3.46) is Faraday's law which describes the generation of electric fields by time-varying magnetic fields, and the fourth (3.47) is Ampère's law which, conversely, describes the generation of magnetic fields by current flows and changing electric fields. Readers who are not familiar with the derivation of these equations from simple physical observations will find it discussed in any standard textbook on electromagnetic theory. This form of Maxwell's equations is valid in the Heaviside–Lorentz system of units and is the microscopic version: the fields $\boldsymbol{D}$ and $\boldsymbol{H}$ which are often used to take approximate account of the properties of dielectric and magnetic materials are not used here.

As far as the classical theory is concerned, I know of no convincing way of arriving at Maxwell's equations other than by inferring them from experimental observations. On the other hand, we shall see in chapter 8 that in quantum mechanics they arise in quite a natural way from geometrical considerations. For now, we shall take them as given and briefly derive some important and elegant properties. Two of the equations, (3.45) and (3.46), are satisfied automatically if we express the fields in terms of an electric scalar potential $\phi(\boldsymbol{x}, t)$ and a magnetic vector potential $\boldsymbol{A}(\boldsymbol{x}, t)$ as

$$\boldsymbol{E} = -\nabla\phi - \frac{1}{c}\frac{\partial \boldsymbol{A}}{\partial t} \tag{3.48}$$

$$\boldsymbol{B} = \nabla \times \boldsymbol{A} \tag{3.49}$$

which follows from the identities $\nabla \times \nabla\phi \equiv 0$ and $\nabla \cdot (\nabla \times \boldsymbol{A}) \equiv 0$. The two remaining equations take on a much more compact appearance if we express them in 4-vector notation. The potentials can be assembled into a contravariant 4-vector A^μ with components $(\phi, \boldsymbol{A})$ or its covariant version A_μ with components $(\phi, -\boldsymbol{A})$. The electric and magnetic fields then form the components of an antisymmetric *field strength 4-tensor*

$$F_{\mu\nu} = \partial_\mu A_\nu - \partial_\nu A_\mu \tag{3.50}$$

whose contravariant form may be written explicitly as

$$F^{\mu\nu} = \begin{pmatrix} 0 & -E^1 & -E^2 & -E^3 \\ E^1 & 0 & -B^3 & B^2 \\ E^2 & B^3 & 0 & -B^1 \\ E^3 & -B^2 & B^1 & 0 \end{pmatrix}. \tag{3.51}$$

In terms of this tensor, (3.44) and (3.47) are simply

$$\partial_\mu F^{\mu\nu} = \frac{1}{c} j_{\mathrm{e}}^\nu \tag{3.52}$$

where j_{e}^ν is the electric 4-vector current with components $(c\rho_{\mathrm{e}}, \boldsymbol{j}_{\mathrm{e}})$.

These equations can be derived from an action principle in more or less the same way as the equations of motion for particles. Because we are now dealing with electromagnetic fields which exist at each point of spacetime rather than with the trajectories of particles, the action must be written as the integral over all space and time of a *Lagrangian density* $\mathscr{L}$:

$$S = \frac{1}{c} \int \mathrm{d}^4x \mathscr{L}(x) \tag{3.53}$$

where

$$\mathscr{L}(x) = -\frac{1}{4} F_{\mu\nu}(x) F^{\mu\nu}(x) - \frac{1}{c} j_{\mathrm{e}}^\mu(x) A_\mu(x). \tag{3.54}$$

The factor $1/c$ in (3.53) arises from the fact that $x^0 = ct$. Readers may easily verify that the Euler–Lagrange equations obtained by varying A^μ are (3.52). To obtain a complete theory of charged particles, we must add to (3.53) the action (3.35) for the particles themselves.

Consider the case of a single particle with charge e. The current is given by (3.41) with $A = e$ and, on substituting this into (3.53), the spacetime integral in the $j^\mu A_\mu$ term can be carried out. Thus, the total action is given by

$$S = \int \mathrm{d}\tau \frac{1}{2} m \eta_{\mu\nu} \frac{\mathrm{d}x^\mu}{\mathrm{d}\tau} \frac{\mathrm{d}x^\nu}{\mathrm{d}\tau} - \frac{e}{c} \int \mathrm{d}\tau \frac{\mathrm{d}x^\mu}{\mathrm{d}\tau} A_\mu(x(\tau)) - \frac{1}{4c} \int \mathrm{d}^4x F_{\mu\nu} F^{\mu\nu}. \tag{3.55}$$

By varying the path of the particle, we find the equation of motion

$$m\,\frac{\mathrm{d}^2x^\mu}{\mathrm{d}\tau^2} = \frac{e}{c}\,\eta_{\nu\sigma}\,\frac{\mathrm{d}x^\nu}{\mathrm{d}\tau}\,F^{\sigma\mu}. \tag{3.56}$$

Its zeroth component can be written as

$$\frac{\mathrm{d}}{\mathrm{d}t}\left(\frac{mc^2}{(1-u^2/c^2)^{1/2}}\right) = -e\boldsymbol{u}\cdot\boldsymbol{E} \tag{3.57}$$

which asserts that the rate of change of energy of the particle is the rate at which work is done on it by the electric field, while the spatial components reproduce the usual Lorentz force

$$\frac{\mathrm{d}\boldsymbol{p}}{\mathrm{d}t} = e\left(\boldsymbol{E} + \frac{\boldsymbol{u}}{c}\times\boldsymbol{B}\right). \tag{3.58}$$

The momentum $\boldsymbol{p}$ here is that written in (3.34). However, it is now not equal to the canonical momentum conjugate to the coordinates of the particle which is

$$p^\mu_{\text{can}} = m\,\frac{\mathrm{d}x^\mu}{\mathrm{d}\tau} - \frac{e}{c}\,A^\mu(x(\tau)). \tag{3.59}$$

The canonical structure of electrodynamics is explored further in the exercises.

Electromagnetism possesses an important symmetry known as *gauge invariance*. In the classical theory, it seems to appear more or less by accident but, as we shall see in chapter 8, it has a deep-seated significance in quantum mechanics and underlies most of our present understanding of the fundamental forces of nature. Let $\theta(x)$ be any function and consider redefining the 4-vector potential according to

$$A'_\mu(x) = A_\mu(x) - \partial_\mu\theta(x). \tag{3.60}$$

The field strengths given by (3.50) are the same functions of A'_μ as they were of A_μ, because the $\partial_\mu\partial_\nu\theta$ terms cancel. This clearly has to do with the antisymmetry of $F_{\mu\nu}$. This antisymmetry also has the consequence that the electric current must be conserved (it must obey (3.40)), as we see by differentiating (3.52). Suppose we demand that the action (3.53) with Lagrangian density (3.54) should be *gauge invariant,* that is that its form should be preserved after the change of variable (3.60), which is called a *gauge transformation*. The change in the action is $-(1/c)\int \mathrm{d}^4x j^\mu_{\mathrm{e}}\partial_\mu\theta$ so, after integrating by parts, we see that this vanishes provided that the current is conserved. Therefore, the quantity whose conservation is associated with the symmetry of gauge invariance is electric charge. If there is no mechanism whereby charged particles can be created or destroyed, then electric charge will naturally be conserved. If there is such a mechanism, then charge may not be conserved and, if it is not, the presence of electromagnetic forces will not make it so. In the latter case, (3.52) could not be true, and Maxwell's theory would not be self-consistent. Readers will recall (I hope!) that the so-called *displacement current* $\partial \boldsymbol{E}/\partial t$ in (3.47) was introduced by Maxwell precisely in order to make his equations consistent with conservation of electric charge. Experimentally, of course, even though individual charged particles can be created or destroyed, it is always found that these processes occur in such a way that electric charge is conserved overall.

Appendix: A Note on Galilean Transformations

The geometrical structure of Galilean spacetime is shown schematically in figure 2.13. As far as geometry is concerned, there is no particular relation between two points at different times: the spatial 'fibres' corresponding to different times are quite independent. Therefore, there is no particular relation between spatial frames of reference belonging to different fibres, and the idea of a *moving* frame of reference has no clear meaning. On the other hand, some sort of relation between the fibres is implied by the fact that material objects retain their identity through time (I take this to be an obvious fact of experience, although philosophers have sometimes found it perplexing: see, for example, Strawson (1959)). The separation of two particles is a smooth function of time and it is natural to choose the spatial frames of reference in successive fibres in such a way that the coordinates of each particle are smooth functions of time as, indeed, we have been assuming. In this way, we obtain a system of space and time coordinates which extends smoothly through the spacetime. The vector which joins the origins of two of these frames will be a smooth function of time, and it makes sense to speak of the relative motion of the two frames. Then, the further assumption that the equations of motion of particles are invariant under Galilean transformations is essentially equivalent to assuming the existence of privileged *inertial* frames of reference, in which these equations have a particularly simple form.

Exercises

3.1. Express the Lagrangian $L = m\dot{\boldsymbol{x}}^2/2 - V(\boldsymbol{x})$ for a single particle in cylindrical coordinates (r, θ, z) with $x = r\cos\theta$ and $y = r\sin\theta$. Show that the generalized momentum conjugate to θ is the angular momentum $mr^2\dot{\theta}$ about the z axis. If the potential V has cylindrical symmetry, that is it is independent of θ, show, by considering the transformation $\theta \to \theta + \varepsilon$, that the conserved quantity F in (3.12) is the angular momentum. When ε is infinitesimal, find the corresponding transformation of the Cartesian coordinates x and y. Working in Cartesian coordinates, show that if the Lagrangian is invariant under this transformation, then the conserved quantity is the z component of the angular momentum $\boldsymbol{J} = \boldsymbol{x} \times \boldsymbol{p}$. Show that if the potential is spherically symmetric, that is it is a function only of $x^2 + y^2 + z^2$, then all three components of angular momentum are conserved. In cylindrical coordinates, show that the generator of rotations about the z axis is $-\mathrm{i}\partial/\partial\theta$. In Cartesian coordinates, show that the rotation generators are $\mathcal{J} = \mathrm{i}\{\boldsymbol{J},\ \}_{\mathrm{P}} = \boldsymbol{x} \times \mathcal{P}$.

3.2. Consider the Lagrangian $L = m\dot{\boldsymbol{x}}^2/2 - V(\boldsymbol{x})$ and the Hamiltonian $H = \boldsymbol{p}^2/2m + V(\boldsymbol{x})$. Show that Hamilton's equations are equivalent to the Euler–Lagrange equations together with the definition of the canonical momentum. Consider the Lagrangian $L = \boldsymbol{p}\cdot\dot{\boldsymbol{x}} - \boldsymbol{p}^2/2m - V(\boldsymbol{x})$, where $\boldsymbol{x}$, $\dot{\boldsymbol{x}}$ and $\boldsymbol{p}$ are to be treated as independent variables. Show that the two sets of Euler–Lagrange equations reproduce the previous equations of motion, together with the relation $\boldsymbol{p} = m\dot{\boldsymbol{x}}$.

3.3. For a single particle in Minkowski space, show that the Hamiltonian $H = \eta_{\mu\nu}p^\mu\dot{x}^\nu - L$ expressed as a function of the momenta leads to Hamilton's equations which reproduce the correct equation of motion together with the definition (3.33) of the momenta, provided that derivatives with respect to proper time are used. Show that this Hamiltonian is not equal to the total energy of the particle.

3.4. Using elementary kinetic theory for a non-relativistic ideal gas in its rest frame, show that $\langle p^i(\mathrm{d}x^j/\mathrm{d}t)\rangle = (p/n)\delta^{ij}$, where p^i and $\mathrm{d}x^i/\mathrm{d}t$ are the Cartesian components of momentum and velocity, p and n are the pressure and number density and the average $\langle\ldots\rangle$ is taken over all the particles. Assume that the same is true for a relativistic gas if the spatial components of momentum in (3.34) are used. For the relativistic gas in its rest frame, imagine dividing the volume it occupies into cells, each of which is small compared with the total volume but still contains many particles. Define the average of the stress tensor (3.42) for each cell as

$$\langle T^{\mu\nu}\rangle = \int_{\text{cell}} \mathrm{d}^3x\, T^{\mu\nu}(x)/\text{Volume of cell}.$$

Show that this average has the form shown in (3.43). More generally, consider a fluid whose stress tensor field has this form at the point x when measured relative to the rest frame of the fluid element at x. Show that its stress tensor field in any frame of reference is

$$T^{\mu\nu} = c^{-2}(\rho + p)u^\mu u^\nu - pg^{\mu\nu}$$

where $u^\mu(x)$ is the 4-velocity of the fluid element at x and $\rho(x)$ and $p(x)$ are the energy density and pressure as measured in the rest frame of this element.

3.5. Consider the Lagrangian density

$$\mathscr{L} = \tfrac{1}{4}F^{\mu\nu}F_{\mu\nu} - \tfrac{1}{2}F^{\mu\nu}(\partial_\mu A_\nu - \partial_\nu A_\mu) - \frac{1}{c}\, j_e^\mu A_\mu .$$

By treating $F^{\mu\nu}$ and A_μ as independent variables, derive two Euler–Lagrange equations and show that they reproduce (3.50) and (3.52).

3.6. In a particular frame of reference, define the Lagrangian $L = -\frac{1}{4}\int \mathrm{d}^3x F_{\mu\nu}F^{\mu\nu}$. Show that $L = \frac{1}{2}\int \mathrm{d}^3x(\boldsymbol{E}^2 - \boldsymbol{B}^2)$. Define the generalized momentum conjugate to $A_\mu(x)$ as $\Pi^\mu(x) = \delta L/\delta(\partial_0 A_\mu)$, where the derivative is a functional one (see appendix 1). Show that $\Pi^i = E^i$ and $\Pi^0 = 0$. Now define the Hamiltonian $H = \int \mathrm{d}^3x \Pi^\mu \partial_0 A_\mu - L$. Show that H is the integral over space of the energy density $(\boldsymbol{E}^2 + \boldsymbol{B}^2)/2$, provided that Gauss' law $\boldsymbol{\nabla}\cdot\boldsymbol{E} = 0$ holds.

4

General Relativity and Gravitation

We now have at our disposal all the mathematical tools that are needed to understand the general theory of relativity and the account it offers of gravitational phenomena. Chapter 2 ended with the question 'what is the structure of our spacetime?' *A priori*, the possibilities are limitless: for a start, there are infinitely many dimensionalities to choose from. However, because special relativity accounts extremely well for a great many phenomena, it is clear that our spacetime must be quite similar to Minkowski spacetime. (Why this should be so is not known, but we shall touch later on some speculative ideas which may point towards an answer.) Our first task in this chapter will be to use this observation to restrict the range of possibilities that need to be considered in practice, which is more or less equivalent to adopting the principle of equivalence mentioned in chapter 2. The next step will be to find out how a given geometrical structure affects the behaviour of material objects, and this will show us how deviations of this structure from that of Minkowski spacetime can be interpreted in terms of gravitational forces. Finally, we shall investigate how the geometrical structure is determined—or at any rate influenced—by the distribution of gravitating matter.

4.1 The Principle of Equivalence

As we stated it in chapter 2, the principle of equivalence asserts that all gravitational effects can be eliminated within a sufficiently small region of space by adopting a freely falling inertial frame of reference. Near the surface of the Earth, for example, this frame of reference is obviously accelerating relative to one fixed in the Earth and the 'equivalence' is between, on the one hand, the acceleration of the inertial frame relative to an earthbound observer and, on the other, the gravitational forces which appear to this observer to act on falling bodies. Let us now see what this principle asserts in terms of spacetime geometry. We shall assume that the metric tensor field $g_{\mu\nu}(x)$ with its associated metric connection (2.49) is the only geometrical structure possessed by the spacetime manifold. The square matrix formed by its components is symmetric and I shall call it $g(x)$. On transforming to a new coordinate system, the new matrix is

$$g' = \Lambda^{\mathrm{T}} g \Lambda \tag{4.1}$$

where Λ is the transformation matrix whose components were defined in (2.14) as $\Lambda^{\mu}{}_{\mu'} = \partial x^{\mu}/\partial x^{\mu'}$, and Λ^{T} is its transpose. Any symmetric matrix can be diagonalized by a transformation of this kind. Let us therefore consider a given point P and a coordinate system in which g is diagonal at P. Assuming that none of the eigenvalues of g is zero (if one of them does vanish, then P is some sort of singular point at which odd things may happen), it will clearly be possible to adjust the scales of the coordinates so that each eigenvalue is either $+1$ or -1. If the equivalence principle is to hold in the neighbourhood of P, then the resulting $g(\mathrm{P})$ must be a 4×4 matrix with one eigenvalue equal to $+1$ and the other three equal to -1. Then, after renumbering the coordinates if necessary, it has the desired Minkowski space form (2.8): $g_{\mu\nu}(\mathrm{P}) = \eta_{\mu\nu}$.

Although P can be any point, it will not in general be possible to find a coordinate system in which $g_{\mu\nu} = \eta_{\mu\nu}$ at every point. If such a coordinate system does exist then, in that system, g is constant, which means that the connection coefficients (2.49) are zero and so therefore is the Ricci scalar curvature R, defined in (2.50). Being a scalar, R is therefore zero in any coordinate system, but we shall shortly see explicit examples of metrics which have non-zero curvature. (Readers may like to consider in detail why there is enough freedom in coordinate transformations to diagonalize the metric at any single point but not at every point simultaneously.) However, it is always possible to find a system in which both $g_{\mu\nu} = \eta_{\mu\nu}$ and all the first derivatives $\partial_{\sigma} g_{\mu\nu}$ vanish at P (see exercise 4.1). A coordinate system of this kind may be called a *locally inertial system* at P. An observer at P who is at rest in such a system will experience the coordinate direction with the positive eigenvalue as time and the other three as spatial. According to the principle of equivalence, if the laws of physics are expressed in terms of locally inertial coordinates, they will reduce at P to the form they take in Minkowski space in terms of Cartesian coordinates, and they will contain no reference to gravitational forces. This, as we are about to discover, is because gravitational forces are given by connection coefficients (2.49) which vanish at P when expressed in locally inertial coordinates.

4.2 Gravitational Forces

Suppose for now that the metric tensor field is fixed and that it does not reduce to that of Minkowski space in any coordinate system (except locally, as discussed above). Normally, this means that spacetime is curved, and we wish to know what effect the curvature has on the laws of motion of particles. From the point of view of chapter 3, this involves finding an action appropriate to the curved spacetime. The two guiding principles here are the principle of equivalence, which we have just been discussing, and the *principle of general covariance*. In Minkowski space, we concluded that equations of motion should be covariant under Poincaré transformations because these left the metric unchanged. In curved spacetime, there are in general no coordinate transformations which leave the metric unchanged. On the other hand, any coordinate system is merely a theoretical device which enables us to label points of spacetime. The only reason for preferring a particular system would be if it enabled a specific metric tensor field to be described in an especially simple way, as is the case with Cartesian coordinates in Minkowski space. If we do not commit ourselves to a specific metric, then any coordinate system should be as good as any other and, in particular, equations of motion should preserve their form under any coordinate transformation. This is the meaning of general covariance.

Clearly, equations of motion will be generally covariant if they are derived from an action which is *invariant* under all transformations, namely a scalar. Scalars can be formed by contracting all the indices of any tensor with the same covariant and contravariant rank. If we allow any derivatives of the metric tensor field to appear in the Lagrangian, then a great many functions would be possible—for example, any function of the Ricci scalar R. In order to satisfy the principle of equivalence, however, we would like the Lagrangian to reduce to its Minkowski space form in a locally inertial frame, and our previous discussion shows that we must work only with $g_{\mu\nu}$ and its first derivatives. But to form tensors and ultimately scalars, we must use covariant derivatives rather than partial ones, and the first covariant derivative of the metric tensor field is, by definition, equal to zero (equation (2.47)). Thus, for a single particle, the Lagrangian must be a scalar formed from the vector $\dot{x}^\mu = \mathrm{d}x^\mu/\mathrm{d}\tau$ and the metric tensor field itself. Because of (2.43), contracting the indices of two g's gives a trivial result, and we see that the Lagrangian can only be a function of the scalar quantity $X = g_{\mu\nu}\dot{x}^\mu\dot{x}^\nu$. As in Minkowski space, we find that the detailed form of this function is immaterial, and we need only replace $\eta_{\mu\nu}$ in (3.32) by $g_{\mu\nu}$:

$$S = \frac{1}{2} m \int \mathrm{d}\tau g_{\mu\nu}(x(\tau)) \frac{\mathrm{d}x^\mu}{\mathrm{d}\tau} \frac{\mathrm{d}x^\nu}{\mathrm{d}\tau}. \tag{4.2}$$

The equation of motion for a free particle moving in the curved spacetime is the Euler–Lagrange equation obtained by varying (4.2) with respect to the path $x^\mu(\tau)$, namely

$$\frac{\mathrm{d}}{\mathrm{d}\tau}\left(g_{\mu\nu}\frac{\mathrm{d}x^\nu}{\mathrm{d}\tau}\right)-\frac{1}{2}g_{\sigma\nu,\mu}\frac{\mathrm{d}x^\sigma}{\mathrm{d}\tau}\frac{\mathrm{d}x^\nu}{\mathrm{d}\tau}=0. \tag{4.3}$$

As in chapter 2, the comma before the index μ is a shorthand for ∂_μ. After carrying out the differentiation and raising the non-contracted index, this may be written as

$$\frac{\mathrm{d}^2x^\mu}{\mathrm{d}\tau^2}+\Gamma^\mu{}_{\nu\sigma}\frac{\mathrm{d}x^\nu}{\mathrm{d}\tau}\frac{\mathrm{d}x^\sigma}{\mathrm{d}\tau}=0 \tag{4.4}$$

which is the equation of a geodesic curve, introduced in chapter 2 as the curved space analogue of a straight line. The affine connection coefficients are given by (2.49).

If our qualitative discussions of the relativistic theory of gravity are to stand up, it must now be possible to find a set of circumstances under which (4.4) can be reinterpreted as the equation of a particle moving through Minkowski or Galilean spacetime under the influence of a gravitational field. I will now show what these circumstances are. An obvious requirement is that the metric should be only slightly different from the $\eta_{\mu\nu}$ of Minkowski spacetime, so let us write it as

$$g_{\mu\nu}=\eta_{\mu\nu}+h_{\mu\nu} \tag{4.5}$$

where $h_{\mu\nu}$ is a small correction. If we keep only terms of first order in $h_{\mu\nu}$, then the connection coefficients are

$$\Gamma^\mu{}_{\nu\sigma}=\tfrac{1}{2}\eta^{\mu\lambda}(h_{\lambda\nu,\sigma}+h_{\lambda\sigma,\nu}-h_{\nu\sigma,\lambda})+\mathrm{O}(h^2). \tag{4.6}$$

The second requirement is that the particle should be moving, relative to our chosen coordinate system, very slowly compared with the velocity of light. This is normally true in those practical situations which appear to support the Newtonian account of gravity: for example, the orbital velocity of the Earth around the Sun is about $10^{-4}c$. The element of proper time along the particle's path is given, according to (4.5), by $c^2\mathrm{d}\tau^2=(\eta_{\mu\nu}+h_{\mu\nu})\mathrm{d}x^\mu\mathrm{d}x^\nu$ and since, for a slowly moving particle, $\mathrm{d}\boldsymbol{x}/\mathrm{d}\tau$ is negligible compared with $\mathrm{d}t/\mathrm{d}\tau$, we have approximately

$$\frac{\mathrm{d}t}{\mathrm{d}\tau}\simeq 1-\frac{1}{2c^2}h_{00}. \tag{4.7}$$

By the same token, the spatial components of (4.4) can be written (using the convention that Latin indices $i, j, k, \ldots$ denote spatial directions) as

$$\frac{\mathrm{d}^2x^i}{\mathrm{d}\tau^2}+\Gamma^i{}_{00}c^2\left(\frac{\mathrm{d}t}{\mathrm{d}\tau}\right)^2\simeq 0. \tag{4.8}$$

The final requirement is that variation of the metric tensor field and hence, as we shall see immediately, of the gravitational field with time is negligible. This has two consequences. First, $\mathrm{d}t/\mathrm{d}\tau$ in (4.7) is approximately a constant, so differentiation with respect to τ in (4.8) can be replaced by differentiation with respect to t. Second, terms in the connection coefficients (4.6) which involve time derivatives can be neglected. In particular, the coefficient which appears in (4.8) is just

$$\Gamma^{i}{}_{00} \simeq \tfrac{1}{2}h_{00,i}. \tag{4.9}$$

So, on multiplying (4.8) by the mass of the particle, we get

$$m\,\frac{\mathrm{d}^2x^i}{\mathrm{d}t^2} \simeq -m\,\frac{\partial}{\partial x^i}\,V \tag{4.10}$$

where V is the gravitational potential of the Newtonian theory, now to be identified as

$$V = \tfrac{1}{2}c^2h_{00}. \tag{4.11}$$

At this point, then, our mathematical account of spacetime geometry begins to make contact with actual observations. If the above requirements are met, we say that the *Newtonian limit* applies. In this limit, we can pretend that Minkowski or Galilean geometry is correct. The small error we incur by doing this is detectable by virtue of the gravitational force on the right-hand side of (4.10), which is related to the true metric through (4.11). Of course, we are not really entitled yet to identify the V in these equations as a *gravitational* potential, rather than a potential of some other kind. We have, certainly, obtained one of the hallmarks of gravity, namely that the force in (4.10) is proportional to the inertial mass of the test particle. The other half of the story is that V should be of the correct form. For example, in the neighbourhood of the Earth, V should be approximately equal to $-GM/r$, where G is Newton's constant, M the mass of the Earth and r the distance from its centre. In the next section, we shall see how this comes about.

4.3 The Field Equations of General Relativity

We have come some way towards answering the question 'what is the structure of our spacetime?'. On empirical grounds, we have seen that it cannot be too far removed from that of Minkowski spacetime. Moreover, we have seen how small deviations from the Minkowski metric can be interpreted in terms of a force field which we would like to identify with gravity. Our basic assumption will now be that the metric tensor field is a physical object whose behaviour is governed, like other physical objects, by an action principle. Although gravity is properly viewed as an 'apparent' force, which disappears when we adopt a truly inertial frame of reference, it is helpful to some extent to think of gravity by analogy with electromagnetism. Thus, the action (4.2), with the metric tensor field decomposed as in (4.5), may be thought of as analogous to the first two terms of (3.55). These lead to the equation of motion (3.56) or (3.58) (analogous to (4.4) or (4.10)) of a charged particle in the presence of given electric and magnetic fields. To find out what electric and magnetic fields are actually present, we have to solve Maxwell's equations (3.52), which relate derivatives of the fields on the left-hand side to the density and currents on the right-hand side. To derive Maxwell's equations, we require the final term in (3.55), which depends on the electromagnetic fields alone.

To find out what the metric tensor field is, for a region of space containing a given distribution of matter, we must solve the gravitational analogues of Maxwell's equations. These are *Einstein's field equations*. The currents on the right-hand side will turn out to be the stress tensor given in (3.42). The left-hand side, analogous to $\partial_\mu F^{\mu\nu}$ in (3.52), is the *Einstein curvature tensor*, which is constructed from the metric tensor field in a manner we have yet to discover. To do this, we must evidently add to the action a term analogous to the last term of (3.55). It must be a scalar quantity, containing just the metric tensor field and its derivatives.

There is one mathematical detail to be sorted out first, namely we need to know how to integrate over spacetime in a covariant manner. Suppose, to take the simplest case, that we have a coordinate system in which the metric tensor field at the point x is diagonal with elements g_{00}, g_{11}, g_{22} and g_{33}. An infinitesimal time interval is $\mathrm{d}t = c^{-1}(g_{00})^{1/2}\mathrm{d}x^0$ and infinitesimal distances are $\mathrm{d}x = (-g_{11})^{1/2}\mathrm{d}x^1$, etc. Therefore, the infinitesimal spacetime volume element is

$$\mathrm{d(spacetime\ volume)} = c^{-1}\mathrm{d}^4x(-g(x))^{1/2} \tag{4.12}$$

where $g(x)$ denotes the determinant of the metric tensor field. On transforming to a new coordinate system, d^4x is multiplied by a Jacobian factor, which is the determinant of the transformation matrix (2.13). Readers should have no difficulty in verifying that this is exactly cancelled by the determinant of the inverse matrix which transforms $g(x)$ according to (4.1). Thus, the volume element (4.12) is a scalar, retaining the same form in all coordinate systems. Correspondingly, we may define a scalar δ function

$$\frac{1}{(-g(x))^{1/2}}\,\delta^4(x-y) \tag{4.13}$$

which has the desired properties when used in conjunction with the scalar volume element (4.12).

Beyond the requirement that the geometrical contribution to the action should be a scalar, there seems to be no *a priori* way of knowing what form it should take. Arguably, the form which has been found to work is the simplest possible one, but simplicity is a somewhat subjective and ill-defined criterion. At any rate, the standard version of general relativity is obtained by taking the total action to be

$$S = \tfrac{1}{2}\int \mathrm{d}^4x\left\{g_{\mu\nu}(x)\sum_n m_n \int \mathrm{d}\tau_n \delta^4(x - x_n(\tau_n))\dot{x}_n^\mu(\tau_n)\dot{x}_n^\nu(\tau_n) + \frac{1}{c\kappa}(-g)^{1/2}[-2\Lambda + R]\right\}. \tag{4.14}$$

The first part is equivalent to a sum of terms like (4.2), representing a collection of particles. Notice that the $(-g)^{1/2}$ factors cancel between the volume element and the δ function. In the second part, R is the Ricci curvature scalar (2.50) and Λ is a constant, called the *cosmological constant*. The overall constant κ determines the strength of coupling between geometry and matter, and consequently the strength of gravitational forces. It must obviously be related to Newton's constant, and we shall shortly derive the exact relationship.

By requiring (4.14) to be stationary against variations in each of the particle trajectories, we obtain an equation of motion of the form (4.4) for each particle. The field equations are obtained by requiring it to be stationary against variations in $g_{\mu\nu}(x)$. In principle, this is no more difficult than obtaining Maxwell's equations from (3.54), but the algebra is considerably more involved. Exercise 4.2 offers guidelines for carrying the calculation through, but here I shall just quote the result: *Einstein's field equations* are

$$R^{\mu\nu} - (\tfrac{1}{2}R - \Lambda)g^{\mu\nu} = \kappa T^{\mu\nu}. \tag{4.15}$$

The terms involving $R^{\mu\nu}$ (defined by (2.36) after raising the indices) and R constitute the Einstein curvature tensor. The cosmological constant Λ is, according to the best astronomical evidence, very close to zero in our universe and may generally be omitted. At the time of writing, there is no understanding of why Λ should be close or equal to zero, and indeed this question is widely regarded as one of the most important mysteries remaining in modern cosmology. The stress tensor on the right-hand side is

$$T^{\mu\nu}(x) = \frac{c}{(-g(x))^{1/2}}\sum_n \int \mathrm{d}\tau_n m_n \frac{\mathrm{d}x_n^\mu}{\mathrm{d}\tau_n}\frac{\mathrm{d}x_n^\nu}{\mathrm{d}\tau_n}\delta^4(x - x_n(\tau_n)). \tag{4.16}$$

It differs from the Minkowski space tensor (3.42) only in so far as the invariant δ function (4.13) has been used.

If the relativistic theory of gravity is to work, it must now be possible to show that the potential $V(x)$ defined in (4.11) reduces to the

Newtonian potential in the appropriate limit. The Newtonian potential of a point mass M at a distance r from it is $V(r) = -GM/r$. Equivalently (as is shown in any textbook on electricity for the analogous Coulomb potential), for a static mass distribution of density $\rho(\boldsymbol{x})$, the potential satisfies Poisson's equation

$$\nabla^2 V = 4\pi G\rho. \tag{4.17}$$

We shall show that this equation follows, in the Newtonian limit, from the (0, 0) component of (4.15). To that end, it is convenient to rewrite (4.15) in the following way. First, define the scalar quantity T by $T = g_{\mu\nu}T^{\mu\nu}$. By contracting (4.15) with $g_{\mu\nu}$, we find that $R = 4\Lambda - \kappa T$ and on substituting this back into (4.15) we get the alternative version

$$R^{\mu\nu} = \kappa(T^{\mu\nu} - \tfrac{1}{2}Tg^{\mu\nu}) + \Lambda g^{\mu\nu}. \tag{4.18}$$

Now assume that a coordinate system can be found in which the matter giving rise to the gravitational potential is at rest and in which the metric tensor field is close to that of Minkowski space, as in (4.5). To the order of accuracy we require, the right-hand side of (4.18) can be evaluated with $h_{\mu\nu} = 0$. For particles at rest, we have $\mathrm{d}x^\mu/\mathrm{d}\tau = (c, 0, 0, 0)$, and this can be used in (3.42) to find the stress tensor. The density is expressed by the $\mu = 0$ component of (3.41) when A is taken to be the mass of a particle, and we find that all components of the stress tensor are zero except for $T^{00} = \rho c^2$, so that $T = \rho c^2$ also. In the Newtonian limit discussed in the last section, the (0, 0) component of the Ricci tensor field is given approximately by

$$R^{00} \simeq -\tfrac{1}{2}\sum_{i=1}^{3} \partial_i\partial_i h_{00}. \tag{4.19}$$

With h_{00} identified as in (4.11), the (0, 0) component of (4.18) now reads

$$\nabla^2 V = (\tfrac{1}{2}\kappa\rho c^2 + \Lambda)c^2. \tag{4.20}$$

This is identical with Poisson's equation (4.17) provided that the cosmological constant is negligibly small and that we identify the constant κ as

$$\kappa = 8\pi G/c^4. \tag{4.21}$$

Equations (4.4) and (4.15) constitute the general relativistic theory of gravity. So long as we have values for the two constants κ and Λ, these equations may in principle be applied to any specific physical situation, their solutions yielding predictions which can be tested against actual observations. The value of κ is determined experimentally by (4.21), but the cosmological constant is, as mentioned above, rather more puzzling. In Einstein's original formulation of the theory, it was zero—which is to say that it did not appear at all. For most purposes, it is assumed to be zero, and this leads to a number of well verified predictions, some of which are discussed in the following section and in chapter 13.

The extent of our knowledge of the actual value of Λ is that it cannot be large enough to invalidate these predictions. We see from (4.20) that the quantity $\Lambda/\kappa c^2 = \Lambda c^2/8\pi G$ is in some respects analogous to a 'mass density of the vacuum', to be considered in addition to the density of real matter, although this does not completely describe its role in the full field equations (4.15). A rough and ready method of placing upper bounds on the value of this quantity is to argue that it must be significantly less than the average density of a system which is well described by the theory with $\Lambda = 0$. For example, the solar system is described by this theory to within the accuracy of observations and of the approximations needed to obtain numerical theoretical predictions. A suitable 'density' might be the mass of the Sun divided by the volume of a sphere which just encloses the orbit of Pluto, which gives about $3 \times 10^{-12}\,\mathrm{g\,cm^{-3}}$, and the agreement of theory with experiment would be upset if the vacuum density were comparable with this. Applying the same argument to much larger systems such as clusters of galaxies (which are, of course, much less precisely understood than the solar system), we obtain a limit of the kind

$$\Lambda c^2/8\pi G \lesssim 10^{-29}\,\mathrm{g\,cm^{-3}}. \tag{4.22}$$

This is roughly the average density of observable matter in the universe and is, of course, vastly smaller than the densities of familiar materials. Whether it is small in an absolute sense depends on our finding some fundamental quantity with the dimensions of a density with which to compare it. We shall see later that such a comparison can be made, which suggests that the smallness of Λ is even more striking than the number quoted in (4.22).

4.4 The Gravitational Field of a Spherical Body

To find out how the general relativistic theory of gravity differs from the Newtonian one, we must, of course, find exact solutions to (4.4) and (4.15), or at least approximate solutions which go beyond the Newtonian approximation. I shall illustrate the nature of general relativistic effects by considering Schwarzschild's solution of the field equations for the metric tensor field associated with a massive spherical body and some of its elementary consequences.

4.4.1 The Schwarzschild solution

The task of finding a general solution to the field equations is too difficult to contemplate, and it is usually possible to find particular solutions only when symmetry or other requirements can be used to reduce the 10 independent components of the metric tensor field to a more manageable number. The solution found by Schwarzschild (1916) rests on several simplifying assumptions. The solution is, however, an exact one: no approximations are required to make the assumptions valid. First, we ask for the gravitational field of a spherically symmetric body and assume that the metric will also be spherically symmetric. Second, since we anticipate that gravitational effects will be extremely weak at large distances from the body, the metric should approach that of Minkowski spacetime at large distances. We therefore use polar coordinates (t, r, θ, ϕ) and expect that for large r the line element will be approximately

$$c^2\,\mathrm{d}\tau^2 \simeq c^2\,\mathrm{d}t^2 - \mathrm{d}r^2 - r^2(\mathrm{d}\theta^2 + \sin^2\theta\,\mathrm{d}\phi^2). \qquad (4.23)$$

It must be borne in mind that these coordinates cannot necessarily be interpreted as time, radial distance and angles in the elementary sense, although these interpretations should become valid in the large r region where (4.23) is valid.

The final assumption, is that, in these coordinates, the components of the metric tensor field are independent of the coordinate t. This implies, in particular, that an observer in the large r region will see a static gravitational field. As a matter of fact, the only assumption which is really needed is that of spherical symmetry. There is a theorem due to G D Birkhoff, explained, for example, by Weinberg (1972), which shows that the only spherically symmetric solution for the metric of a space which is empty apart from a central spherical body is the time-independent Schwarzschild solution. Here, to make matters simpler, I shall assume that the metric is static. With these assumptions, the line element can be written as

$$c^2\,\mathrm{d}\tau^2 = A(r)c^2\,\mathrm{d}t^2 - B(r)\,\mathrm{d}r^2 - r^2(\mathrm{d}\theta^2 + \sin^2\theta\,\mathrm{d}\phi^2). \qquad (4.24)$$

The two functions $A(r)$ and $B(r)$, which should approach the value 1 for large r, remain to be determined. A third unknown function $C(r)$ could have been included in the coefficient of the angular term. However, we could then define a new radial coordinate by $r'^2 = C(r)r^2$, and so recover the form (4.24) with A and B appropriately redefined.

We shall consider only the *exterior* solution, namely the metric as it exists outside the central massive body. In this region there is no matter, so, taking the cosmological constant to be zero, we have to solve (4.15) in the special case that $\Lambda = T^{\mu\nu} = 0$. This is actually a set of 10 equations for the 10 independent components of the metric tensor field. Provided, as is in fact the case, that our assumptions are consistent with the structure of the field equations, it will be possible to find functions $A(r)$ and $B(r)$ such that all 10 equations are satisfied. The task of finding these functions and verifying that all the field equations are satisfied is straightforward but quite lengthy, although the result is a simple one. I shall outline the steps and leave it to sufficiently energetic readers to fill in the details. The components $g_{\mu\nu}$ can be read off from (4.24). We must use them to calculate the connection coefficients (2.49) and thence the Ricci tensor (2.36) and the scalar curvature (2.50). A useful short cut to finding the connection coefficients is to write out the action (4.2) explicitly:

$$S = \tfrac{1}{2}m \int \mathrm{d}\tau [c^2 A(r)\dot{t}^2 - B(r)\dot{r}^2 - r^2(\dot{\theta}^2 + \sin^2\theta\,\dot{\phi}^2)]. \quad (4.25)$$

By varying each of the coordinates, it is easy to find the Euler–Lagrange equations, from which the Γ can be picked out by comparison with (4.4).

There is now nothing for it but to work out the components of $R^{\mu\nu}$ and equate them to zero (by contracting the equation $R^{\mu\nu} - Rg^{\mu\nu}/2 = 0$ with $g_{\mu\nu}$, we find that $R^{\mu\nu}$ and R are separately equal to zero). As it turns out, all the off-diagonal elements vanish identically. The remaining four equations are differential equations for $A(r)$ and $B(r)$ which have the solution $A(r) = 1/B(r) = 1 + \alpha/r$, where α is a constant of integration. To identify the constant, we note that h_{00} in (4.11) is just α/r. For large r, this is indeed small and must equal $2/c^2$ times the Newtonian potential $-GM/r$, where M is the mass of the central body. The Schwarzschild line element is therefore

$$c^2\,\mathrm{d}\tau^2 = \left(1 - \frac{2GM}{c^2 r}\right)c^2\,\mathrm{d}t^2 - \left(1 - \frac{2GM}{c^2 r}\right)^{-1}\mathrm{d}r^2 - r^2(\mathrm{d}\theta^2 + \sin^2\theta\,\mathrm{d}\phi^2). \quad (4.26)$$

It has an obvious peculiarity at the *Schwarzschild radius*

$$r_{\mathrm{S}} = 2GM/c^2 \quad (4.27)$$

which has, for example, values of 0.886 cm for the Earth, 2.95 km for the Sun and 2.48×10^{-52} cm for a proton. As we shall see, this singularity is associated with the possibility of 'black holes'. Remember, however, that (4.26) is the exterior solution for the metric, valid outside the massive body. It does not follow that there is a black hole of radius 0.886 cm lurking at the centre of the Earth! Before discussing this in more detail, we shall take a look at some more prosaic features of the Schwarzschild solution.

4.4.2 Time near a massive body

A normal body, such as the Earth or the Sun, is larger than the Schwarzschild radius calculated from its mass. Let us consider a stationary observer near such a body to be one whose (r, θ, ϕ) coordinates are fixed, for example somebody standing on the surface of the Earth. For such an observer, the flow of proper time is measured by

$$d\tau = \left(1 - \frac{r_S}{r}\right)^{1/2} dt. \tag{4.28}$$

The time experienced by a stationary observer is thus proportional to the coordinate t, but with a factor which changes with r. Two events occurring at the same value of t will appear simultaneous to any stationary observer, and therefore the spacetime can be separated in a meaningful way into three-dimensional spatial slices, each labelled by its own value of t. All stationary observers agree on this splitting, but the time which elapses between two given values of t is different for observers at different radial positions.

The variation of time intervals with radial position can be investigated by the shift it causes in atomic spectral lines. Consider a radiating atom located at r_{at} and an observer at r_{obs}. Suppose a pulse of light is emitted at coordinate time t_e and received at t_r, and a second pulse is emitted at $t_e + \Delta t_e$, being received at $t_r + \Delta t_r$ (see figure 4.1). Since the metric is independent of t, the paths of the two pulses are exactly similar, and therefore the coordinate time interval $t_r - t_e$ between emission and reception of the first pulse is equal to the corresponding interval $(t_r + \Delta t_r) - (t_e + \Delta t_e)$ for the second. It follows that the coordinate time interval Δt_e between the moments at which the two pulses are emitted is equal to the interval Δt_r between the moments at which they are received: $\Delta t_e = \Delta t_r$. The corresponding proper time intervals are therefore different, and the ratio of the observed frequency of the received wave to the frequency of the wave as emitted by the atom follows trivially from (4.28):

$$\frac{\text{frequency at reception}}{\text{frequency at emission}} = \frac{(\Delta\tau_{obs})^{-1}}{(\Delta\tau_{at})^{-1}} = \left(\frac{1 - r_S/r_{at}}{1 - r_S/r_{obs}}\right)^{1/2}. \tag{4.29}$$

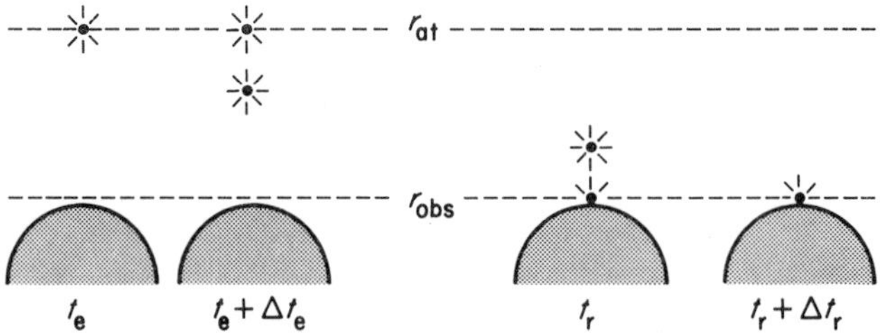

Figure 4.1 Passage of two pulses of light from a radiating atom to an observer in the gravitational field of a spherical body.

This ratio involves only the (0, 0) component of the metric tensor, which we have identified in terms of the gravitational potential. In general, for a static spacetime (i.e. one which can be divided into identical spatial slices), we have

$$\frac{\text{observed frequency}}{\text{frequency at emission}} = \left(\frac{1 + 2V_{\text{at}}/c^2}{1 + 2V_{\text{obs}}/c^2}\right)^{1/2}. \tag{4.30}$$

In a weak gravitational field, the frequency shift $\Delta\nu = \nu_{\text{obs}} - \nu_{\text{at}}$ is given approximately by

$$\frac{\Delta\nu}{\nu} = \frac{V_{\text{at}} - V_{\text{obs}}}{c^2}. \tag{4.31}$$

Although this shift can have either sign, what can normally be observed in practice is light from the atmospheres of stars. The radiating atom is therefore at a lower gravitational potential than an earthbound telescope, so a *gravitational redshift* is observed. Such observations confirm the prediction (4.31) to precisions of a few per cent. A method of measuring frequency shifts in the Earth's gravitational field was devised by Pound and Rebke (1960), who used the Mössbauer effect to determine the change in frequency of γ rays from ^{57}Fe nuclei on travelling a vertical distance of some 22 m. It should be clear that these observations test only the principle of equivalence, with the gravitational potential identified as in (4.11). Provided that the Newtonian potential is used in (4.31), the derivation of the frequency shift does not require the field equations or Schwarzschild's solution of them.

4.4.3 Distances near a massive body

Within an equal-time slice of the Schwarzschild spacetime, distances are measured by the spatial part of the line element:

$$\mathrm{d}l^2 = \left(1 - \frac{2GM}{c^2 r}\right)^{-1} \mathrm{d}r^2 + r^2(\mathrm{d}\theta^2 + \sin^2\theta\,\mathrm{d}\phi^2). \tag{4.32}$$

This is a non-Euclidean space, and the departure from Euclidean geometry may be illustrated by the fact that the circumference of a circle is not equal to 2π times its radius. Consider a circle concentric with the central body in the equatorial plane $\theta = \pi/2$ at a fixed radial coordinate r. Its circumference is

$$\text{circumference} = \int_0^{2\pi} \frac{\mathrm{d}l}{\mathrm{d}\phi}\,\mathrm{d}\phi = 2\pi r. \tag{4.33}$$

Its radius cannot be determined exactly, because (4.32) is valid only outside the central body. We can, however, compare two circles of radii r_1 and r_2. In Euclidean geometry, the difference between their circumferences in 2π times the difference between their radii. In the Schwarzschild space, the difference in circumference is $2\pi(r_2 - r_1)$, but the radial distance between them is

$$\text{radial distance} = \int_{r_1}^{r_2} \frac{\mathrm{d}l}{\mathrm{d}r}\,\mathrm{d}r = \int_{r_1}^{r_2} \frac{\mathrm{d}r}{(1 - r_S/r)^{1/2}} = r_2 f(r_2) - r_1 f(r_1) \quad (4.34)$$

where the function $f(r)$ is

$$f(r) = \left(1 - \frac{r_S}{r}\right)^{1/2} + \left(\frac{r_S}{r}\right) \ln\left\{\left(\frac{r}{r_S}\right)^{1/2}\left[1 + \left(1 - \frac{r_S}{r}\right)^{1/2}\right]\right\}. \quad (4.35)$$

When r is much greater than r_S, this may be approximated as

$$f(r) \simeq 1 + \left(\frac{r_S}{r}\right) \ln\left[2\left(\frac{r}{r_S}\right)^{1/2}\right] \quad (4.36)$$

and for two circles satisfying this condition, we find

$$\frac{\text{difference in circumference}}{\text{radial distance}} \simeq 2\pi\left[1 - \frac{1}{2}\left(\frac{r_S}{r_2 - r_1}\right) \ln\left(\frac{r_2}{r_1}\right)\right] \quad (4.37)$$

provided that $r_2 - r_1$ is also larger than r_S. As an example, if r_S is the Schwarzschild radius of the Sun, r_1 is the radius of the Sun (6.96×10^8 m) and r_2 is the semi-latus rectum of the orbit of Mercury (5.5×10^{10} m), then the correction term is about 10^{-7}. For many purposes, therefore, the solar system can adequately be described in terms of Euclidean geometry.

4.4.4 Particle trajectories near a massive body

The analogy drawn above between the field equations and Maxwell's equations may be misleading in one important respect: the field strength tensor (3.51) is linear in the electromagnetic fields, while the curvature tensors are non-linear in the metric tensor field. Suppose, for example, that we wish to calculate, according to classical mechanics, the orbit of an electron near a positive nucleus, which we take to remain stationary. The linearity of the field strength tensor allows us to express the total electric field as the sum of fields due to the nucleus and the electron. The field due to the electron exerts no force on the electron itself. It can be subtracted from the total field, and we simply regard the electron as moving in the field of the nucleus. In general, this cannot be done with gravity. Given, say, a star and a single planet, the true metric cannot be expressed as the sum of two Schwarzschild metrics. If we wish to find the metric and the relative motion of the two bodies, it is necessary to solve the whole problem in one go: since we do not know the metric, we cannot immediately find the orbits and, not knowing these, we cannot write down any explicit form for the stress tensor which appears in the field equations we must solve for the metric. In fact, the exact solution of this two-body problem is not known.

What we can do without too much trouble is to work out the trajectories of 'test particles' in the Schwarzschild spacetime—or at least we can write down their equations of motion and solve these by numerical means if necessary. A test particle is one whose effect on the metric is negligible, and its equations of motion are the geodesic equations (4.4) with the connection coefficients calculated in this case from the Schwarzschild metric. I shall write out explicitly only the form of these equations which applies to motion in the equatorial plane: this can, of course, be any plane passing through the centre of the massive body if we choose our coordinates appropriately. With θ fixed at $\pi/2$, the equations are

$$\frac{\mathrm{d}}{\mathrm{d}\tau}\left[\left(1 - \frac{r_\mathrm{S}}{r}\right)\dot{t}\right] = 0 \tag{4.38}$$

$$\frac{\mathrm{d}}{\mathrm{d}\tau}(r^2\dot{\phi}) = 0 \tag{4.39}$$

$$\left(1 - \frac{r_\mathrm{S}}{r}\right)^{-1}\ddot{r} + \frac{1}{2}c^2\left(\frac{r_\mathrm{S}}{r^2}\right)\dot{t}^2 - \frac{1}{2}\left(1 - \frac{r_\mathrm{S}}{r}\right)^{-2}\left(\frac{r_\mathrm{S}}{r^2}\right)\dot{r}^2 - r\dot{\phi}^2 = 0. \tag{4.40}$$

As in previous equations, the overdot denotes $\mathrm{d}/\mathrm{d}\tau$.

The derivation of the equations of motion (4.4) was valid for massive particles. For photons, or other massless particles, the action (4.2) vanishes. To deal with this case, we simply define a new parameter λ such that $\mathrm{d}\tau = m\,\mathrm{d}\lambda$. The mass then disappears from the action and can be set to zero. The equations of motion (4.4) then follow as before, but with τ replaced by λ. The trajectories for massless particles are still geodesics, but are not parametrized by proper time. Clearly, indeed, they are *null geodesics*, along which $\mathrm{d}\tau = 0$.

These equations lead to a number of interesting predictions when applied to the solar system. Light passing close to the Su is predicted to be deflected by 1.75 seconds of arc, and the expeditions of Dyson, Eddington and Davidson to observe this effect during a total eclipse in 1919 resulted in one of the earliest confirmations of Einstein's theory. (Their measurements were actually not precise enough to justify the confirmation which was claimed at the time, but later, more accurate, observations do confirm the theoretical result.) When the planets are treated as test particles, it is found that their orbits are not elliptical as in the simple Newtonian theory, but can be described as ellipses whose perihelia (distances of closest approach to the Sun) precess slowly. The largest precession rate, that for Mercury, is predicted to be some 43 seconds of arc per century, in good agreement with observations. Planetary orbits have, of course, been studied for centuries and are known with extreme precision. Even within Newtonian theory, the approximation of treating the planets as test particles is far too crude, and their perturbing influence on each other must be taken into account. These perturbations themselves cause precessions, to which the general relativistic effect is a small correction. In order to apply general relativity to the solar system in a meaningful way, systematic methods of obtaining corrections to the detailed Newtonian theory must be devised. These techniques, known as post-Newtonian approximations, are discussed in specialized textbooks, but are well beyond the scope of this one. Finally, as first worked out by Shapiro, radar signals reflected from a neighbouring planet are slightly delayed by comparison with their round-trip time according to Newtonian theory. The simpler aspects of the theory of these phenomena are explored in the exercises.

4.5 Black and White Holes

So far, we have considered the spacetime near a massive body whose radius is larger than its Schwarzschild radius r_S. In this section, we shall consider the case of an object which is smaller than its Schwarzschild radius. First, let us see whether it is possible to make sense of the metric (4.26) all the way down to $r = 0$. This metric is valid only outside the central body, so physically we will want to know what has happened to the said body. This question will be addressed in due course: for now, let us take it to be an idealized point particle, which nevertheless has a substantial mass M.

To simplify matters, I shall discuss only the paths of free particles moving in the radial direction, which are described by the two functions $r(\tau)$ and $t(\tau)$. Remember that while τ is by definition the time experienced by the particle, the coordinates r and t have no unique interpretation as distances or times. In the region where r is very large, however, they are, to a good approximation, the radial distance and time as experienced by a stationary observer. The paths of radially moving particles can be found by solving (4.38) and the equation

$$\dot{r}^2 = \left(1 - \frac{r_S}{r}\right)^2 c^2 \dot{t}^2 - c^2\left(1 - \frac{r_S}{r}\right) \tag{4.41}$$

which follows from the line element (4.26) with $d\theta = d\phi = 0$. Of course, (4.40) must also be satisfied (with $\dot{\phi} = 0$), but readers may easily check that it follows from (4.41) and (4.38). By differentiating (4.41) and using (4.38), we find the radial equation of motion

$$\ddot{r} = -\frac{1}{2} c^2\left(\frac{r_S}{r^2}\right). \tag{4.42}$$

In view of the definition (4.27) of r_S, this is precisely the equation satisfied by a particle in the Newtonian potential. Two particular solutions are those in which the particle passes through the point r_0, at time $\tau = 0$, with the corresponding escape velocity $(2GM/r_0)^{1/2} =$

$c(r_S/r_0)^{1/2}$, in either the outward or the inward direction. They are

$$r(\tau) = (r_0^{3/2} \pm \tfrac{3}{2}cr_S^{1/2}\tau)^{2/3} \tag{4.43}$$

where the positive sign corresponds to an outgoing particle and the negative sign to an ingoing one. In either case, the particle can apparently pass through the point $r = r_S$ without encountering anything unusual.

Suppose that r_0 is greater than r_S. The solution for $t(\tau)$ is most easily written in terms of $r(\tau)$ and is given by

$$r_S^{1/2}ct = \pm\left[\tfrac{2}{3}r^{3/2} + 2r_Sr^{1/2} + r_S^{3/2}\ln\left(\frac{r^{1/2} - r_S^{1/2}}{r^{1/2} + r_S^{1/2}}\right)\right] \tag{4.44}$$

to which we could add a constant of integration to specify the time at which the particle passes through r_0. We see that, as an ingoing particle approaches r_S, its coordinate $t(\tau)$ approaches ∞, although the time interval it experiences while travelling from r_0 to r_S is finite, being equal to $2(r_0^{3/2} - r_S^{3/2})/3cr_S^{1/2}$. This means that in the neighbourhood of r_S, the coordinate t is no longer useful as a measure of physical time. Correspondingly, the metric given by (4.26) does not give a useful description of the geometry near r_S, because one of its components becomes infinite.

Although we have considered only one special kind of particle trajectory, much the same thing happens for any trajectory passing through r_S. Mathematically, we have to say that the spacetime manifold described by (4.26) does not include the spherical surface at $r = r_S$. Strictly speaking, this metric applies to two separate spacetimes, namely the two regions $r > r_S$ and $r < r_S$. In that case, what becomes of our ingoing particle when it reaches the edge of the first region in which it started? There are two possibilities. One is that the singularity at $r = r_S$ is a genuine singularity of the geometrical structure. If so, the particle would have reached the end of the space and time available to it. We would have to investigate whether it could be reflected, remain trapped on the 'edge of the universe' or simply disappear from the universe altogether. In view of the fact that the radial component of its path (4.43) passes perfectly smoothly past r_S, it seems unlikely that such measures should be necessary. The other possibility is that the singularity is merely a 'coordinate singularity'. That is to say, the particle has not reached the end of spacetime, but merely the end of that region of spacetime for which t serves as a useful coordinate. This second possibility is in fact the correct one. Nevertheless, from a mathematical point of view, we have at hand only the region $r > r_S$. We must add on to it a second region, in which $r < r_S$, which is an extension of the same geometrical structure. This will be possible if we can trade in t for a new coordinate which will describe a smooth join between the two regions.

This means that when we express the line element (4.26) in terms of this new coordinate, all the components of the metric tensor field will be smooth at r_S.

Let us call the region $r > r_S$ region I. This region covers most of the universe, although it is a universe populated only by 'test particles' and therefore cannot describe the whole of our actual universe. Region I has in fact two 'edges' at $r = r_S$ and $t = +\infty$ or $-\infty$. At these two edges we can join on two regions. That which joins on at $t = \infty$, called region II, is the one into which ingoing particles fall; that which joins on at $t = -\infty$, called region II′, is the one from which outgoing particles can emerge. Each of these regions has the same geometrical structure as the region $r < r_S$ of the original Schwarzschild solution: the trick is to find a way of smoothly joining the various regions together. The join between regions I and II can be described in terms of the Eddington–Finkelstein coordinate v, defined by

$$v = ct + r + r_S \ln\left(\frac{r}{r_S} - 1\right). \tag{4.45}$$

If we substitute for t the expression (4.44) with the $-$ sign to represent the path of an ingoing particle, we see that v remains finite as the particle passes through r_S. Moreover, when rewritten in terms of v, the line element becomes

$$c^2\,\mathrm{d}\tau^2 = \left(1 - \frac{r_S}{r}\right)\mathrm{d}v^2 - 2\,\mathrm{d}v\,\mathrm{d}r \tag{4.46}$$

which is perfectly smooth at the boundary between regions I and II. To describe the boundary with region II′, we can use instead the coordinate w, defined by

$$w = ct - r + r_S \ln\left(\frac{r}{r_S} - 1\right) \tag{4.47}$$

in terms of which the line element takes the form (4.46) with $\mathrm{d}v$ replaced by $-\mathrm{d}w$.

The boundary between regions I and II can be crossed only by ingoing particles and, in fact, only by ingoing light rays also. Nothing ever crosses from region II into region I, for which reason region II is called a *black hole*. Conversely, particles and light rays may cross from region II′ to region I, but not in the opposite direction, so region II′ is sometimes called a *white hole*. It turns out that regions II and II′ each have a second boundary, to which can be joined a fourth region I′. This is an exact replica of region I. Particles can pass out of region II′ into either of regions I and I′, or out of I or I′ into region II. However, there is no route by which a particle can pass from region I to region I′ or vice versa. Each of regions II and II′ has a real singularity at $r = 0$, which cannot be removed by any coordinate transformation. The one in region II is discussed below. The collection of four regions is called the *maximal extension* of the Schwarzschild solution. A description of the whole of this spacetime can be given by trading in both t and r for the two coordinates v and w. For further details, the reader is referred to more specialized books, for example Hawking and Ellis (1973).

So far in this section, our discussion has been purely mathematical: we have asked only about the geometrical structure implied by the Schwarzschild solution. We must now consider whether the curious phenomena associated with black and white holes can be brought about by known physical processes. Although the geometry described above represents an entire universe, this universe has to satisfy the symmetry assumptions which went into the Schwarzschild solution in the first place. This is obviously not true of our universe which, for example, contains more than one massive body. The most we hope for in practice is that some fair-sized region in the neighbourhood of, say, a star, is very similar to a corresponding region of the Schwarzschild spacetime.

The structure of a star is supported by its internal pressure and the flow of energy from nuclear reactions in its core. When its nuclear fuel is exhausted, the star collapses and, if it shrinks to a size equal to its own Schwarzschild radius, the conditions exist for the formation of a black hole. It appears, indeed, that once a mass is contained within its Schwarzschild radius, the gravitational attraction between its constituent parts cannot be counteracted by the outward pressure of any known force, and the mass is inevitably compressed to a single point—a singularity at $r = 0$. What becomes of this matter is not clear and readers should bear in mind that our whole discussion ignores any quantum-mechanical considerations which might profoundly affect the fate of the collapsing star.

From the point of view of the collapsing matter, this occurs within a finite time although, as we shall see, the collapse appears to an external observer to take an infinite time. Theorems of Hawking and Penrose (discussed, for example, by Hawking and Ellis (1973)) show that this phenomenon is rather general: for example, it does not depend on the exact spherical symmetry assumed by Schwarzschild. On the other hand, it seems likely that the geometry of these black holes will usually not be of the Schwarzschild type, but rather will correspond to a Kerr solution, in which axial symmetry but not full spherical symmetry is assumed. This allows for the angular momentum possessed by a rotating star. Here, however, I shall consider only black holes of the simpler Schwarzschild type, which illustrate many of the same qualitative features. Notice that prior to the stellar collapse, the exterior Schwarzschild solution we have considered is valid only outside the star, and therefore only for $r > r_S$. There is therefore no boundary at $r = r_S$ and

$t = -\infty$ to which we might attach a region of type II$'$, and the question of forming a white hole does not arise. In fact there is not, to my knowledge, any physical process which is known to give rise to a white hole, and discussions of such objects are largely confined to the more speculative popular literature.

In Minkowski space, the line element (2.6) implies that $|dt/d\tau| > |d\boldsymbol{x}/d\tau|$ along the path of any particle which is not massless, and that $|d\boldsymbol{x}/dt| = c$ for a light ray. As illustrated in figure 4.2, this implies that all the possible light rays passing through a given point P lie on a cone, and the path of a particle passing through P must be contained in this cone. This is expressed by saying that the path is *time-like* or, since the path is directed forwards in time, that it lies in the *forward light cone* of P. This is true both for freely falling particles and for those accelerated by some non-gravitational force. The familiar result of special relativity that no body can be accelerated past the speed of light is of course a direct consequence of this. Since any sufficiently small region of spacetime looks like a small region of Minkowski space, the same is true of particle trajectories in every spacetime of physical interest.

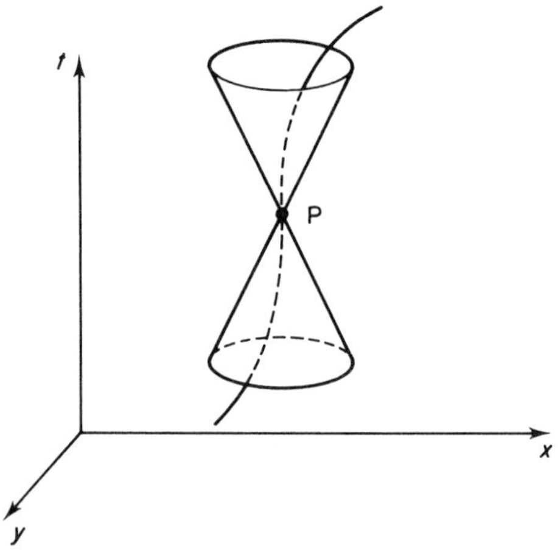

Figure 4.2 The light cone of a spacetime point P and a possible trajectory of a particle through P.

The qualitative effect of black-hole geometry on the paths of particles can be understood by plotting out the paths of light rays and imagining particle trajectories to thread through the cones they produce. Using r and v to describe radial motion, we see from (4.46) that light rays, for which $d\tau = 0$, satisfy

$$v = \text{constant} \tag{4.48}$$

or

$$\frac{dr}{dv} = \frac{1}{2}\left(1 - \frac{r_S}{r}\right). \tag{4.49}$$

Readers may verify that these curves are indeed geodesics. By differentiating (4.45) with v held constant, we find that when r is very large, so that t provides a measure of time, $dr/dt \approx -c$. Thus, the set of curves corresponding to (4.48) represent inward-falling light rays. In figure 4.3, these curves are represented by diagonal lines from bottom right to top left. Vertical lines are lines of constant r. The peculiarities of the geometry arise from the other set of light-ray curves (4.49). One of these is the line $r = r_S$, namely a light ray which remains stationary at the Schwarzschild radius. Outside this radius, rays governed by (4.49) are outgoing; inside, both sets of light rays fall inwards, terminating at the singularity at $r = 0$. Inside the Schwarzschild radius, therefore, all light rays and all particles fall inwards. Events in region II are invisible to an outside observer, and the spherical surface at $r = r_S$ (obtained by reinstating the angular coordinates) is called the *event horizon*.

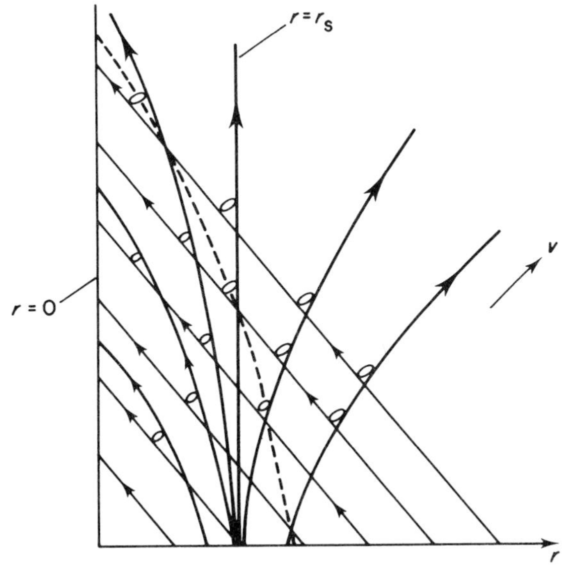

Figure 4.3 Trajectories of light rays (full lines) and an inward-falling particle (broken line) moving radially near a black hole.

The broken line in figure 4.3 represents the path of a particle falling from outside the event horizon. Suppose that it radiates light as it falls, so that a distant observer can follow its progress. It is apparent from the paths of the outgoing light rays that this observer will have to wait an infinite time (measured for him by t) before receiving the signal emitted by the particle as it crosses the horizon. If light energy is radiated at a constant rate as measured by the proper time of the particle, then the finite amount of energy emitted in a short period before the particle reaches the horizon is received over an infinite period of time by the observer. To him, therefore, the signal becomes ever fainter, and disappears entirely as the particle reaches the horizon. Also at this point, the interval between successive crests of a light wave becomes, for the observer, infinitely long, so the light is infinitely redshifted.

Obviously, a black hole is, in itself, difficult to detect. On the other hand, if large amounts of matter are drawn in by the strong gravitational field which surrounds it, this matter may be expected to become very hot, giving rise to intense x- and γ radiation. This may happen, for example, in a binary star system, one of whose stars collapses to a black hole which can then accrete matter from its companion. At the time of writing, there are one or two objects known to astronomers which might contain black holes, but none has been unambiguously identified as such. Nevertheless, it is thought that most stars which are bigger than a few solar masses will eventually collapse, unless they first explode into smaller fragments. In addition, clusters of large stars such as are found at the cores of galaxies appear to stand a good chance of coalescing and eventually collapsing to form very large black holes. In theory, therefore, black holes ought not to be particularly scarce.

Exercises

4.1. In a system of coordinates x^μ, let the coordinates of the point P be x^μ_P. If the connection coefficients are given by (2.49), show that, in a new coordinate system given by

$$x^{\mu'} = \delta^{\mu'}_\mu (x^\mu - x^\mu_\mathrm{P}) + \delta^{\mu'}_\mu \Gamma^\mu_{\nu\sigma}(x_\mathrm{P})(x^\nu - x^\nu_\mathrm{P})(x^\sigma - x^\sigma_\mathrm{P})$$

all first derivatives of the new components of the metric tensor field vanish at P.

4.2. The object of this exercise is to derive the field equations (4.15). Some of the results given in appendix 1 will be needed. The overall strategy is to make a small change in the metric, $g_{\mu\nu} \rightarrow g_{\mu\nu} + \delta g_{\mu\nu}$, and require that the first-order change in the action (4.14) should vanish. The change in the gravitational part is

$$\delta S_\mathrm{grav} = \frac{1}{2c\kappa} \int \mathrm{d}^4x[(\delta(-g)^{1/2})(R - 2\Lambda) + (-g)^{1/2}(\delta g^{\mu\nu} R_{\mu\nu} + g^{\mu\nu}\delta R_{\mu\nu})].$$

(a) Show that, to first order, $\delta g^{\mu\nu} = -\delta g_{\mu\nu}$, where indices are raised using the original metric.

(b) Show that $\delta((-g)^{1/2}) = \frac{1}{2}(-g)^{1/2} g^{\mu\nu}\delta g_{\mu\nu}$.

(c) Show that the difference between two connections, such as $\Gamma(g)$ and $\Gamma(g + \delta g)$, is a tensor field.

(d) Show that

$$g^{\mu\nu}\delta R_{\mu\nu} = g^{\mu\nu}[(\delta\Gamma^{\lambda}{}_{\mu\nu})_{;\lambda} - (\delta\Gamma^{\lambda}{}_{\mu\lambda})_{;\nu}] = [g^{\mu\nu}\delta\Gamma^{\lambda}{}_{\mu\nu} - g^{\mu\lambda}\delta\Gamma^{\nu}{}_{\mu\nu}]_{;\lambda}.$$

Hence show that this term contributes to δS only a surface integral, which does not affect the field equations.

(e) Find the change in the matter action and complete the derivation of the field equations.

4.3. This exercise investigates the bending of light by the Sun, by considering the path of a light ray in the equatorial plane of the Schwarzschild spacetime, with coordinates (r, ϕ). First note that, in Euclidean space, the equation $r = r_0/\sin\phi$ describes a straight line, with $\phi = 0$ or π at $r = \infty$, whose distance of closest approach to the origin is r_0. In the Schwarzschild spacetime, define $u = 1/r$ and let r_0 be the coordinate distance of closest approach.

(a) Show that (4.38) and (4.39) are valid for a null geodesic, if $\mathrm{d}/\mathrm{d}\tau$ is replaced by differentiation with respect to a suitable parameter λ. Use these and (4.26) to derive the equation

$$\left(\frac{\mathrm{d}u}{\mathrm{d}\phi}\right)^2 + u^2(1 - r_S u) = (r_0 - r_S)/r_0^3.$$

(b) Treating r_S/r_0 as a small parameter, show that the solution to this equation for which $u = 0$ when $\phi = 0$ is

$$r_0 u = \sin\phi + (r_S/2r_0)[(1 - \cos\phi)^2 - \sin\phi] + \mathrm{O}((r_S/r_0)^2).$$

(c) To first order in r_S/r_0, show that the angle α, such that $u = 0$ when $\phi = \pi + \alpha$, is $\alpha = 2r_S/r_0$. Taking r_0 to be the solar radius 6.96×10^5 km (why is this allowed?), show that a light ray which just grazes the surface of the Sun is deflected by an angle of 1.75 seconds of arc.

4.4. Suppose that Mercury and the Earth could be frozen in their orbits at coordinate distances r_M and r_E in a direct line from the centre of the Sun. The distance between them can be found from (4.34) with r_S the Schwarzschild radius of the Sun. If space were Euclidean, what would be the round-trip time for a radar signal reflected from the surface of Mercury? What is the coordinate time for the round trip in Schwarzschild spacetime? What is the actual round-trip time as measured by an observer on the Earth? Take r_E and r_M to be much smaller than r_S. Show that, compared with the Euclidean result, the return of the signal to Earth appears to be delayed by a time interval given approximately by

$$\Delta\tau \approx \frac{r_S}{c}\left[\ln\left(\frac{r_E}{r_M}\right) + \left(\frac{r_M}{r_E}\right) - 1\right].$$

Estimate the magnitude of this effect by taking $r_M = 5.5 \times 10^7$ km and $r_E = 1.5 \times 10^8$ km.

4.5. A planet orbits a star whose Schwarzschild radius is r_S along a circular path with radial coordinate r. Verify that this orbit is a geodesic of the Schwarzschild metric. Show that the coordinate time for one revolution is the same as the period of an orbit of radius r in the Newtonian theory. Show that a proper time interval experienced by inhabitants of the planet is $(1 - 3r_S/2r)^{1/2}$ times the corresponding coordinate time interval.

4.6. Suppose that a photon of frequency ν can be considered as having kinetic energy $h\nu$ and the same gravitational potential energy as a particle of mass $h\nu/c^2$. Deduce the expression (4.31) for the frequency shift in a weak gravitational field. Do you think that such an interpretation could be rigorously justified? (Photons are discussed in chapter 5 and subsequent chapters.)

5

Quantum Theory

Much of the remainder of this book will concern itself with those aspects of theoretical physics which seek to understand the nature of matter. Such understanding as we have has mainly been achieved by probing the structure of successively smaller constituents and, at least on the face of things, the regions of space and time we need to consider are far too small for spacetime curvature to be of any significance. Most of our considerations will therefore be restricted to Minkowski or, as in the present chapter, Galilean spacetime. Paradoxically, it seems that gravity and the structure of space and time may have a vital role to play in our understanding of matter on the very smallest scales, and the ways in which this comes about will be hinted at in later chapters.

In chapter 3, we studied some general theoretical aspects of classical or Newtonian mechanics which at one time appeared to provide a firm basis for understanding the properties and behaviour of material objects. As I hope the reader is aware, it became apparent towards the end of the nineteenth century that a number of experimental observations could not be accommodated in this framework. As it turned out, a radical revision of both the mathematical and the conceptual foundations of mechanics is required to give an adequate account of these and subsequent observations, which arise most importantly in connection with atomic and subatomic phenomena. While the mathematical developments which constitute quantum mechanics have been outstandingly successful in describing all manner of observed properties of matter, it is fair to say that the conceptual basis of the theory is still somewhat obscure. I myself do not properly understand what it is that quantum theory tells us about the nature of the physical world, and by this I mean to imply that I do not think anybody else understands it either, though there are respectable scientists who write with confidence on the subject. This need not worry us unduly. There does exist a canon of generally accepted phrases which, if we do not examine them too critically, provide a reliable means of extracting from the mathematics well defined predictions for the outcome of any experiment we can conduct (apart, that is, from the difficulty of solving the mathematical equations, which can be very great). I will generally use these without comment, and readers must choose for themselves whether or not to accept them at face value.

This chapter deals with non-relativistic quantum mechanics, and I am going to assume that readers are already familiar with the more elementary aspects of the subject. The following section outlines the reasons why classical mechanics has proved inadequate and reviews the elementary ideas of wave mechanics. Although the chapter is essentially self-contained, readers who have not met this material before are urged to consult a textbook on quantum mechanics for a fuller account. The remaining sections develop the mathematical theory in somewhat more general terms, and this provides a point of departure for the quantum field theories to be studied in later chapters.

5.0 Wave Mechanics

The observations which led to the quantum theory are often summarized by the notion of *particle–wave duality*. Phenomena which might normally be regarded as wave motions turn out to have particle-like aspects, while particles behave in some respects like waves.

The phenomena in question are basically of three kinds. First, there is evidence that electromagnetic radiation, which for many purposes is described in terms of waves, behaves for other purposes like a stream of particles, called *photons*. (It is interesting to recall that Newton believed in a 'corpuscular' theory of light, propounded in his *Opticks*, but for quite erroneous reasons.) In the photoelectric effect, for example, light incident on the surface of a metal causes electrons to be ejected. Contrary to what might have been expected, the energy of one of these electrons is found to be quite independent of the intensity of the radiation, although the number ejected per unit time does increase with the intensity. On the other hand, the energy of an electron increases linearly with the frequency of the radiation. As Einstein was the first to realize, this can be understood if the radiation is considered to consist of photons, each carrying a definite amount of energy

$$E = h\nu \tag{5.1}$$

where ν is the frequency and $h = 6.6256 \times 10^{-34}$ J s is Planck's constant. The energy of a single photon is transferred to a single electron, and the observed kinetic energy of the electron is this *quantum* of energy less a certain amount, the *work function*, required to release the electron from the metallic surface. Planck himself had been concerned with understanding the spectrum of black-body radiation, namely the way in which energy is distributed over frequencies. The analogous question of the distribution of molecular velocities in a gas could be well understood from a statistical analysis based on Newton's laws of motion, but this method failed when applied to electromagnetic waves. Planck discovered that, if the statistical analysis were to be modified by assuming that the energy carried by a wave of frequency v could only be some multiple of the quantum (5.1), then the correct spectrum could be obtained. Finally, the picture of radiation as a stream of photons is directly corroborated by the Compton effect, in which x rays scattered from electrons are found to undergo an increase in wavelength. According to electromagnetic theory which, as we have seen, is consistent with special relativity, a wave carrying energy E also carries a momentum $p = E/c$. If Compton scattering is viewed as a collision between a photon and an electron, then the change in wavelength is correctly found simply by requiring conservation of energy and momentum in each such collision. Since for electromagnetic radiation wavelength is related to frequency by $\lambda = c/v$, the momentum of a photon can be expressed as

$$p = h/\lambda \tag{5.2}$$

though as far as photons are concerned, this amounts merely to rewriting (5.1).

The second kind of evidence is that which shows that objects normally conceived of as particles have some wave-like properties. It was first suggested by de Broglie that the motion of a particle of energy E and momentum p might have associated with it a wave, whose frequency and wavelength would be given by (5.1) and (5.2). These would now be two independent equations, since the wave velocity would not, in general, be that of light. Celebrated experiments by Thompson and by Davisson and Germer showed that indeed electrons could be diffracted by a crystal lattice, just as light is by a diffraction grating, and confirmed the relation (5.2) between momentum and wavelength.

Lastly, there is the fact that atoms have definite ionization energies and radiate discrete rather than continuous spectra. This suggests that electrons in atoms occupy certain preferred orbits with definite allowed energies. If the electrons have waves associated with them, then the preferred states of motion can be envisaged as standing wave patterns, from which discrete energy levels arise in the same way as notes of a definite pitch from any musical instrument.

This talk of particle–wave duality may well strike readers as a leap in the dark. Indeed, it is undoubtedly the case that the elementary constituents of matter are neither particles nor waves, but rather entities of some other kind, for which our everyday experience provides no reliable analogy. Nevertheless, the de Broglie relations (5.1) and (5.2) point the way towards a quantitative theory which has become extraordinarily successful. I shall develop the essential points of this theory in more or less the traditional way, which should be made plausible, though it certainly is not justified in detail, by the experimental facts we have discussed.

Consider first a free particle, with energy E and 3-vector momentum $\boldsymbol{p}$. Classically, it would move in a straight line with constant velocity. With this motion, we must somehow associate a *wavefunction* $\Psi(\boldsymbol{x}, t)$, and since, according to (5.2), it must have a definite wavelength, the most reasonable guess for the nature of this wave is that it should be a plane wave. It turns out that wavefunctions must in general be complex, and a suitable guess is

$$\Psi(\boldsymbol{x}, t) = \exp[\mathrm{i}(\boldsymbol{k}\cdot\boldsymbol{x} - \omega t)]. \tag{5.3}$$

In terms of the angular frequency $\omega = 2\pi\nu$ and wavevector $\boldsymbol{k}$, with $|\boldsymbol{k}| = 2\pi/\lambda$, we have $E = \hbar\omega$ and $\boldsymbol{p} = \hbar\boldsymbol{k}$, where $\hbar = h/2\pi$. We see at once that, since this wave exists everywhere in space, there is nothing to tell us where the particle is. The accepted interpretation is that, in general, the quantity

$$P(\boldsymbol{x}, t)\mathrm{d}^3x = |\Psi(\boldsymbol{x}, t)|^2\mathrm{d}^3x \tag{5.4}$$

is the probability of finding the particle in an infinitesimal region d^3x surrounding the point $\boldsymbol{x}$. Alternatively, we can refer to $P(\boldsymbol{x}, t)$ itself as the *probability density* for finding the particle in the neighbourhood of $\boldsymbol{x}$. This means that the integral over all space of P should be 1. Therefore, (5.3) is not quite satisfactory as it stands, since it gives the value 1 for P itself. One method of modifying (5.3) is to suppose that the particle is confined to some large region of space and to divide the right-hand side of (5.3) by the square root of this volume.

More generally, if we wish to predict the result of a measurement of some quantity, say A, given that the state of motion of our system is described by a known wavefunction Ψ, it may well be that Ψ does not yield any particular value for A. In that case, we must be content with calculating probabilities for the measurement to yield various possible values of A. How such probabilities are obtained will be discussed in the next section. Clearly, however, we must have some means of extracting from the wavefunction whatever information it contains about the quantity A. To this end, we associate with every physical quantity a *differential operator* which can act on any wavefunction. For the cases of energy and momentum, these are taken to be

$$\text{energy operator: } \mathrm{i}\hbar\frac{\partial}{\partial t} \tag{5.5}$$

$$\text{momentum operator: } -\mathrm{i}\hbar\boldsymbol{\nabla}. \tag{5.6}$$

Obviously, acting with these on the plane wave (5.3) is equivalent to multiplying the wavefunction by E or $\boldsymbol{p}$ respectively. Other wavefunctions, corresponding to states in which the particle does not have a uniquely defined energy or momentum, can be written as a superposition of waves of the form (5.3) by Fourier transformation. If we act with the above operators on this wavefunction, we obtain a new wavefunction, in which each component of the superposition has been multiplied by its own energy or momentum. In a manner which will become clear below, we can compare the new wavefunction with the old one, or with plane waves, and by making these comparisons we obtain all the information quantum mechanics allows us to have about the energy or momentum of the particle in the given state of motion.

To find out how the state of motion of a system evolves with time, we can, in simple cases at least, make use of the fact that its energy can be expressed in terms of other quantities. For example, if we have a single particle of mass m moving in a potential $V(\boldsymbol{x})$, then its energy is $E = (\boldsymbol{p}^2/2m) + V(\boldsymbol{x})$. If we substitute the operators (5.5) and (5.6) into this equation, and allow each side to act on the wavefunction, we obtain *Schrödinger's equation*

$$i\hbar \frac{\partial}{\partial t} \Psi(\boldsymbol{x}, t) = \left(-\frac{\hbar^2}{2m} \nabla^2 + V(\boldsymbol{x})\right)\Psi(\boldsymbol{x}, t). \tag{5.7}$$

With these preliminary ideas in mind, we can proceed to develop the mathematical theory in detail. One of our main concerns will be to show how equations (5.5) and (5.6), which we obtained more or less by guesswork, can be justified at a deeper level in terms of the symmetries we studied in chapter 3.

5.1 The Hilbert Space of State Vectors

In order to develop the theory of classical mechanics, we had first to decide how a unique state of a physical system could be specified, and this question must now be reconsidered. We have already seen that, if a quantum-mechanical particle has a definite momentum, then it cannot also have a definite position. More generally, there will be *maximal sets* of observable quantities, say $\{A, B, C, \ldots\}$, such that every quantity in the set can, at the same time, have a definite value, while any other quantity is either forbidden to have a definite value at the same time, or has a value which is determined by the values of $A, B, C, \ldots$. For a single free particle whose only properties are position and momentum, $\{\boldsymbol{x}\}$ and $\{\boldsymbol{p}\}$ are examples of such maximal sets. The energy $E = \boldsymbol{p}^2/2m$ does not count, because it can be expressed in terms of $\boldsymbol{p}$. We shall say that a system is specified to be in a *pure quantum state* when all the values $\{a, b, c, \ldots\}$ of quantities belonging to some maximal set have been given. The criterion for deciding which sets of observables actually are maximal sets will emerge later on.

The first crucial assumption we made in chapter 3 for classical mechanics was that every instantaneous state could be specified in terms only of the positions and velocities of all the particles in the system. We now need a corresponding assumption for quantum mechanics, which again can ultimately be justified only by the fact that it leads to successful predictions about experimental observations. It consists in the following enigmatic statement:

> all possible instantaneous states of the system can be represented by vectors in a Hilbert space.

The precise definition and properties of Hilbert spaces are discussed in many mathematical textbooks (see, for example, Simmons (1963)) but I propose not to burden readers with them here. For many purposes in physics, the state vectors can be thought of as a straightforward generalization of ordinary Euclidean 3-vectors and I will follow this line of thought, ignoring a number of rigorous mathematical subtleties. The main generalizations are:

(i) The Hilbert space can have any number of dimensions, and we usually need an infinite number to accommodate all possible states.

(ii) A 3-vector can be multiplied by any positive real number α, the effect being to multiply its length by α, leaving its direction unchanged, or by a negative number which reverses the direction. A state vector may be multiplied by any complex number.

(iii) We denote a state vector by $|\Psi\rangle$, the Ψ being simply a label for identification. The scalar product $\boldsymbol{u}\cdot\boldsymbol{v}$ of two 3-vectors generalizes to a complex number $\langle\Phi|\Psi\rangle$ which has the property

$$\langle\Psi|\Phi\rangle = \langle\Phi|\Psi\rangle^*. \tag{5.8}$$

Suppose for the moment that each observable quantity in the maximal set $\{A, B, C, \ldots\}$ can assume only a discrete set of values. The state in which these values are $a, b, c, \ldots$ will be represented by a vector $|a, b, c, \ldots\rangle$ normalized so that $\langle a, b, c, \ldots|a, b, c, \ldots\rangle = 1$. Each of the vectors obtained by multiplying this vector by any non-zero complex number represents the same physical state, and the set of all such vectors is called a *ray*. Thus, each physical state corresponds to a ray or, in other words, a *direction* in the Hilbert space. The relationship between the quantum state of a system and physical measurements performed upon it is the subject of the following basic postulate of the theory. Suppose the actual state is represented by a normalized vector $|\Psi\rangle(\langle\Psi|\Psi\rangle = 1)$ and a measurement is made of all the quantities in some maximal set. Then the probability of obtaining the set of results $\{a, b, c, \ldots\}$ is

$$P(a, b, c, \ldots|\Psi) = |\langle a, b, c, \ldots|\Psi\rangle|^2. \tag{5.9}$$

Clearly, the goal of quantum-mechanical calculations will be to find these scalar products, though we do not yet know how to set about this. Readers who have understood chapter 2 will realize that the existence of scalar products implies that the Hilbert space possesses a metric, and that this metric gives a unique correspondence between a vector $|\Psi\rangle$ and a one-form $\langle\Psi|$ which is the other half of the scalar product symbol. Other readers may take it for granted that a given state can be represented by either of these two types of vector. A whimsical terminology due to P A M Dirac calls $|\;\rangle$ a 'ket' and $\langle\;|$ a 'bra' so that the scalar product becomes a bra(c)ket. It is convenient to represent the correspondence between bra and ket vectors by writing

$$\langle\Psi| = |\Psi\rangle^\dagger \quad \text{and} \quad |\Psi\rangle = \langle\Psi|^\dagger \tag{5.10}$$

although the † symbol is more properly reserved for use with operators as described below. The property (5.8) of the scalar product implies that, if α is a complex number, then

$$(\alpha|\Psi\rangle)^\dagger = \alpha^*\langle\Psi| \quad \text{and} \quad (\alpha\langle\Psi|)^\dagger = \alpha^*|\Psi\rangle. \tag{5.11}$$

If $|\Psi\rangle$ is the state $|a', b', c', \ldots\rangle$, then the probability in (5.9) must be equal to 1 if the two sets of values are the same and zero otherwise. This implies that two state vectors associated with the same maximal set of observables are *orthonormal*, which means

$$\langle a, b, c, \ldots|a', b', c', \ldots\rangle = \delta_{aa'}\delta_{bb'}\delta_{cc'}\ldots. \tag{5.12}$$

On the other hand, the total probability of getting *some* set of values from the measurement is found by summing (5.9) over all possible values of $a, b, c, \ldots$ and must be equal to 1. This will be true if every state vector can be expressed as a sum of the form

$$|\Psi\rangle = \sum_{a,b,c,\ldots} \psi_{abc\ldots}|a, b, c, \ldots\rangle. \tag{5.13}$$

If $|\Psi\rangle$ is normalized, the complex coefficients in this expression satisfy

$$\langle\Psi|\Psi\rangle = \sum_{a,b,c\ldots} |\psi_{abc\ldots}|^2 = 1 \tag{5.14}$$

and readers may verify, using (5.12), that the sum of the probabilities (5.9) is indeed 1. If $|\Psi\rangle$ is not normalized, then the right-hand side of (5.9) must be divided by $\langle\Psi|\Psi\rangle$.

The fact that every state vector can be expressed in the form (5.13) means that the set of vectors $|a, b, c, \ldots\rangle$ associated with a maximal set of observables is a set of orthonormal basis vectors for the Hilbert space. Choosing a new set of basis vectors, corresponding to a different maximal set of observables, is like rotating the coordinate axes in Euclidean geometry.

If one of the observables, say A, can assume a continuous range of values, then the $\delta_{aa'}$ in (5.12) must be replaced by the Dirac function $\delta(a - a')$ and the sums in (5.13) and (5.14) must be replaced by integrals. As far as A is concerned, the probability (5.9) then becomes a probability density, in the sense discussed in the last section. Consider, for example, a single particle, and choose the maximal set to be $\{\boldsymbol{x}\}$. Although a state vector is not the same thing as a wavefunction, a given state of motion of the particle can be represented either by a state vector $|\Psi\rangle$ or by a wavefunction $\psi(\boldsymbol{x})$. In fact, if $|\boldsymbol{x}\rangle$ represents the state in which the particle is instantaneously at the point $\boldsymbol{x}$, then the wavefunction is simply the coefficient of $|\boldsymbol{x}\rangle$ in the expansion

$$|\Psi\rangle = \int \mathrm{d}^3x\,\psi(\boldsymbol{x})|\boldsymbol{x}\rangle. \tag{5.15}$$

Since the orthonormality condition is now $\langle \boldsymbol{x}|\boldsymbol{x}'\rangle = \delta^3(\boldsymbol{x} - \boldsymbol{x}')$, we get

$$\psi(\boldsymbol{x}) = \langle \boldsymbol{x}|\Psi\rangle \tag{5.16}$$

and for the probability density we find

$$P(\boldsymbol{x}|\Psi) = |\psi(\boldsymbol{x})|^2. \tag{5.17}$$

Apart from the fact that $\psi(\boldsymbol{x})$ describes only an instantaneous state, this agrees exactly with (5.4).

5.2 Operators and Observable Quantities

Suppose we have a rule which enables us to associate with any given vector $|\Psi\rangle$ another vector $|\Psi'\rangle$. We say that an *operator* $\hat{O}$ acts on $|\Psi\rangle$ to produce $|\Psi'\rangle$:

$$|\Psi'\rangle = \hat{O}|\Psi\rangle. \tag{5.18}$$

I shall usually use the circumflex to indicate operators. The rule which defines an operator may be specified in various ways, and sometimes rather indirect means are necessary since it is impractical to consider each vector in the Hilbert space individually. The simplest operator of all is the *identity* operator $\hat{I}$, which leaves every vector unchanged. Almost all the operators used in quantum theory are *linear*. This means that, for any two vectors $|\Phi\rangle$ and $|\Psi\rangle$ and any two complex numbers α and β, $\hat{O}(\alpha|\Phi\rangle + \beta|\Psi\rangle) = \alpha\hat{O}|\Phi\rangle + \beta\hat{O}|\Psi\rangle$. All operators in this book are linear unless otherwise stated.

Observable quantities can be represented by operators in the following way. Let A belong to a maximal set $\{A, B, C, \ldots\}$. If the state of the system is one of the corresponding basis vectors $|a, b, c, \ldots\rangle$ then A has the definite value a, and we define the action of an operator $\hat{A}$ on each basis vector to be that of multiplying it by a:

$$\hat{A}|a, b, c, \ldots\rangle = a|a, b, c, \ldots\rangle. \tag{5.19}$$

An equation of this form, in which the action of an operator is just to multiply the vector by a number, is called an *eigenvalue equation*. We say that $|a, b, c, \ldots\rangle$ is an *eigenvector* of $\hat{A}$ with *eigenvalue* a. Since any vector can be expanded as in (5.13), this tells us how $\hat{A}$ acts on every vector. The probability $P(a|\Psi)$ of getting the result a from a measurement of A, irrespective of the values of any other quantities, is found by summing (5.9) over all values of $b, c, \ldots$. Readers should be able to verify that the *expectation value* $\langle A\rangle$, which denotes the average value of A obtained from many measurements, is

$$\langle A\rangle = \sum_a aP(a|\Psi) = \langle\Psi|\hat{A}|\Psi\rangle. \tag{5.20}$$

The expression on the right-hand side means the scalar product of $\langle\Psi|$ with the vector $\hat{A}|\Psi\rangle$.

In view of the symmetrical appearance of expressions like this, it is useful to define the action of operators on bra vectors also. The new bra vector $\langle\Phi|\hat{A}$ is defined by requiring that, for any $\langle\Phi|$ and any $|\Psi\rangle$, the expression $\langle\Phi|\hat{A}|\Psi\rangle$ has the same value, whether we regard it as the scalar product of $\langle\Phi|\hat{A}$ and $|\Psi\rangle$ or of $\langle\Phi|$ and $\hat{A}|\Psi\rangle$. This quantity is called a *matrix element* of $\hat{A}$. The reason is discussed in exercise 5.1, which readers may like to study before proceeding. There is a second method by which an operator may be used to obtain a new bra vector. If $\langle\Psi|$ is the bra whose corresponding ket is $|\Psi\rangle$, we can first form the new ket vector $\hat{A}|\Psi\rangle$ and then find its corresponding bra. The new bras formed from $\langle\Psi|$ by these two methods are not necessarily the same. We can describe the second method in terms of the action of an operator $\hat{A}^\dagger$, which is called the *adjoint* or the *Hermitian conjugate* of $\hat{A}$:

$$(\hat{A}|\Psi\rangle)^\dagger = \langle\Psi|\hat{A}^\dagger. \tag{5.21}$$

Using (5.8), we find that for any two vectors

$$\langle\Psi|\hat{A}|\Phi\rangle = \langle\Phi|\hat{A}^\dagger|\Psi\rangle^*. \tag{5.22}$$

An operator which equals its own adjoint

$$\hat{A}^\dagger = \hat{A} \tag{5.23}$$

is called *self-adjoint* or *Hermitian*. Strictly speaking, these two terms have slightly different meanings, but the distinction will not concern us.

In (5.22), let us take $\hat{A}$ to be Hermitian, $|\Psi\rangle$ to be an eigenvector of $\hat{A}$ with eigenvalue a_1 and $|\Phi\rangle$ an eigenvector with eigenvalue a_2. We find

$$(a_1 - a_2^*)\langle\Phi|\Psi\rangle = 0. \tag{5.24}$$

In the case that $|\Phi\rangle = |\Psi\rangle$, we have, of course, $a_1 = a_2$, and we see that this eigenvalue must be real. On the other hand, if the two eigenvalues are different, then the vectors must be orthogonal. These two properties are just what we need if $\hat{A}$ is to represent a measurable quantity, since its eigenvalues are possible results of measurements and therefore real numbers, and we want its eigenvectors to satisfy (5.12). We therefore assume that all observable quantities are represented by Hermitian operators.

The sum of any two operators is defined so as to be consistent with the addition of two vectors. That is, to act with $(\hat{A} + \hat{B})$ on a vector $|\Psi\rangle$, we first act with $\hat{A}$ and $\hat{B}$ separately and then add the resulting vectors: $(\hat{A} + \hat{B})|\Psi\rangle = \hat{A}|\Psi\rangle + \hat{B}|\Psi\rangle$.

The product $\hat{A}\hat{B}$ of two operators represents the combined effect of acting on a ket vector with $\hat{B}$ and then acting on the resulting vector with $\hat{A}$: $\hat{A}\hat{B}|\Psi\rangle = \hat{A}(\hat{B}|\Psi\rangle)$. The product $\hat{B}\hat{A}$, in which $\hat{A}$ acts before $\hat{B}$ does not necessarily have the same effect. The difference between these two products is another operator, called the *commutator* of $\hat{A}$ and $\hat{B}$ and written as

$$[\hat{A}, \hat{B}] = \hat{A}\hat{B} - \hat{B}\hat{A} \tag{5.25}$$

In practice, most of the information we have about operators derives from *commutation relations* which express commutators in terms of other operators. This is largely because of the role played by commutators in the symmetry operations discussed in the next section. We can use the definition of a commutator to express the criterion for building the maximal sets of observables with which our discussion started. If A and B belong to the same set, then acting on one of the associated basis vectors $|a, b, c, \ldots\rangle$ with $\hat{A}$ and $\hat{B}$ in either order gives the same result, namely multiplying it by ab. Since this is true for every basis vector, the result of acting with $\hat{A}$ and $\hat{B}$ in either order on any vector is the same. Therefore, their commutator is zero and they are said to *commute*. ('Zero' here means the operator which acts on any vector to give the vector whose length is zero.) Thus, a maximal set of observables is such that all the corresponding operators commute with each other, and no other independent operator commutes with all of them (except $\hat{I}$, which commutes with everything).

We shall often need to consider operators which are functions of other operators. To illustrate what is involved, consider the expression

$\hat{A} = \exp(\alpha\hat{B})$. Since we know how to multiply operators, we can make sense of this by using the power series expansion

$$\hat{A} = \hat{I} + \alpha\hat{B} + \tfrac{1}{2}\alpha^2\hat{B}^2 + \ldots. \qquad (5.26)$$

For some purposes, we can treat this function as if $\hat{B}$ were a number. For example, the inverse operator $\hat{A}^{-1}$ is defined by $\hat{A}^{-1}\hat{A} = \hat{A}\hat{A}^{-1} = \hat{I}$. It is equal to $\exp(-\alpha\hat{B})$, as may readily be verified by multiplying the two series together. On the other hand, readers may verify in the same way that the product $\exp(\hat{B})\exp(\hat{C})$ is not equal to $\exp(\hat{B} + \hat{C})$ unless $\hat{B}$ and $\hat{C}$ commute. Obviously, functions of operators must be handled with care. A power series is often the best way of resolving doubts as to whether a particular manipulation is permissible. By using (5.22), we find that the adjoint of $\hat{A} = \exp(\alpha\hat{B})$ is $\hat{A}^\dagger = \exp(\alpha^*\hat{B}^\dagger)$. If $\alpha = \mathrm{i}$ and $\hat{B}$ is Hermitian, this implies that

$$\hat{A}^\dagger = \hat{A}^{-1} \qquad (5.27)$$

in which case A is said to be *unitary*.

5.3 Spacetime Translations and the Properties of Operators

In order to make use of the formalism we have developed so far, we obviously need information about the specific properties of operators which represent particular physical quantities. The only way to acquire this information is to make informed guesses and see whether they lead to a successful theory. Our only guide in this enterprise is classical mechanics, and I propose to make the required guesses as plausible as possible by drawing analogies with the discussions in chapter 3. We begin with time translations.

There are several different ways of describing the evolution in time of the state of a system. The most obvious, which we consider first, is called the *Schrödinger picture*. Each vector in the Hilbert space is associated with an instantaneous state of the system, so we denote by $|\Psi(t)\rangle$ its state at time t. If we suppose that the initial state $|\Psi(0)\rangle$ at $t = 0$ is known, then the relation between these two states can be described by a *time evolution operator* $\hat{U}(t)$:

$$|\Psi(t)\rangle = \hat{U}(t)|\Psi(0)\rangle. \qquad (5.28)$$

In order to preserve the probabilistic interpretation of $|\Psi(t)\rangle$ in a systematic way, we require its normalization to remain constant:

$$\langle\Psi(t)|\Psi(t)\rangle = \langle\Psi(0)|\hat{U}^\dagger(t)\hat{U}(t)|\Psi(0)\rangle = \langle\Psi(0)|\Psi(0)\rangle. \qquad (5.29)$$

Evidently, $\hat{U}(t)$ must be a unitary operator with $\hat{U}(0) = \hat{I}$ and, according to our discussion at the end of the last section, it can be written as

$$\hat{U}(t) = \exp(-i\hat{\mathscr{H}}t) \tag{5.30}$$

where $\hat{\mathscr{H}}$ is an Hermitian operator. If we assume that $\hat{\mathscr{H}}$ is independent of time, insert (5.30) into (5.28) and differentiate, we get

$$i\frac{d}{dt}|\Psi(t)\rangle = \hat{\mathscr{H}}|\Psi(t)\rangle \tag{5.31}$$

which has the same form as equation (3.22) for the evolution of the state in classical mechanics. Now, $\mathscr{H}$ in (3.22) was a differential operator constructed from the Hamiltonian function, which is usually the same as the total energy. The quantum-mechanical operator $\hat{\mathscr{H}}$ is Hermitian and therefore suitable for representing an observable quantity. A reasonable guess, therefore, is that $\hat{\mathscr{H}}$ is proportional to the quantum-mechanical Hamiltonian or total energy operator $\hat{H}$. Since the argument of the exponential in (5.30) must be dimensionless, our guess is

$$\hat{\mathscr{H}} = \hat{H}/\hbar \tag{5.32}$$

where $\hbar$ is a fundamental constant with dimensions of energy × time. The value of this constant must eventually be determined experimentally, and it is, of course, none other than Planck's constant divided by 2π.

A different view of time evolution, called the *Heisenberg picture*, comes about when we realize that $|\Psi(t)\rangle$ is not itself an observable quantity. The expectation value of an observable quantity at time t can be written without reference to $|\Psi(t)\rangle$ as

$$\langle\Psi(t)|\hat{A}|\Psi(t)\rangle = \langle\Psi|\hat{A}(t)|\Psi\rangle \tag{5.33}$$

where $|\Psi\rangle$ means $|\Psi(0)\rangle$ and

$$\hat{A}(t) = \hat{U}^\dagger(t)\hat{A}\hat{U}(t). \tag{5.34}$$

The two operators $\hat{A}$ and $\hat{A}(t)$ have their analogues in classical mechanics where, as we have seen, a function $A(p, q)$ defines the meaning of a given dynamical quantity in terms of momenta and coordinates, whereas $A(t) = A(p(t), q(t))$ gives the value of this quantity when we substitute for p and q the actual solutions of the equations of motion. In fact, we can easily derive an equation of motion for $\hat{A}(t)$, analogous to (3.17), by differentiating (5.34). Since $\hat{\mathscr{H}}$ obviously commutes with $\hat{U}$ and $\hat{U}^\dagger$, these can be differentiated as if $\hat{\mathscr{H}}$ were a number. But, since $\hat{\mathscr{H}}$ does not necessarily commute with $\hat{A}$, we must be careful about the order of operators in the result, which is

$$\frac{d}{dt}\hat{A}(t) = -i[\hat{A}(t), \hat{\mathscr{H}}] = -\frac{i}{\hbar}[\hat{A}(t), \hat{H}]. \tag{5.35}$$

An immediate consequence of this is that any quantity whose associated operator commutes with $\hat{H}$ is conserved. In particular, $\hat{H}$ commutes with itself and is conserved. The assumption which went into this result was that $\hat{\mathcal{H}}$, and therefore the quantum-mechanical law of motion, did not depend explicitly on time. In view of our discussion in §3.2, we would expect conservation of energy to be an automatic consequence of this assumption, which reinforces our interpretation of $\hat{H}$ as representing the total energy.

In chapter 3, we constructed from the total momentum an operator $\mathcal{P}$ (equation (3.24)) which generates translations in space just as $\mathcal{H}$ does in time. From considerations similar to those above, we can ascertain the properties of the quantum-mechanical momentum operator. Comparing (5.35) with (3.17), we observe a correspondence of the form

$$[\hat{A}, \hat{B}] = i\hbar\{\hat{A}, \hat{B}\}_{\mathrm{P}}, \tag{5.36}$$

where the right-hand side means that we first evaluate the Poisson bracket in classical mechanics and then substitute the quantum operators for their corresponding classical variables. While some care must be exercised in making this entirely general, I shall show that it can be established for the position and momentum operators with the same degree of plausibility as in the preceding discussion. Considering for simplicity the case of a single particle with position operator $\hat{\boldsymbol{x}}$ and momentum $\hat{\boldsymbol{p}}$, (5.36) asserts that

$$[\hat{x}_\alpha, \hat{p}_\beta] = i\hbar\delta_{\alpha\beta} \tag{5.37}$$

where α and β label Cartesian components. On the right-hand side of this equation and in similar contexts we understand a complex number to mean the operator which multiplies a vector by that number. This commutation relation comprises the whole of our knowledge about momentum and position operators. The entire theory of quantum mechanics rests upon (5.37) and (5.35).

If $|\hat{\boldsymbol{x}}\rangle$ is any position eigenvector ($\hat{\boldsymbol{x}}|\boldsymbol{x}\rangle = \boldsymbol{x}|\boldsymbol{x}\rangle$) and $\hat{\mathcal{P}}$ is the quantum-mechanical generator of space translations, then for any 3-vector $\boldsymbol{a}$

$$\exp(-i\boldsymbol{a}\cdot\hat{\mathcal{P}})|\boldsymbol{x}\rangle = |\boldsymbol{x} + \boldsymbol{a}\rangle. \tag{5.38}$$

This implies that $\hat{\boldsymbol{x}}$ and $\hat{\mathcal{P}}$ satisfy the relation

$$\exp(i\boldsymbol{a}\cdot\hat{\mathcal{P}})\hat{\boldsymbol{x}}\exp(-i\boldsymbol{a}.\hat{\mathcal{P}}) = \hat{\boldsymbol{x}} + \boldsymbol{a}. \tag{5.39}$$

If we expand the exponentials to first order in $\boldsymbol{a}$ and identify

$$\hat{\mathcal{P}} = \hat{\boldsymbol{p}}/\hbar \tag{5.40}$$

then we obtain (5.37). Exercise 5.3 shows that (5.39) will then be true to all orders. (We also need to know that the Cartesian components of

$\hat{p}$ commute with each other, and readers should satisfy themselves that this is indeed so.) If the Hamiltonian for the single particle is translationally invariant, it is independent of $\hat{x}$. It is therefore a function of $\hat{p}$ only, and so commutes with $\hat{p}$. Thus $\hat{p}$ is conserved in a translationally invariant system, which reinforces our confidence in the interpretation (5.40). More generally, for a system containing several particles, the generator $\hat{\mathcal{P}}$ will be the sum of their momentum operators divided by $\hbar$, and a translationally invariant Hamiltonian may depend on any of the coordinate differences $(\hat{x}_i - \hat{x}_j)$. The results of exercise 5.3 show that the total momentum is then conserved.

5.4 Quantization of a Classical System

Until we have some experience of quantum-mechanical systems, the only sensible way we have of specifying such a system is to model it upon a classical one. Given the formal correspondences we have seen to exist between classical and quantum mechanics, it is not difficult to give a prescription for 'quantizing' a classical system. It is called the *canonical quantization* scheme. Usually, the classical system can be specified by giving its Lagrangian as a function of generalized coordinates $\{q_i\}$ and their velocities. The momentum $p_i = \partial L/\partial \dot{q}_i$ conjugate to each coordinate can be found and the velocities eliminated in favour of the momenta. The Hamiltonian can then be obtained in the standard way. The quantum-mechanical system is then defined by replacing coordinates and momenta by the corresponding operators and requiring these operators to satisfy the commutation relations

$$[\hat{q}_i, \hat{p}_j] = i\hbar\delta_{ij}. \tag{5.41}$$

These relations apply to Schrödinger-picture operators or to Heisenberg-picture operators *at the same time*. The commutator $[\hat{q}(t), \hat{p}(t')]$, for example, is not equal to $i\hbar$ unless $t' = t$. Its value depends upon how the system has evolved between t and t', and it is different for systems with different Hamiltonians. In most cases, no simple expression can be derived for it.

When implementing this procedure, ambiguities can arise from the ordering of quantum-mechanical operators which, unlike the corresponding classical variables, do not commute. Quite often, the general correspondence (5.36) will be found to apply, but the safest course is to impose (5.41) on Cartesian coordinates and momenta, and work out other commutators from these. When the classical Hamiltonian contains products of variables whose operators do not commute, the quantum Hamiltonian is not unambiguously prescribed. A possible course is to replace, say, $\hat{A}\hat{B}$ by the symmetrized product $(\hat{A}\hat{B} + \hat{B}\hat{A})/2$, but other solutions may be desirable in specific cases. A further difficulty arises if the time derivative of some coordinate does not appear in the Lagrangian. The momentum conjugate to this coordinate is identically zero, and (5.41) obviously cannot hold. This is not likely to happen when the classical Lagrangian describes a system of particles, since the kinetic energy term involves all the velocity components. It does happen, however, when we try to extend the formalism to treat the electromagnetic field as a quantum system.

A point worth noting is that velocities do not, in general, have a well defined meaning in quantum mechanics. We have seen that, if a particle has a definite momentum, its position is completely undetermined. To assign it a velocity would require two exact measurements of its position, separated by an infinitesimal time interval, which does not make good quantum-mechanical sense, even as an idealized limiting process. The momenta which appear in (5.41) are always the canonically defined ones. In the presence of electromagnetic forces, for example, they correspond to classical quantities of the kind shown in (3.59) (though we have not yet given a proper account of the quantum mechanics of relativistic particles) rather than to just $m\dot{x}$.

Although the formulation of quantum theory in terms of state vectors and operators acting on them is more general than wave mechanics, the solution of specific problems is often most conveniently achieved in terms of wavefunctions. Let us therefore see how the algebra of operators acting on state vectors can be reinterpreted in terms of differential operators on wavefunctions. The wavefunction corresponding to a state vector $|\Psi\rangle$ is given by (5.16). The wavefunction corresponding to $\hat{x}|\Psi\rangle$ is

$$\langle x|\hat{x}|\Psi\rangle = x\langle x|\Psi\rangle = x\psi(x) \tag{5.42}$$

and so the action of the Schrödinger-picture position operator corresponds to multiplication of the wavefunction by the coordinate. Similarly, using (5.38), with the generator identified as in (5.40), we can write

$$\psi(x + a) = \exp(a\cdot\nabla)\psi(x) = \langle x|\exp(\mathrm{i}a\cdot\hat{p}/\hbar)|\Psi\rangle. \tag{5.43}$$

As in (3.23), the exponential of the gradient operator represents a Taylor series. Clearly, the action of $\hat{p}$ on $|\Psi\rangle$ corresponds to that of $-\mathrm{i}\hbar\nabla$ on the wavefunction. Readers should be able to satisfy themselves that, given any operator which can be expressed as a function $A(\hat{x}, \hat{p})$, the wavefunction corresponding to the vector $A(\hat{x}, \hat{p})|\Psi\rangle$ is $A(x, -\mathrm{i}\hbar\nabla)\psi(x)$. In particular, if $\hat{A}$ is the Hamiltonian for a particle moving in the potential $V(x)$, we see from (5.31) that the time-dependent wavefunction $\Psi(x, t) = \langle x|\Psi(t)\rangle$ obeys Schrödinger's equation (5.7). To complete the correspondence between state vectors and wavefunctions, we note first that the operators x and $-\mathrm{i}\hbar\nabla$ satisfy the same commutation relation (5.37) as $\hat{x}$ and $\hat{p}$ and second, as readers may show, that any matrix element may be expressed as

$$\langle\Phi|A(\hat{x}, \hat{p})|\Psi\rangle = \int d^3x\phi^*(x)A(x, -i\hbar\nabla)\psi(x). \tag{5.44}$$

The extension of these considerations to systems containing more than one particle, with wavefunctions $\psi(x_1, x_2, \ldots) = \langle x_1, x_2, \ldots|\Psi\rangle$ should be obvious.

5.5 An Example: The One-dimensional Harmonic Oscillator

The harmonic oscillator provides a standard illustration of the mathematical ideas we have developed. It also serves to introduce the idea of raising and lowering operators which are of fundamental importance for second quantization and field theory which we study in the following chapter. The classical system from which we start consists of a single particle of mass m, moving in one dimension in the potential $V(x) = m\omega^2x^2/2$, and the classical trajectories are, of course, sinusoidal oscillations of angular frequency ω. The Lagrangian is

$$L = \tfrac{1}{2}m\dot{x}^2 - \tfrac{1}{2}m\omega^2x^2. \tag{5.45}$$

The momentum conjugate to x is $p = m\dot{x}$, and the Hamiltonian is

$$H = \frac{1}{2m}p^2 + \tfrac{1}{2}m\omega^2x^2. \tag{5.46}$$

None of the above-mentioned difficulties occurs here, and we are free to impose the commutation relation $[\hat{x}, \hat{p}] = i\hbar$.

We developed the mathematics of state vectors and operators by assuming that a Hilbert space describing all possible states of motion of our system was given, and enquiring about the properties of operators acting on it. Now, however, we see that the practical problem of theoretical physics is the reverse: our physical principles supply us with operators having definite properties, and we have to construct a Hilbert space by finding the states of motion which are permitted by these properties. This problem will be solved if we can find a set of basis vectors and if we know how any operator acts on each basis vector. A set of basis vectors will be associated with some maximal set of observables, and the most useful sets are $\{x\}$, $\{p\}$ and $\{H\}$. The description in terms of a particular set of basis vectors is called a *representation*, and the representations associated with the above maximal sets are called, logically enough, the coordinate, momentum and energy representations.

We shall first construct the basis vectors of the energy representation.

These are eigenvectors of the Hamiltonian, labelled by an integer n, with energy eigenvalues ε_n:

$$\hat{H}|n\rangle = \varepsilon_n|n\rangle. \tag{5.47}$$

They are of particular physical interest, since they are *stationary states*. Time-dependent vectors of the form $\exp(-\mathrm{i}\varepsilon_n t)|n\rangle$ are solutions of (5.31), and the expectation value in such a state of any operator which is defined in a time-independent manner is constant. If, for example, the oscillator is regarded as a model for the vibrations of a diatomic molecule, then the observed spectral lines arise from transitions between these states, caused by external forces which are not included in our description. The fact that the allowed energy levels have discrete values rather than a continuous range is at present a matter of assumption, but will be verified in due course. It is advantageous to exchange the position and momentum for two new operators $\hat{a}$ and $\hat{a}^\dagger$, defined by

$$\hat{a} = \left(\frac{\omega m}{2\hbar}\right)^{1/2}\left[\hat{x} + \left(\frac{1}{\omega m}\right)\mathrm{i}\hat{p}\right] \tag{5.48}$$

$$\hat{a}^\dagger = \left(\frac{\omega m}{2\hbar}\right)^{1/2}\left[\hat{x} - \left(\frac{1}{\omega m}\right)\mathrm{i}\hat{p}\right] \tag{5.49}$$

in terms of which, the Hamiltonian can be written as

$$\hat{H} = (\hat{a}^\dagger\hat{a} + \tfrac{1}{2})\hbar\omega. \tag{5.50}$$

Using the commutation relation for $\hat{x}$ and $\hat{p}$, we find that these operators satisfy

$$[\hat{a}, \hat{a}^\dagger] = 1 \tag{5.51}$$

$$[\hat{a}, \hat{H}] = \hbar\omega\hat{a} \tag{5.52}$$

$$[\hat{a}^\dagger, \hat{H}] = -\hbar\omega\hat{a}^\dagger. \tag{5.53}$$

From (5.52) and (5.53), it is easy to show that if $|n\rangle$ is an energy eigenvector with energy ε_n, then $\hat{a}|n\rangle$ is an eigenvector with energy $(\varepsilon_n - \hbar\omega)$ and $\hat{a}^\dagger|n\rangle$ is an eigenvector with energy $(\varepsilon_n + \hbar\omega)$. For this reason, $\hat{a}$ and $\hat{a}^\dagger$ are called energy lowering and raising operators.

Written in terms of $\hat{x}$ and $\hat{p}$, the Hamiltonian is a sum of squares of Hermitian operators. Therefore, none of its eigenvalues can be negative, and there must be a *ground state* of minimum energy, which we denote by $|0\rangle$. Since $\hat{a}|0\rangle$ cannot be a state with lower energy, the only way to satisfy (5.52) when it acts on $|0\rangle$ is to have $\hat{a}|0\rangle = 0$. Then, acting on $|0\rangle$ with (5.50) shows that $\varepsilon_0 = \hbar\omega/2$ and so, by acting n times on $|0\rangle$ with $\hat{a}^\dagger$, we generate an infinite series of energy eigenvectors with energies

$$\varepsilon_n = (n + \tfrac{1}{2})\hbar\omega. \tag{5.54}$$

Furthermore, there cannot be any states with energies between these values. If there were, then by acting enough times with $\hat{a}$, we could generate a state with energy between 0 and $\hbar\omega$, but not equal to $\hbar\omega/2$. Acting once more with $\hat{a}$ would have to produce zero, but we know that a state with this property has an energy of exactly $\hbar\omega/2$, which is a contradiction. Thus, the states $|n\rangle$, with energy eigenvalues given by (5.54), constitute the complete set of basis vectors for the energy representation. We require these basis vectors to be normalized so that $\langle n|n'\rangle = \delta_{nn'}$. I leave it as an exercise for readers to establish (by induction) that they are given by

$$|n\rangle = (n!)^{-1/2}(\hat{a}^\dagger)^n|0\rangle \tag{5.55}$$

and that

$$\hat{a}^\dagger|n\rangle = (n+1)^{1/2}|n+1\rangle \quad \text{and} \quad \hat{a}|n\rangle = n^{1/2}|n-1\rangle. \tag{5.56}$$

This is, essentially, the solution to our problem. Any observable property of the oscillator can be expressed in terms of $\hat{x}$ and $\hat{p}$, and it is a trivial matter to express these in terms of $\hat{a}$ and $\hat{a}^\dagger$ by solving (5.48) and (5.49). Any state vector can be expressed in terms of the $|n\rangle$, and so (5.56) tells us how any operator acts on any vector. A particularly useful operator is $\hat{n} = \hat{a}^\dagger\hat{a}$, which has the property

$$\hat{n}|n\rangle = \hat{a}^\dagger\hat{a}|n\rangle = n|n\rangle. \tag{5.57}$$

It is called the *number operator*, because it counts the number of *quanta* $\hbar\omega$ of energy in the state.

These results may be translated into the coordinate representation by finding the wavefunctions $\psi_n(x)$ of the energy eigenstates. The two sets of basis vectors are related by

$$|n\rangle = \int_{-\infty}^{\infty} dx\,\psi_n(x)|x\rangle \quad \text{and} \quad |x\rangle = \sum_{n=0}^{\infty} \psi_n^*(x)|n\rangle. \tag{5.58}$$

To find the wavefunctions, we rewrite the raising and lowering operators in terms of x and $-i\hbar\partial/\partial x$. $\psi_0(x)$ is the solution of the equation $a(x, -i\hbar\partial/\partial x)\psi_0(x) = 0$, and the others are found by applying the raising operator to it. The result may be written as

$$\psi_n(x) = \left[n!\left(\frac{\pi\hbar}{\omega m}\right)^{1/2}\left(\frac{2\omega m}{\hbar}\right)^n\right]^{-1/2} \exp\left(\frac{\omega m x^2}{2\hbar}\right)\left(-\frac{d}{dx}\right)^n \exp\left(\frac{-\omega m x^2}{\hbar}\right) \tag{5.59}$$

where the normalizing factor ensures that

$$\int_{-\infty}^{\infty} |\psi_n(x)|^2 dx = 1. \tag{5.60}$$

A further translation into the momentum representation is simply a matter of Fourier transformation. It can easily be found that the relations

$$|x\rangle = (2\pi\hbar)^{-1/2} \int \mathrm{d}p \exp(-\mathrm{i}px/\hbar)|p\rangle \tag{5.61}$$

$$|p\rangle = (2\pi\hbar)^{-1/2} \int \mathrm{d}x \exp(\mathrm{i}px/\hbar)|x\rangle \tag{5.62}$$

are uniquely determined by (5.38) and the orthonormality requirements $\langle x|x'\rangle = \delta(x - x')$ and $\langle p|p'\rangle = \delta(p - p')$. Consequently, the energy eigenvectors may be expressed as

$$|n\rangle = \int \mathrm{d}p \pi_n(p)|p\rangle \tag{5.63}$$

where the *momentum space wavefunction* is

$$\pi_n(p) = (2\pi\hbar)^{-1/2} \int \mathrm{d}x \exp(-\mathrm{i}px/\hbar)\psi_n(x). \tag{5.64}$$

Obviously, this method of solving the problem works only for the particular case of the harmonic oscillator. For single particles in other potentials, the most practical method of constructing the Hilbert space is to use the coordinate representation. The eigenvalue equation (5.47) becomes the *time-independent Schrödinger equation*

$$\left[-\frac{\hbar^2}{2m}\nabla^2 + V(\boldsymbol{x})\right]\psi_\varepsilon(\boldsymbol{x}) = \varepsilon\psi_\varepsilon(\boldsymbol{x}). \tag{5.65}$$

In the case of the harmonic oscillator, the boundary conditions on the solutions of this equation are that the wavefunction must vanish sufficiently fast as $|x| \to \infty$ for the integral in (5.60) to converge to a finite value, which can be normalized to 1. This is possible only when ε has one of the values in (5.54), so it is these boundary conditions which lead to the energy of the oscillator being quantized in a set of discrete levels. In all these states, the probability density (5.17) vanishes rapidly when x becomes sufficiently large. In this sense, the particle is constrained by the parabolic potential to remain close to the origin, and the states are known as *bound states*.

In almost every physical problem, the potential approaches a finite value, which might as well be zero, at infinity: the Coulomb potential seen by the electron in a hydrogen atom is an archetypical example. If the potential possesses a well, then there may be bound states of negative energy, in which the particle is most probably to be found in the well. The spectrum of bound state energy levels is always a discrete one. In positive energy states, however, the particle can escape to infinity, where the wavefunction becomes similar to (5.3). These are called *scattering states*. The energies of scattering states form a continuous spectrum, because different boundary conditions apply to them.

The exact nature of these boundary conditions is slightly complicated, because the wavefunctions cannot be made to satisfy (5.60) or its three-dimensional equivalent. In fact, if the particle is not bound by the potential, the usefulness of the energy eigenfunctions associated with the potential alone is limited, and a different kind of description is appropriate. I shall return briefly to this question in chapter 9.

The use of wavefunctions to solve both bound state and scattering problems is of the utmost importance in many areas of physics. The practical techniques available are described in any respectable textbook on quantum mechanics, but they are not part of the subject matter of this book and I must ask readers to look elsewhere for further details.

Exercises

5.1. The object of this exercise is to show that manipulation of state vectors and operators is entirely analogous to the algebra of complex matrices and is in fact identical in the case of a Hilbert space of finite dimension. Readers are invited to satisfy themselves of this, and to gain some further insight, by considering the various assertions made below. Little or no detailed working may be needed. Let $|\psi\rangle$ stand for the column matrix $(\psi_1, \ldots, \psi_N)^{\mathrm{T}}$, where the ψ_i are complex numbers and T denotes the transpose. An orthonormal basis is given by the vectors $|i\rangle$, where $|1\rangle = (1, 0, 0, \ldots, 0)^{\mathrm{T}}$, $|2\rangle = (0, 1, 0, \ldots, 0)^{\mathrm{T}}$ and so on.

(a) Any column matrix $|\psi\rangle$ can be expressed as a linear combination of the $|i\rangle$ with coefficients ψ_i.

(b) For any complex number α, $\alpha|\psi\rangle = (\alpha\psi_1, \ldots, \alpha\psi_N)^{\mathrm{T}}$.

(c) If $\langle\psi|$ is the row matrix $(\psi_1^*, \ldots, \psi_N^*)$, and $\langle\psi|\phi\rangle$ is taken to be the usual matrix product, then (5.8) and (5.11) are true.

(d) Multiplication by any $N \times N$ square matrix $\hat{A}$ provides a rule for converting any column matrix into another column matrix.

(e) Any square matrix can be multiplied on the left by a row matrix, and the elements of $\hat{A}$ are $\hat{A}_{ij} = \langle i|\hat{A}|j\rangle$.

(f) If the elements of $\hat{A}^\dagger$ are $(\hat{A}^\dagger)_{ij} = \hat{A}_{ji}^*$, then (5.21) and (5.22) are true.

(g) If $\hat{A}|i\rangle = a_i|i\rangle$ for each basis vector, then $\hat{A}$ is a diagonal matrix with diagonal elements a_i.

(h) If $\hat{A}$ is a diagonal matrix, $\hat{B}$ is a square matrix such that $[\hat{A}, \hat{B}] = 0$, and $a_i \neq a_j$, then $\hat{B}_{ij} = 0$.

(i) If $\{\hat{A}, \hat{B}, \hat{C}, \ldots\}$ is a maximal set of operators (square matrices) in the sense discussed following (5.25), and the basis vectors $|i\rangle$ are their simultaneous eigenvectors, then $\hat{A}, \hat{B}, \hat{C}, \ldots$ are all diagonal and, for any pair of indices i and j, there is at least one member of the set whose ith and jth eigenvalues are not equal.

(j) If $\hat{A}$ is a diagonal matrix with diagonal elements a_i, then $f(\hat{A})$ is the diagonal matrix whose elements are $f(a_i)$.

5.2. For any set of operators $\hat{A}, \hat{B}, \hat{C}, \ldots$ show that $(\hat{A}\hat{B}\hat{C}\ldots)^\dagger = \ldots \hat{C}^\dagger\hat{B}^\dagger\hat{A}^\dagger$ and $(\hat{A}\hat{B}\hat{C}\ldots)^{-1} = \ldots \hat{C}^{-1}\hat{B}^{-1}\hat{A}^{-1}$.

5.3. For any function f, show that $\hat{p}f(\hat{x}) = f(\hat{x})\hat{p} - \mathrm{i}\hbar f'(\hat{x})$ and that $\hat{x}f(\hat{p}) = f(\hat{p})\hat{x} + \mathrm{i}\hbar f'(\hat{p})$. Hence verify (5.39) and find an analogous relation in which the roles of $\hat{\boldsymbol{x}}$ and $\hat{\boldsymbol{p}}$ are reversed. Show that if the potential energy of a system of particles depends only on the relative coordinates of pairs of particles then the total momentum is conserved.

5.4. The symbol $|\psi\rangle\langle\psi|$ represents a *projection operator*, which acts on any ket vector $|\phi\rangle$ to produce the new ket vector $(\langle\psi|\phi\rangle)|\psi\rangle$ and analogously on any bra vector. Show that the probability (5.9) is the expectation value of a projection operator. If $|a, b, c, \ldots\rangle$ are a complete set of basis vectors, show that their projection operators form a *resolution of the identity*, which means that

$$\sum_{a,b,c,\ldots} |a, b, c, \ldots\rangle\langle a, b, c, \ldots| = \hat{I}.$$

Show that the operator $\hat{A}$, for which $\hat{A}|a, b, c, \ldots\rangle = a|a, b, c, \ldots\rangle$, can be expressed as

$$\hat{A} = \sum_{a,b,c,\ldots} |a, b, c, \ldots\rangle a \langle a, b, c, \ldots|.$$

How can this be generalized to represent an operator which is not diagonal in this representation?

5.5 If $f'(x)$ denotes the derivative $\mathrm{d}f(x)/\mathrm{d}x$ when x is an ordinary number, show that $\mathrm{d}f(\alpha\hat{A})/\mathrm{d}\alpha = \hat{A}f'(\alpha\hat{A})$.

5.6. Let $|i\rangle$ and $|\alpha\rangle$ be two sets of orthonormal basis vectors such that

$$|i\rangle = \sum_\alpha u_{i\alpha}|\alpha\rangle.$$

Show that the complex coefficients $u_{i\alpha}$ are the components of a unitary matrix.

5.7. Let $\hat{A}$ and $\hat{B}$ be two operators such that the commutator $\hat{C} = [\hat{A}, \hat{B}]$ commutes with both $\hat{A}$ and $\hat{B}$, and let $:\ldots:$ denote an ordering of operators such that $\hat{A}$'s always stand to the left of $\hat{B}$'s, so that

$$:(\hat{A} + \hat{B})^n: = \sum_{m=0}^{n} \binom{n}{m} \hat{A}^m \hat{B}^{n-m}$$

where $\binom{n}{m}$ is the binomial coefficient.

(a) Show that

$$(\hat{A} + \hat{B})^{n+1} = \hat{A}(\hat{A} + \hat{B})^n + (\hat{A} + \hat{B})^n\hat{B} - n\hat{C}(\hat{A} + \hat{B})^{n-1}.$$

(b) Show that

$$(\hat{A} + \hat{B})^n = \sum_{m=0}^{[n/2]} \alpha_{nm}\hat{C}^m :(\hat{A} + \hat{B})^{n-2m}:$$

where $[n/2]$ equals $n/2$ if n is even or $(n-1)/2$ if n is odd, and the expansion coefficients satisfy the recursion relation

$$\alpha_{n+1,m+1} = \alpha_{n,m+1} - n\alpha_{n-1,m}.$$

(c) Verify that these coefficients are given by

$$\alpha_{nm} = \left(-\frac{1}{2}\right)^m \frac{n!}{(n-2m)!m!}$$

and hence derive the *Baker–Campbell–Hausdorff* formula

$$\exp(\hat{A} + \hat{B}) = \exp(\hat{A})\exp(\hat{B})\exp(\hat{C}/2).$$

6

Second Quantization and Quantum Field Theory

Up to a point, the quantum theory developed in chapter 5 was quite general. However, the systems we had in mind were non-relativistic ones consisting of either a single particle or a fixed number of particles. In this chapter, we extend the theory to deal with systems in which the number of particles can change. There are several reasons for wanting to do this. The most obvious is that we need a method of describing high-energy scattering and decay processes in which particles are observed to be created and destroyed. A second is that, when we try to make quantum theory consistent with relativity, we encounter difficulties of interpretation which can be resolved only in this more general setting. These difficulties will be discussed in chapter 7. The final reason is that, even for systems with a fixed number of non-relativistic particles, the mathematics rapidly becomes intractable as the number of particles increases. A useful device for dealing with large systems is, roughly speaking, to imagine adding an extra particle, which serves as theoretical probe of the state of the system. To put the matter another way, the method of *second quantization* developed in this chapter provides a means of dealing with the entire system by considering only a few particles at a time.

The term 'second quantization' is an unfortunate one, in so far as it suggests a theory which is 'twice as quantum-mechanical' as the one we started with. This is emphatically not the case: all we are doing is developing a convenient mathematical technique for dealing with the original theory. The origin of the term will become clear, but briefly it is this. Addition or subtraction of particles to or from the system is represented by creation and annihilation operators which are closely analogous to the raising and lowering operators of the harmonic oscillator. From these we can construct *field operators* which, in the absence of interactions, satisfy the same Schrödinger equation as single-particle wavefunctions. By turning a wavefunction, which is acted on by operators representing physical quantities, into an operator which itself acts on state vectors, we might appear to be adding a further layer of quantumness, but readers who follow the development carefully will realize that this is not a good description of what is actually taking place.

6.1 The Occupation Number Representation

Consider a system containing a fixed number N of identical particles. For the moment, we shall assume that they do not interact with each other. Some, though not all, states of the system can be specified by giving the state of motion of each particle. I shall label a complete set of single-particle states by the symbol k. Quite often, it will be convenient to take these single-particle states to be momentum eigenstates, in which case k will represent the value of the momentum. Other sets of states, such as the Bloch states which describe the motion of electrons in a crystal lattice, may be more useful in particular circumstances. Also, if the particles have spin, then the spin state of the particle is included in k. (Readers who are not familiar with spin will find a brief discussion in appendix 2; those who are not familiar with Bloch states may consult a book on solid state physics, but need not do so for the purposes of this book.) Thus, if we choose to specify the momentum (k_x, k_y, k_z) and spin s of an electron, then k is a shorthand for this set of four numbers.

Using these single-particle states, we can choose a set of basis vectors for the whole system of the form $|k_1, k_2, \ldots, k_N\rangle$, where the nth label in the list refers to the nth particle. Because quantum-mechanical particles do not follow definite trajectories, it is impossible in principle to distinguish two identical particles. Therefore, the two vectors $|k_1, k_2, \ldots\rangle$ and $|k_2, k_1, \ldots\rangle$ must be taken as referring to the same physical state and can only differ by a phase factor. That is, $|k_2, k_1, \ldots\rangle = \alpha|k_1, k_2, \ldots\rangle$, where α is a complex number of unit magnitude. On interchanging the particles a second time, we get back to the original vector, so $\alpha^2 = 1$. The same is, of course, true for any pair of particles. The state is said to be *symmetric* if $\alpha = 1$ or *antisymmetric* if $\alpha = -1$. It is found that particles with integral spin can exist only in symmetric states. They are said to obey *Bose–Einstein statistics* and are called *bosons*. Particles with half-integer spin exist only in antisymmetric states. They obey *Fermi–Dirac statistics* and are called *fermions*. The only known explanation for this state of affairs (the *spin–statistics theorem*) comes from relativistic field theories and will be touched on in chapter 7.

For the moment, we deal only with bosons. The order of k labels in a basis vector is immaterial: the same set of labels in any order identify the same vector. It is a simple matter to allow for variable numbers of particles to be present. We simply include in the Hilbert space state vectors with arbitrary numbers of k labels. The orthonormality condition for these vectors is a bit cumbersome to write down correctly. I shall exhibit an expression for it, and then explain its meaning. The expression is

$$\langle k_1, k_2, \ldots, k_N | k'_1, k'_2, \ldots, k'_{N'} \rangle = C\delta_{NN'} \sum_P \delta(k_1 - k'_{P(1)})\delta(k_2 - k'_{P(2)}) \ldots \delta(k_N - k'_{P(N)}). \tag{6.1}$$

We want this scalar product to be zero unless the two vectors represent the same physical state. They must first of all have the same number of particles, which accounts for $\delta_{NN'}$. Then, we need delta function constraints to ensure that each vector represents the same set of single-particle states. Each delta function in (6.1) stands for a product of deltas for each variable represented by k: a Kronecker symbol for discrete variables and a Dirac function for continuous ones. If we list the k labels of a given vector in a different order, we still have the same vector. Therefore, we must arrange matters so that the constraints will be satisfied if any permutation of the labels $k'_1, \ldots, k'_N$ matches the set $k_1, \ldots, k_N$. In (6.1), the set of numbers $P(1), \ldots, P(N)$ is a permutation of $1, \ldots, N$, and we achieve the desired effect by summing over all permutations. If, say, n of the k's are equal, then $n!$ of the terms in this sum will simultaneously be satisfied and, to get the correct normalization, we divide by $n!$. If there are several sets of equal k's, then we divide by the $n!$ for each set, and this normalization factor is denoted by C.

If at least one of the variables represented by k is continuous, it will be extremely rare for two k's to have exactly the same values, and C is almost always equal to 1. In mathematical terms, the Dirac delta function makes good sense only when it appears inside an integral and, for readers who understand such matters, 'almost always' means 'except on a set of zero measure'. It often happens that all the variables in k have only a discrete set of values. For example, if the particles are confined to a cubical box of side L, then each momentum component can have only the values $2\pi\hbar n/L$, where n is an integer. In that case, it is possible to use a different notation in which $k_1, k_2, \ldots$ are the allowed values of k, rather than the k associated with different particles. The basis vectors can then be denoted by $|n_1, n_2, \ldots\rangle$, where n_i is the number of particles in the state k_i. This is called the *occupation number representation,* n_i being the occupation numbers of single-particle states. The orthonormality condition can be written much more straightforwardly as

$$\langle n_1', n_2', \ldots, | n_1, n_2, \ldots \rangle = \delta_{n_1 n_1'} \delta_{n_2 n_2'} \ldots . \tag{6.2}$$

At this point, it is interesting to note the greater generality of the formulation of quantum theory in terms of state vectors as opposed to wavefunctions. In the Schrödinger picture, the time-dependent state vector is some linear combination of basis vectors:

$$|\Psi(t)\rangle = \sum_{n_1, n_2, \ldots} \Psi_{n_1 n_2 \ldots}(t) | n_1, n_2, \ldots \rangle . \tag{6.3}$$

In a quite natural way, this represents in general a superposition of states in which the system contains different numbers of particles. If the system does contain a fixed number N of particles, then only those coefficients for which the occupation numbers add to N will be non-zero. If the Hamiltonian does not allow for processes in which particles are created or destroyed, then this number will be conserved. If such processes are possible, then even if we start with a definite number of particles, the number remaining after some period of time will be uncertain, and the superposition will contain states with different numbers of remaining particles.

It is now possible to introduce creation and annihilation operators which convert a given basis vector into one with an extra particle or one with a particle missing. In the occupation number representation, the process is precisely analogous to changing the number of quanta of energy in the state of an harmonic oscillator. For each single-particle state k, we define operators $\hat{a}_k$ and $\hat{a}_k^\dagger$ by

$$\hat{a}(k_i) | n_1, n_2, \ldots, n_i, \ldots \rangle = n_i^{1/2} | n_1, n_2, \ldots, (n_i - 1), \ldots \rangle \tag{6.4}$$

$$\hat{a}^\dagger(k_i) | n_1, n_2, \ldots, n_i, \ldots \rangle = (n_i + 1)^{1/2} | n_1, n_2, \ldots, (n_i + 1), \ldots \rangle \tag{6.5}$$

Since each operator affects only one of the occupation numbers, it is easy to show that operators for different k's commute, while those for the same k satisfy (5.51). In summary, the commutation relations are

$$[\hat{a}(k_i), \hat{a}(k_j)] = [\hat{a}^\dagger(k_i), \hat{a}^\dagger(k_j)] = 0 \tag{6.6}$$

$$[\hat{a}(k_i), \hat{a}^\dagger(k_j)] = \delta_{ij} . \tag{6.7}$$

If some of the k variables are continuous, we must revert to the previous representation. The commutation relation (6.7) becomes

$$[\hat{a}(k), \hat{a}^\dagger(k')] = \delta(k - k') . \tag{6.8}$$

If we restrict attention to basis vectors whose k arguments are all different, then the action of the creation and annihilation operators is

$$\hat{a}^{\dagger}(k)|k_1, k_2, \ldots, k_N\rangle = |k_1, k_2, \ldots, k_N, k\rangle \tag{6.9}$$

$$\hat{a}(k)|k_1, k_2, \ldots, k_N\rangle = \sum_{n=1}^{N} \delta(k - k_n)|k_1, k_2, \ldots, (k_n), \ldots, k_N\rangle \tag{6.10}$$

where, in the second equation (k_n) denotes a label which is missing from the original list. By acting with $[\hat{a}, \hat{a}^{\dagger}]$ on an arbitrary basis vector, it is easily verified that (6.9) and (6.10) imply the relation (6.8). Readers will also find it instructive to verify from the above equations that $\hat{a}^{\dagger}(k)$ is indeed the adjoint of $\hat{a}(k)$.

The entire set of basis vectors can be constructed by the method we used in the case of the harmonic oscillator. We start from the vacuum state $|0\rangle$, which contains no particles, and use the creation operator to populate it:

$$|k_1, k_2, \ldots, k_N\rangle = \hat{a}^{\dagger}(k_N) \ldots \hat{a}^{\dagger}(k_2)\hat{a}^{\dagger}(k_1)|0\rangle. \tag{6.11}$$

The Hilbert space constructed in this way is called a *Fock space*. A subtle point is worth noting. When particles interact with each other, it is still possible to form state vectors in terms of single-particle states, but these will not, in general, be energy eigenstates. It is not necessarily true that every possible state of the system can be represented as a superposition of the Fock basis vectors, so the Fock space constructed according to (6.11) may be only a part of the whole Hilbert space. For most purposes, though, it will not be necessary to worry about this.

6.2 Field Operators and Observables

From now on, we always take the single-particle states to be momentum eigenstates. For the moment, we consider only spinless particles, so k stands just for the three momentum components, or rather for the wavevector $\boldsymbol{k} = \boldsymbol{p}/\hbar$. The wavefunction for a single particle in the state $|\Psi(t)\rangle$ can be written as

$$\begin{aligned}\Psi(\boldsymbol{x}, t) = \langle \boldsymbol{x}|\Psi(t)\rangle &= (2\pi)^{-3/2} \int \mathrm{d}^3k\, \mathrm{e}^{\mathrm{i}\boldsymbol{k}\cdot\boldsymbol{x}} \langle \boldsymbol{k}|\Psi(t)\rangle \\ &= (2\pi)^{-3/2} \int \mathrm{d}^3k\, \mathrm{e}^{\mathrm{i}\boldsymbol{k}\cdot\boldsymbol{x}} \langle 0|\hat{a}(k)|\Psi(t)\rangle.\end{aligned} \tag{6.12}$$

The annihilation operator $\hat{a}$ creates the one-particle bra vector from the vaccum because it is the adjoint of $\hat{a}^{\dagger}$. In the non-relativistic theory, we define the Schrödinger-picture *field operators* by

$$\hat{\psi}(\boldsymbol{x}) = (2\pi)^{-3/2} \int \mathrm{d}^3k\, \mathrm{e}^{\mathrm{i}\boldsymbol{k}\cdot\boldsymbol{x}}\, \hat{a}(k) \tag{6.13}$$

$$\hat{\psi}^{\dagger}(\boldsymbol{x}) = (2\pi)^{-3/2} \int \mathrm{d}^3k\, \mathrm{e}^{-\mathrm{i}\boldsymbol{k}\cdot\boldsymbol{x}}\, \hat{a}^{\dagger}(k). \tag{6.14}$$

Obviously, these create or annihilate a particle at a definite point $\boldsymbol{x}$, rather than in a definite momentum state. In relativistic theories, we shall find that the situation is a little more complicated because of the need to maintain Lorentz covariance. The commutation relations for the field operators follow from those of $\hat{a}(k)$ and $\hat{a}^\dagger(k)$, and are

$$[\hat{\psi}(\boldsymbol{x}), \hat{\psi}(\boldsymbol{x}')] = [\hat{\psi}^\dagger(\boldsymbol{x}), \hat{\psi}^\dagger(\boldsymbol{x}')] = 0 \tag{6.15}$$

$$[\hat{\psi}(\boldsymbol{x}), \hat{\psi}^\dagger(\boldsymbol{x}')] = \delta(\boldsymbol{x} - \boldsymbol{x}'). \tag{6.16}$$

The operators which represent observable properties of many-particle systems are constructed from the creation and annihilation operators or from the field operators. The operator $\hat{n}(k) = \hat{a}^\dagger(k)\hat{a}(k)$ is a number operator, which counts the number of particles in state k, if k is discrete, or $\hat{n}(k)\mathrm{d}^3k$ counts the number of particles in the momentum range d^3k. The total number of particles is counted by the operator

$$\hat{N} = \int \mathrm{d}^3k\, \hat{a}^\dagger(k)\hat{a}(k) = \int \mathrm{d}^3x\, \hat{\psi}^\dagger(\boldsymbol{x})\hat{\psi}(\boldsymbol{x}) \tag{6.17}$$

and, by summing $\hbar\boldsymbol{k}$ times the number of particles having that momentum, we find that the following operator represents the total momentum:

$$\hat{\boldsymbol{P}} = \int \mathrm{d}^3k\, \hbar\boldsymbol{k}\, \hat{a}^\dagger(k)\hat{a}(k) = \int \mathrm{d}^3x\, \hat{\psi}^\dagger(\boldsymbol{x})(-\mathrm{i}\hbar\boldsymbol{\nabla})\hat{\psi}(\boldsymbol{x}). \tag{6.18}$$

The number and total momentum are *one-body* operators, in the sense that they represent the total for the system of a property possessed by individual particles. Kinetic energy, mass, electric charge and the potential energy due to an externally applied field are examples of other properties of the same kind. There is clearly a general rule for constructing one-body operators. If $A(\boldsymbol{x}, -\mathrm{i}\hbar\boldsymbol{\nabla})$ is the wave-mechanical operator which represents some property of a single particle, then the total property for the whole system is represented by

$$\hat{A} = \int \mathrm{d}^3x\, \hat{\psi}^\dagger(\boldsymbol{x})A(\boldsymbol{x}, -\mathrm{i}\hbar\boldsymbol{\nabla})\hat{\psi}(\boldsymbol{x}). \tag{6.19}$$

We may also consider operators which depend for their definition on two or more particles at a time. An example of a *two-body* operator is a potential $V(\boldsymbol{x}, \boldsymbol{x}')$, such as the Coulomb potential, which acts between two particles. In a state with particles at the points $\boldsymbol{x}_1, \ldots, \boldsymbol{x}_N$, the total potential energy is

$$\begin{aligned} V &= \tfrac{1}{2}\sum_{i,j=1}^{N} V(\boldsymbol{x}_i, \boldsymbol{x}_j) \\ &= \tfrac{1}{2}\int \mathrm{d}^3x\, \mathrm{d}^3x'\, V(\boldsymbol{x}, \boldsymbol{x}') \sum_{i,j=1}^{N} \delta(\boldsymbol{x} - \boldsymbol{x}_i)\delta(\boldsymbol{x}' - \boldsymbol{x}_j) \end{aligned} \tag{6.20}$$

the terms with $i = j$ being excluded from the sum. This will be correctly represented if we can find an operator which, when acting on any state of the form $|\boldsymbol{x}_1, \ldots, \boldsymbol{x}_N\rangle$ gives the same state multiplied by the sum of delta functions in (6.20). The action of the field operators on this state is exactly analogous to (6.9) and (6.10), and I leave it as an exercise for readers to verify that the total potential energy is represented by

$$\hat{V} = \tfrac{1}{2}\int \mathrm{d}^3x\,\mathrm{d}^3x'\,\hat{\psi}^\dagger(\boldsymbol{x})\hat{\psi}^\dagger(\boldsymbol{x}')V(\boldsymbol{x}, \boldsymbol{x}')\hat{\psi}(\boldsymbol{x}')\hat{\psi}(\boldsymbol{x}). \tag{6.21}$$

6.3 Equation of Motion and Lagrangian Formalism for Field Operators

We have dealt so far only with Schrödinger-picture field operators. In the Heisenberg picture, time-dependent operators are defined by the usual method through (5.34):

$$\hat{\psi}(\boldsymbol{x}, t) = \mathrm{e}^{\mathrm{i}\hat{H}t/\hbar}\hat{\psi}(\boldsymbol{x})\mathrm{e}^{-\mathrm{i}\hat{H}t/\hbar}. \tag{6.22}$$

For a system of free particles, the Hamiltonian is just the kinetic energy operator. Because the Hamiltonian commutes with itself, it can be expressed in terms of either Schrödinger-picture or Heisenberg-picture fields:

$$\hat{H} = \int \mathrm{d}^3x\,\hat{\psi}^\dagger(\boldsymbol{x})\left(-\frac{\hbar^2}{2m}\nabla^2\right)\hat{\psi}(\boldsymbol{x}) = \int \mathrm{d}^3x\,\hat{\psi}^\dagger(\boldsymbol{x}, t)\left(-\frac{\hbar^2}{2m}\nabla^2\right)\hat{\psi}(\boldsymbol{x}, t). \tag{6.23}$$

Readers to whom this is not obvious should verify it by substituting (6.22) into the second expression. The same is true if $\hat{H}$ contains a potential energy of the form (6.21), or a one-body external potential. The Heisenberg-picture operators satisfy the commutation relations (6.15) and (6.16), provided that all operators are evaluated at the same time; these are called *equal-time commutation relations*. By using them in (5.35), we can find the equation of motion for $\hat{\psi}(\boldsymbol{x}, t)$. If we include the potential (6.21) and an external potential $U(\boldsymbol{x})$, the result is

$$\begin{aligned}\mathrm{i}\hbar\frac{\partial}{\partial t}\hat{\psi}(\boldsymbol{x}, t) = &-\frac{\hbar^2}{2m}\nabla^2\hat{\psi}(\boldsymbol{x}, t) + U(\boldsymbol{x})\hat{\psi}(\boldsymbol{x}, t)\\ &+ \int \mathrm{d}^3x'\,\hat{\psi}^\dagger(\boldsymbol{x}', t)V(\boldsymbol{x}, \boldsymbol{x}')\hat{\psi}(\boldsymbol{x}', t)\hat{\psi}(\boldsymbol{x}, t).\end{aligned} \tag{6.24}$$

When the two-body potential is absent, this is the same as the Schrödinger equation satisfied by the wavefunction. This is just as well, since the wavefunction (6.12) can be written in the Heisenberg picture as $\Psi(\boldsymbol{x}, t) = \langle 0|\hat{\psi}(\boldsymbol{x}, t)|\Psi\rangle$, and it must obey the Schrödinger equation.

I shall now show that the whole structure of second quantization can be obtained from a Lagrangian formalism, by means of the canonical quantization prescription described in chapter 5. For brevity, I shall give the derivation just for the free-particle theory whose Hamiltonian is (6.23), but readers should be able to extend it without difficulty to the case of particles interacting via a two-body potential. Consider the action defined by

$$S = \int \mathrm{d}t\,\mathrm{d}^3x\,\psi^*(\boldsymbol{x}, t)\left(\mathrm{i}\hbar\frac{\partial}{\partial t} + \frac{\hbar^2}{2m}\nabla^2\right)\psi(\boldsymbol{x}, t) \tag{6.25}$$

where $\psi(\boldsymbol{x}, t)$ is a complex function, not, for the moment, a field operator. In chapter 3, we saw that Maxwell's equations for the electromagnetic field could be obtained by finding the Euler–Lagrange equations for an action somewhat akin to this. In this case, the real and imaginary parts of ψ are independent functions, but it is more convenient to treat ψ and ψ^* as the independent variables. By varying $\psi^*(\boldsymbol{x}, t)$ we obtain Schrödinger's equation for $\psi(\boldsymbol{x}, t)$, and by varying $\psi(\boldsymbol{x}, t)$ itself we get the complex conjugate of the same equation. The values of $\psi(\boldsymbol{x}, t)$ at each point $\boldsymbol{x}$ form an infinite set of generalized coordinates, and there is an infinite set of conjugate momenta, which form a function $\Pi(\boldsymbol{x}, t)$. This function is found by functional differentiation (which is explained in appendix 1 for readers who are not familiar with it):

$$\Pi(\boldsymbol{x}, t) = \frac{\delta S}{\delta\dot{\psi}(\boldsymbol{x}, t)} = \mathrm{i}\hbar\psi^*(\boldsymbol{x}, t). \tag{6.26}$$

In (6.25), the time derivative acts only on ψ or, if we integrate by parts and take ψ to vanish at the boundaries of our system, it acts only on ψ^*. Therefore, we cannot define conjugate momenta for ψ and ψ^* at the same time. This kind of difficulty was mentioned in chapter 5, and the solution which works in this case is to ignore ψ^* as an independent variable, except for the purpose of deriving the equation of motion. Then the Hamiltonian is

$$H = \int \mathrm{d}^3x\,\Pi\dot{\psi} - L = \int \mathrm{d}^3x\,\psi^*\left(-\frac{\hbar^2}{2m}\nabla^2\right)\psi. \tag{6.27}$$

To get back to our quantum theory, we simply follow the canonical quantization scheme, replacing $\psi(\boldsymbol{x}, t)$ with the field operator $\hat{\psi}(\boldsymbol{x}, t)$ and its complex conjugate with $\hat{\psi}^\dagger(\boldsymbol{x}, t)$. The Hamiltonian (6.27) becomes identical to (6.23). In the canonical commutator (5.41), the coordinate $\hat{q}_i$ becomes $\hat{\psi}(\boldsymbol{x}, t)$, the momentum $\hat{p}_j$ is replaced by the momentum $\hat{\Pi}(\boldsymbol{x}', t)$ obtained from (6.26), and the Kronecker symbol is replaced by $\delta(\boldsymbol{x} - \boldsymbol{x}')$. The result is none other than the commutator (6.16) for the field operators. For the kind of theory we have been studying, this new bit of formalism provides no new information, since we have just returned to our starting point. Suppose, however, that we wish to treat the electromagnetic field as a quantum system. The analysis we have just been through shows us how to do this, although there is an added difficulty to be overcome, as will be discussed in chapter 9. The vector potential A^μ becomes a field operator, which obeys Maxwell's equations rather than the Schrödinger equation, and its commutation relations will again be given by the canonical prescription. In the light of our experience in this chapter, we may anticipate that this field operator can be interpreted in terms of creation and annihilation operators for particles, namely *photons*, which are quanta of electromagnetic energy. In fact, the Lagrangian formalism provides the most convenient basis for most relativistic field theories.

6.4 Second Quantization for Fermions

Many of the most important applications of non-relativistic field theory concern electronic systems. Electrons have spin $\frac{1}{2}$, and are therefore fermions. Although the consequences of this are far reaching, the modifications needed in the basic theory are quite simple. We must take the label k of single-particle states to include the variable s, which measures the component of spin along a chosen quantization axis and has the values $\pm\frac{1}{2}$. Slightly more tricky is the antisymmetry of multi-particle states. For simplicity, let us consider two-particle states, for which $|k, k'\rangle = -|k', k\rangle$. I shall follow the common practice of using $\hat{b}(k)$ and $\hat{b}^\dagger(k)$ to denote fermionic annihilation and creation operators, to distinguish them from bosonic ones. It is now important to keep track of the ordering of k labels in a state vector. A sensible convention when using $\hat{b}^\dagger$ to add a particle is to place the label for the added particle at the end of the list. Thus

$$\hat{b}^\dagger(k')\hat{b}^\dagger(k)|0\rangle = |k, k'\rangle = -|k', k\rangle = -\hat{b}^\dagger(k)\hat{b}^\dagger(k')|0\rangle. \quad (6.28)$$

Similarly, the annihilation operator can be regarded as removing the last particle in the list. It can, of course, remove any particle in the state, so to write down the result we first move the particle to the end, if necessary, incurring a minus sign for each interchange. For a two-particle state:

$$\hat{b}(k)|k_1, k_2\rangle = \delta(k - k_2)|k_1\rangle - \delta(k - k_1)|k_2\rangle = -\hat{b}(k)|k_2, k_1\rangle. \quad (6.29)$$

More generally, (6.10) is modified to read

$$\hat{b}(k)|k_1, k_2, \ldots, k_N\rangle = \sum_{n=1}^{N} (-1)^{N-n}\delta(k - k_n)|k_1, k_2, \ldots, (k_n), \ldots, k_N\rangle. \quad (6.30)$$

Evidently, $\hat{b}$ and $\hat{b}^\dagger$ cannot obey the commutation relations (6.6)–(6.8). In fact, as it is not difficult to see, the relations consistent with the antisymmetry of the state vectors are the *anticommutation relations*

$$\{\hat{b}(k), \hat{b}(k')\} = \{\hat{b}^\dagger(k), \hat{b}^\dagger(k')\} = 0 \tag{6.31}$$

$$\{\hat{b}(k), \hat{b}^\dagger(k')\} = \delta(k - k') \tag{6.32}$$

where the anticommutator is defined by $\{\hat{A}, \hat{B}\} = \hat{A}\hat{B} + \hat{B}\hat{A}$. In the occupation number representation, (6.4) and (6.5) are still valid for fermions, but each occupation number can be only 0 or 1. Thus, if we act twice with the same creation operator, (6.31) shows that we get zero instead of two particles in the same state. This, of course, is the second-quantization version of the Pauli exclusion principle. In the same way, the anticommutation relations which replace (6.15) and (6.16) for the field operators are

$$\{\hat{\psi}_s(\boldsymbol{x}), \hat{\psi}_{s'}(\boldsymbol{x}')\} = \{\hat{\psi}_s^\dagger(\boldsymbol{x}), \hat{\psi}_{s'}^\dagger(\boldsymbol{x}')\} = 0 \tag{6.33}$$

$$\{\hat{\psi}_s(\boldsymbol{x}), \hat{\psi}_{s'}^\dagger(\boldsymbol{x}')\} = \delta_{ss'}\delta(\boldsymbol{x} - \boldsymbol{x}'). \tag{6.34}$$

After the Fourier transformation (6.13), we get, of course, two field operators, labelled by s, corresponding to the two spin polarizations.

These are all the changes we need to make, in order to accommodate fermions. In equations (6.19), (6.21) and (6.24), it is only necessary to add spin labels to the fields and include a sum over these labels with each space integration. I have ordered the operators in these expressions so as to make them correct for both fermions and bosons.

Exercises

6.1. Let $A(\boldsymbol{x}, -i\hbar\boldsymbol{\nabla})$, $B(\boldsymbol{x}, -i\hbar\boldsymbol{\nabla})$ and $C(\boldsymbol{x}, -i\hbar\boldsymbol{\nabla})$ be wave-mechanical operators with the commutation relation $[A, B] = C$. Show that the corresponding second-quantized one-body operators $\hat{A}$, $\hat{B}$ and $\hat{C}$ satisfy the same commutation relation, if the field operators have either commutation or anticommutation relations.

6.2. Using time-independent field operators, show that the Hamiltonian (6.23) can be expressed as

$$\hat{H} = \int \mathrm{d}^3k \hbar\omega(\boldsymbol{k})\hat{a}^\dagger(\boldsymbol{k})\hat{a}(\boldsymbol{k})$$

where $\omega(\boldsymbol{k}) = \hbar k^2/2m$. Show that for any n, $H^n a(\boldsymbol{k}) = a(\boldsymbol{k})[H - h\omega(\boldsymbol{k})]^n$, and hence that the time-dependent field operator in (6.22) is

$$\hat{\psi}(\boldsymbol{x}, t) = (2\pi)^{-3/2} \int \mathrm{d}^3k \exp[\mathrm{i}\boldsymbol{k}\cdot\boldsymbol{x} - \mathrm{i}\omega(\boldsymbol{k})t]\hat{a}(k).$$

Check that this works for both bosons and fermions.

7

Relativistic Wave Equations and Field Theories

Up to this point, our study of quantum theory has concerned itself with the behaviour of particles which inhabit a Galilean spacetime. For many purposes, in atomic, molecular and condensed matter physics, this theory is quite adequate. We saw in earlier chapters, however, that our actual spacetime has a structure which is much closer to that of the Minkowski spacetime of special relativity and that more general geometrical structures than this must be considered when gravitational phenomena are significant. From a purely theoretical point of view, it is therefore important to formulate quantum theory in a way which is consistent with these more general spacetimes. The benefits of constructing a relativistic quantum theory actually go far beyond the aesthetic satisfaction of making our geometrical and quantum-mechanical reasoning compatible. For one thing, we shall discover that the relativistic theory provides a deeper understanding of the nature of spin and the distinction between fermions and bosons, which in the non-relativistic theory appear simply as facts of life which we must strive to accommodate. Also, of course, there are many situations in which relativistic effects become observable, for which non-relativistic theory provides no explanation. The most obvious are high-energy scattering experiments, in which particles acquire kinetic energies comparable with or greater than their rest energies mc^2, and the correct 4-momentum (3.34) must be used. There are, however, more subtle effects, such as the spin–orbit coupling which is essential for interpreting atomic spectra, which are also of relativistic origin.

For the most part, I shall deal only with quantum theory in Minkowski spacetime, which is well understood. At the end of the chapter, I shall allude briefly to the question of setting up quantum theories in curved spacetimes, which involves some surprising difficulties and is not understood quite so well. If our world is thoroughly quantum-mechanical (and the prevailing view is that it must be), then we ought to treat the geometrical structures themselves in quantum-mechanical terms, which means constructing a quantum theory of gravity. How this can be done is the subject of intensive and highly technical debate, and I can do no more than hint (in chapter 12) at some of the ideas which have been put forward.

I shall write all equations having to do with relativistic theories in terms of *natural units*, which are defined so that $c = \hbar = 1$. This leaves us free to define one fundamental unit, which is normally taken to be energy, measured in MeV. In these units, length and time have the same dimensions and are measured in $(\mathrm{MeV})^{-1}$. Mass, momentum and energy have the same dimensions, being measured in MeV. Appendix 3 discusses these units in more detail and gives some conversion factors between natural and laboratory units.

7.1 The Klein–Gordon Equation

If we wish to invent a Minkowski space version of wave mechanics, the first problem to be overcome is that the Schrödinger equation (5.7) expresses the non-relativistic relationship between energy and momentum. The relationship in special relativity is that implied by (3.34), which may be written in various ways as

$$E^2 - \boldsymbol{p}\cdot\boldsymbol{p} = (p^0)^2 - \boldsymbol{p}\cdot\boldsymbol{p} = p_\mu p^\mu = m^2. \qquad (7.1)$$

At least for free particles, it is a simple matter to convert this into a relativistic wave equation, called the *Klein–Gordon equation*. We just substitute (5.5) and (5.6) and let the resulting operator act on a wavefunction ϕ:

$$\left[\frac{\partial^2}{\partial t^2} - \nabla^2 + m^2\right]\phi(\boldsymbol{x}, t) = [\partial_\mu\partial^\mu + m^2]\phi(x) = [\Box + m^2]\phi(x) = 0. \qquad (7.2)$$

The d'Alembertian operator $\Box$ defined here is the Minkowski space version of the Laplacian ∇^2; it is sometimes written as $\Box^2$. We should certainly expect the Klein–Gordon equation to be valid for free relativistic particles, but whether it should be regarded as a generalization of the Schrödinger equation is a moot point, since it is not related in a simple way to a time evolution of the form (5.31).

It is important to ask how the wavefunction is to be interpreted in a relativistic context. If (7.2) is to have a Lorentz covariant meaning, then ϕ must be some kind of 4-tensor, as discussed in §3.5. For now, we shall consider spinless particles whose wavefunctions have only a single component, so ϕ must be a scalar. This implies that the probability density is not correctly given by (5.4). Like the number density in

(3.36), it must be the time-like component of a conserved 4-vector, whose other components are the probability current density. In a loose, intuitive manner, we can think of the probability density as a kind of number density. In the non-relativistic theory, the current density is

$$\boldsymbol{j}(x) = \frac{1}{m}\,\mathrm{Re}[\Psi^*(-\mathrm{i}\boldsymbol{\nabla})\Psi] = \frac{\mathrm{i}}{2m}\,\Psi^*\overleftrightarrow{\boldsymbol{\nabla}}\Psi \tag{7.3}$$

where the notation $A\overleftrightarrow{\boldsymbol{\nabla}}B$ means $A\boldsymbol{\nabla}B - (\boldsymbol{\nabla}A)B$. Intuitively, this expression is rather like (velocity × density), as in (3.37). More precisely, the reason for (7.3) is that it satisfies the equation of continuity $\partial P/\partial t + \boldsymbol{\nabla}\cdot\boldsymbol{j} = 0$, as may easily be verified by using the Schrödinger equation. In the present case, it may similarly be verified using the Klein–Gordon equation that the equation of continuity in the form $\partial_\mu j^\mu = 0$ is satisfied, provided that we identify the 4-vector probability current density as

$$j^\mu(x) = \frac{\mathrm{i}}{2m}\,\phi^*\overleftrightarrow{\partial}^\mu\phi. \tag{7.4}$$

This is fortunate, in so far as (7.4) is manifestly a 4-vector, so that the equation of continuity is Lorentz covariant. The unfortunate thing about (7.4) is that j^0, which we want to identify as the probability density, is, unlike $|\Psi|^2$, not necessarily positive. This is one of two problems which afflict all relativistic wave equations.

The second problem emerges when we look at plane-wave solutions of the Klein–Gordon equation. Evidently, the function

$$\phi_k(x) = \exp(-\mathrm{i}k\cdot x) \tag{7.5}$$

where $k\cdot x = k_\mu k^\mu = k^0 t - \boldsymbol{k}\cdot\boldsymbol{x}$, is a solution of (7.2) and also an energy–momentum eigenfunction, provided that $k^0 = \pm(\boldsymbol{k}^2 + m^2)^{1/2}$. The negative energy solutions are a severe embarrassment, because they imply the existence of single-particle states with energy less than that of the vacuum. Intuitively, this is nonsensical. In fact, there is no lower limit to the energy spectrum. This means that the vacuum is unstable, since an infinite amount of energy could be released from it by the spontaneous creation of particles in negative energy states. We can see that this problem is related to the first one, because it is the negative energy states which give rise to a negative probability density.

Because of these problems, the Klein–Gordon equation does not lead to a tenable wave-mechanical theory of relativistic particles. It is, indeed, impossible to construct such a theory. We shall see shortly that it does lead to a perfectly sensible quantum field theory. To develop this field theory, we follow the canonical quantization procedure explained in §6.3, but the requirement of Lorentz covariance leads to some minor changes. Like the Schrödinger equation, the Klein–Gordon equation can be obtained as an Euler–Lagrange equation from an action. Assuming that ϕ is a complex function, the action is

$$S = \int d^4x \mathscr{L}(\phi) \tag{7.6}$$

where the Lagrangian density is given by

$$\mathscr{L}(\phi) = (\partial_\mu \phi^*)(\partial^\mu \phi) - m^2 \phi^* \phi. \tag{7.7}$$

This action is manifestly a scalar quantity, as we require for a Lorentz covariant theory. In contrast to the non-relativistic theory, it contains the time derivatives of both ϕ and ϕ^*, so two independent canonical momenta can be defined:

$$\Pi(x) = \partial^0 \phi^*(x) \qquad \Pi^*(x) = \partial^0 \phi(x). \tag{7.8}$$

The general solution of the Klein–Gordon equation can be written in terms of energy–momentum eigenfunctions. To ensure that it is a scalar, we first write it in a form which does not distinguish between space and time components:

$$\phi(x) = \int \frac{d^4k}{(2\pi)^3} \delta(k^2 - m^2)\alpha(k)e^{-ik\cdot x}. \tag{7.9}$$

The energy (k^0) integral can be carried out using the delta function. We get two terms, corresponding to the positive and negative energy solutions, with $k^0 = \pm\omega(\boldsymbol{k})$, where $\omega(\boldsymbol{k}) = (\boldsymbol{k}^2 + m^2)^{1/2}$. For reasons which will become apparent, we write the coefficient $\alpha(k)$ as

$$\alpha(k) = \begin{cases} a(\boldsymbol{k}) & \text{for } k^0 = +\omega(\boldsymbol{k}) \\ c^*(-\boldsymbol{k}) & \text{for } k^0 = -\omega(\boldsymbol{k}). \end{cases} \tag{7.10}$$

Then, after changing the sign of $\boldsymbol{k}$ in the negative energy term, we get

$$\phi(x) = \int \frac{d^3k}{[(2\pi)^3 2\omega(\boldsymbol{k})]} [a(\boldsymbol{k})e^{-ik\cdot x} + c^*(\boldsymbol{k})e^{ik\cdot x}]. \tag{7.11}$$

In each term, k^0 stands for $+\omega(\boldsymbol{k})$. The $2\omega(\boldsymbol{k})$ in the denominator appears for the reason explained in appendix 1. Because of this, the coefficients $a(\boldsymbol{k})$ and $c(\boldsymbol{k})$ cannot be obtained from $\phi(x)$ by a simple Fourier transformation. Instead, we have

$$a(\boldsymbol{k}) = i \int d^3x e^{ik\cdot x} \overleftrightarrow{\partial}^0 \phi(x) \qquad c(\boldsymbol{k}) = i \int d^3x e^{ik\cdot x} \overleftrightarrow{\partial}^0 \phi^*(x). \tag{7.12}$$

We are now ready to develop the second-quantized description.

7.2 Scalar Field Theory for Free Particles

As in the non-relativistic case, we carry out second quantization by replacing complex functions with field operators. Because a relativistic theory treats space and time on much the same footing, these are initially given as time-dependent Heisenberg-picture operators. Nevertheless, we are still free only to specify the equal-time commutators. First of all, we have

$$\begin{aligned}[\hat{\phi}(\boldsymbol{x}, t), \hat{\Pi}(\boldsymbol{x}', t)] &= \mathrm{i}\delta(\boldsymbol{x} - \boldsymbol{x}') \\ [\hat{\phi}(\boldsymbol{x}, t), \hat{\phi}(\boldsymbol{x}', t)] &= [\hat{\Pi}(\boldsymbol{x}, t), \hat{\Pi}(\boldsymbol{x}', t)] = 0.\end{aligned} \tag{7.13}$$

Taking the adjoints of these equations, we find that $\hat{\phi}^\dagger$ and $\hat{\Pi}^\dagger$ satisfy exactly the same relations. The two sets of operators $(\hat{\phi}, \hat{\Pi})$ and $(\hat{\phi}^\dagger, \hat{\Pi}^\dagger)$ are to be treated as independent variables, so we also have

$$[\hat{\phi}(\boldsymbol{x}, t), \hat{\phi}^\dagger(\boldsymbol{x}', t)] = [\hat{\Pi}(\boldsymbol{x}, t), \hat{\Pi}^\dagger(\boldsymbol{x}', t)] = [\hat{\phi}(\boldsymbol{x}, t), \hat{\Pi}^\dagger(\boldsymbol{x}', t)] = 0. \tag{7.14}$$

The operator version of (7.12) can be written, in a manner reminiscent of the definition (5.48) of the harmonic oscillator lowering operator, as

$$\hat{a}(\boldsymbol{k}) = \int \mathrm{d}^3x \exp\{\mathrm{i}[\omega(\boldsymbol{k})t - \boldsymbol{k}\cdot\boldsymbol{x}]\}[\omega(\boldsymbol{k})\hat{\phi}(\boldsymbol{x}, t) + \mathrm{i}\hat{\Pi}^\dagger(\boldsymbol{x}, t)] \tag{7.15}$$

$$\hat{c}(\boldsymbol{k}) = \int \mathrm{d}^3x \exp\{\mathrm{i}[\omega(\boldsymbol{k})t - \boldsymbol{k}\cdot\boldsymbol{x}]\}[\omega(\boldsymbol{k})\hat{\phi}^\dagger(\boldsymbol{x}, t) + \mathrm{i}\hat{\Pi}(\boldsymbol{x}, t)]. \tag{7.16}$$

So, by using (7.13) and (7.14), we can work out the commutation relations for the $\hat{a}$ and the $\hat{c}$. The result is that

$$[\hat{a}(\boldsymbol{k}), \hat{a}^\dagger(\boldsymbol{k}')] = [\hat{c}(\boldsymbol{k}), \hat{c}^\dagger(\boldsymbol{k}')] = (2\pi)^3 2\omega(\boldsymbol{k})\delta(\boldsymbol{k} - \boldsymbol{k}') \tag{7.17}$$

while all other commutators between these operators are zero. Apart from a normalization factor, we recognize these as two independent sets of creation and annihilation operators. The effect of the normalization factor for single-particle states is that

$$\langle \boldsymbol{k}|\boldsymbol{k}'\rangle = (2\pi)^3 2\omega(\boldsymbol{k})\delta(\boldsymbol{k} - \boldsymbol{k}'). \tag{7.18}$$

This means that the integral of the probability density over a volume of $1/2\omega(\boldsymbol{k})$ is 1 or, loosely, that this is the volume occupied by the particle. If we make a Lorentz transformation to a new frame of reference, the energy $\omega(\boldsymbol{k})$ transforms, of course, as the time component of a 4-vector. As readers may verify, the volume occupied by the particle changes by the appropriate Fitzgerald contraction factor, so (7.18) is a natural Lorentz covariant normalization.

The fact that we have two sets of creation and annihilation operators leads to the resolution of the problems of negative energies and probability densities. The field theory we have constructed actually describes two species of particles. Particles of one species are called the *antiparticles* of the other. For the sake of argument, I will refer to the particles created by $\hat{a}^\dagger$ as 'particles' and to those created by $\hat{c}^\dagger$ as 'antiparticles'. The solution to the problem of negative energies is apparent from (7.11), when we reinterpret it as a field operator. The coefficient of the positive energy term is, as in the non-relativistic theory, the annihilation operator for particles. However, the coefficient of the negative energy term is not an annihilation operator for particles in negative energy states, but rather a creation operator for positive energy antiparticles. We can construct the Hamiltonian operator by the usual canonical method. It is

$$\begin{aligned}\hat{H} &= \int \mathrm{d}^3x\left[\frac{\partial\hat{\phi}^\dagger}{\partial t}\frac{\partial\hat{\phi}}{\partial t} + (\boldsymbol{\nabla}\hat{\phi}^\dagger)\cdot(\boldsymbol{\nabla}\hat{\phi}) + m^2\hat{\phi}^\dagger\hat{\phi}\right]\\ &= \int \frac{\mathrm{d}^3k}{[(2\pi)^3 2\omega(\boldsymbol{k})]}\,\omega(\boldsymbol{k})[\hat{a}^\dagger(\boldsymbol{k})\hat{a}(\boldsymbol{k}) + \hat{c}^\dagger(\boldsymbol{k})\hat{c}(\boldsymbol{k}) + (2\pi)^3 2\omega(\boldsymbol{k})\delta(0)].\end{aligned} \tag{7.19}$$

In the second expression, the last term comes from rewriting $\hat{c}(\boldsymbol{k})\hat{c}^\dagger(\boldsymbol{k})$ by means of the commutator. If we act on the vacuum state, which contains no particles or antiparticles, the first two terms give zero. The last term is an infinite constant. It may be dropped on the usual grounds that the total energy of a system is only defined up to an arbitrary constant, and the most sensible value for the energy of the vacuum is zero. (If we allow the structure of spacetime to be determined by Einstein's equations (4.15), however, the energy of the vacuum contributes to $T^{\mu\nu}$ and must be considered more carefully.) Another way of looking at this is to remember that the ordering of operators is not unambiguously prescribed by the quantization procedure. We can regard the vanishing of the vacuum energy as a criterion for ordering operators such that annihilation operators appear to the right of creation operators. This is called *normal ordering*. Bearing in mind the normalization in (7.17), we recognize (7.19) as summing the quantity (energy of state $\boldsymbol{k}$) $\times$ (number of particles and antiparticles in state $\boldsymbol{k}$) over positive energy states. Thus, the total energy of any state is positive.

The solution of the problem of negative probabilities is quite similar. We define a number operator $\hat{N}$ by integrating over all space the operator corresponding to the probability density in (7.4):

$$\begin{aligned}\hat{N} = \mathrm{i}\int \mathrm{d}^3x :\hat{j}^0: &= \mathrm{i}\int \mathrm{d}^3x :\hat{\phi}^\dagger(\boldsymbol{x}, t)\overleftrightarrow{\partial}^0\hat{\phi}(\boldsymbol{x}, t):\\ &= \int \frac{\mathrm{d}^3k}{[(2\pi)^3 2\omega(\boldsymbol{k})]}[\hat{a}^\dagger(\boldsymbol{k})\hat{a}(\boldsymbol{k}) - \hat{c}^\dagger(\boldsymbol{k})\hat{c}(\boldsymbol{k})]\end{aligned} \tag{7.20}$$

where the colons :(. . .): denote normal ordering of the creation and annihilation operators. Again, the denominator $(2\pi)^3 2\omega(\boldsymbol{k})$ appears just because of the covariant normalization, and we see that $\hat{N}$ represents the quantity (number of particles − number of antiparticles). A negative value for this quantity simply indicates a state with more antiparticles than particles and presents no difficulty. Another way of expressing this is to assign to each particle a *particle number* $n = 1$ and to each antiparticle a particle number $n = -1$. Then $\hat{N}$ may be said to represent the particle number, rather than the number of particles. In the field operator obtained from (7.11), both terms act on a given state to reduce the particle number by one unit, either by annihilating a particle or by creating an antiparticle. This rule applies to any other properties the particles may possess. For example, if particles carry an electric charge, then their antiparticles carry exactly the opposite charge, and the same is true of all the other *quantum numbers* (lepton number, baryon number, isospin, strangeness, etc) which are required to classify the observed particles. Historically, the existence of antiparticles was first predicted by Dirac (1928, 1929) on the basis of his relativistic wave equation for electrons discussed in the next section, and the antielectron, or positron, was discovered experimentally by Anderson (1933) in cosmic ray showers. All observed particles are indeed found to have antiparticles. However, particles and antiparticles may in some cases be identical. Mathematically, this will be so if $\hat{c}(\boldsymbol{k}) = \hat{a}(\boldsymbol{k})$, which means that the wavefunction (7.11) is real and the corresponding field operator is Hermitian. In that case, the number operator (7.20) is identically zero, and the particle number must be taken as $n = 0$. Clearly, only a restricted range of properties are available to particles which are their own antiparticles; for example they must be electrically neutral. Examples are the photon and the neutral pion. In the case of the photon, the space and time derivatives of its Hermitian field operators are observable quantities, namely electric and magnetic fields.

7.3 The Dirac Equation and Spin-$\frac{1}{2}$ Particles

7.3.1 The Dirac equation

The problems of negative energies and probabilities encountered in connection with the Klein–Gordon equation evidently have something to do with the fact that this equation involves a second time derivative. Dirac attempted to solve these problems by inventing a new wave equation containing only the first time derivative, which is more closely analogous to the non-relativistic Schrödinger equation. As we shall see, it is in fact not possible to solve the problems in this way, and Dirac's theory also makes sense only as a second-quantized field theory. The Dirac equation is nevertheless of vital importance because it predicts the existence of particles with intrinsic angular momentum or *spin* of magnitude $\hbar/2$. Such particles are indeed observed, electrons being perhaps the most familiar, and the Dirac theory is the proper Lorentz covariant means of describing them.

Since special relativity treats time and space on more or less the same footing, an equation which contains only the first time derivative can also contain only first spatial derivatives. The equation must therefore be of the form

$$(\mathrm{i}\gamma^\mu \partial_\mu - m)\psi(x) = 0 \tag{7.21}$$

where the four coefficients γ^μ are constants. We shall see immediately that these coefficients cannot commute with each other. They must therefore be square matrices rather than simple numbers, so the wavefunction $\psi(x)$ must be a column matrix. This wavefunction must also satisfy the Klein–Gordon equation, which simply expresses the relationship between energy and momentum and, indeed, this should be an automatic consequence of the Dirac equation (7.21). Obviously, we get an equation bearing some resemblance to the Klein–Gordon equation if we act twice with the operator $(\mathrm{i}\gamma^\mu \partial_\mu - m)$:

$$(\mathrm{i}\gamma^\mu \partial_\mu - m)^2\psi(x) = (-\gamma^\mu\gamma^\nu\partial_\mu\partial_\nu - 2\mathrm{i}m\gamma^\mu\partial_\mu + m^2)\psi(x) = 0. \tag{7.22}$$

Using the original equation (7.21) and the fact that $\partial_\mu\partial_\nu = \partial_\nu\partial_\mu$, we can rewrite this as

$$(\tfrac{1}{2}\{\gamma^\mu, \gamma^\nu\}\partial_\mu\partial_\nu + m^2)\psi(x) = 0. \tag{7.23}$$

In order for this to be the same as the Klein–Gordon equation (7.2), the γ matrices must satisfy the condition

$$\{\gamma^\mu, \gamma^\nu\} \equiv \gamma^\mu\gamma^\nu + \gamma^\nu\gamma^\mu = 2g^{\mu\nu} \tag{7.24}$$

where $g^{\mu\nu}$ is the (μ, ν) component of the Minkowski space metric tensor (2.8) and is understood to be multiplied by the unit matrix. This is called a *Clifford algebra*.

The smallest matrices which can be made to obey the Clifford algebra are 4×4 matrices, and we shall consider only these. Even so, there are infinitely many representations of the algebra, that is infinitely many sets of four 4×4 matrices which satisfy the condition (7.24). Each representation gives a different, but equivalent, mathematical description of the same physical situation. For this reason, it is possible to derive all the physical consequences of the theory from the fact that the γ matrices satisfy (7.24), without ever specifying exactly what these matrices are. Nevertheless, it is often helpful to have in mind at least one possible set of such matrices. A standard representation is

$$\gamma^0 = \begin{pmatrix} I & 0 \\ 0 & -I \end{pmatrix} \qquad \gamma^i = \begin{pmatrix} 0 & \sigma^i \\ -\sigma^i & 0 \end{pmatrix} \tag{7.25}$$

where each entry is itself a 2×2 matrix, I being the unit matrix and σ^i

the *Pauli matrices*, given by

$$\sigma^1 = \begin{pmatrix} 0 & 1 \\ 1 & 0 \end{pmatrix} \qquad \sigma^2 = \begin{pmatrix} 0 & -\mathrm{i} \\ \mathrm{i} & 0 \end{pmatrix} \qquad \sigma^3 = \begin{pmatrix} 1 & 0 \\ 0 & -1 \end{pmatrix}. \quad (7.26)$$

Readers who are familiar with the non-relativistic theory of spin-$\frac{1}{2}$ particles, or who have studied appendix 2, will recognize these matrices as the operators which represent the three components of the particle's intrinsic angular momentum. We shall shortly see that this is no coincidence.

7.3.2 Lorentz covariance and spin

As we have discussed in some detail in previous chapters, the equations which express laws of physics are expected to take the same form when referred to any frame of reference. In Minkowski space, it is usually convenient to consider only Cartesian frames, in which the metric tensor has the simple form (2.8). Because we have done this, the metric tensor does not appear explicitly in most of our equations, and we expect the form of these equations to remain the same only when we make Lorentz transformations of the form (3.25) (or, more generally, Poincaré transformations which include spacetime translations). In classical physics, this property of Lorentz covariance is guaranteed if all equations can be expressed in terms of 4-tensors. A Lorentz transformation rearranges the components of a tensor amongst themselves, in such a way that the form of the tensor equations is preserved. In (7.2), we assumed that the wavefunction $\phi(x)$ was the simplest kind of tensor, namely a scalar. The detailed meaning of this is as follows. Suppose that the state of the particle is described by two observers, using sets of coordinates x and x', related by (3.25). The same state will be described by these observers in terms of two wavefunctions $\phi(x)$ and $\phi'(x')$. In general, ϕ and ϕ' are different functions, but if x and x' are the coordinates in the two frames of the same spacetime point, then $\phi(x) = \phi'(x')$. Since $\partial_\mu \partial^\mu = \partial_{\mu'} \partial^{\mu'}$, each wavefunction also satisfies the Klein–Gordon equation written in its own set of coordinates.

The Dirac wavefunction is a four-component column matrix, so we may expect that, on transforming to a new frame of reference, not only will the components be different functions of the new coordinates, but they will also be rearranged amongst themselves. It turns out that this rearrangement is not the same as those specified by any of the tensor transformation laws (2.19). Although ψ has four components, these do not refer to spacetime directions as do the components of a 4-vector. They actually refer, as we shall see, to different states in which the particle can exist. I shall label these components as ψ_α, where α has the values 1, . . ., 4. Thus, ψ is a geometrical object of a kind we have not previously met. It is called a *spinor*, and its transformation law can be written as

$$\psi'_\alpha(x') = S_{\alpha\beta}(\Lambda)\psi_\beta(x). \tag{7.27}$$

The two sets of coordinates are again related by (3.25). The new transformation matrix S is usually represented as a function of the matrix Λ as I have done here, but it is probably clearer to think of S and Λ as different matrices which both depend on the same parameters, namely rotation angles and boost velocities such as those in (3.26) and (3.27).

If the Dirac equation is to be covariant, then the transformed wavefunction must satisfy the equation

$$(\mathrm{i}\gamma^{\mu'}\partial_{\mu'} - m)\psi'(x') = 0. \tag{7.28}$$

The transformed derivative is $\partial_{\mu'} = \Lambda^{\mu}{}_{\mu'}\partial_\mu$, and it might appear that the γ matrices should transform as a contravariant 4-vector. This is not correct, though. A constant 4-vector singles out a special direction in spacetime, and the whole point of covariance is that no such special direction exists. The new equation (7.28) is supposed to have the same form as the original equation (7.21), and this means that both observers are entitled to use the same set of γ matrices. Thus, if the old matrices have the numerical values in (7.25) and (7.26), then so do the new ones. But the index μ' indicates that they are associated with the x' coordinate axes. From this requirement, we can work out what the spinor transformation matrix must be. We substitute (7.27) into (7.28), rewrite $\partial_{\mu'}$ in terms of ∂_μ, and multiply by S^{-1} to get

$$(\mathrm{i}S^{-1}(\Lambda)\gamma^{\mu'}\Lambda^{\mu}{}_{\mu'}S(\Lambda)\partial_\mu - m)\psi(x) = 0. \tag{7.29}$$

Remembering that $\Lambda^{\mu}{}_{\mu'}\Lambda^{\mu'}{}_{\nu} = \delta^{\mu}{}_{\nu}$, we see that this is the same as (7.21) provided that

$$S^{-1}(\Lambda)\gamma^{\mu'}S(\Lambda) = \Lambda^{\mu'}{}_{\mu}\gamma^{\mu}. \tag{7.30}$$

Only if a matrix S with this property can be found will the Dirac equation be Lorentz covariant.

It is sufficient to find S for the case of infinitesimal transformations. This will give us the generators for Lorentz transformations, and the matrix for finite transformations can be built up by exponentiation, in just the same way as for spacetime translations. By expanding (3.26) and (3.27) in powers of the rotation angle or boost velocity, we see that Λ can be written as

$$\Lambda^{\mu'}{}_{\mu} = \delta^{\mu'}{}_{\mu} + g^{\mu'\nu'}\omega_{\nu'\mu} + \ldots \tag{7.31}$$

where $\omega_{\nu\mu}$ is antisymmetric in its two indices, each of its components being proportional to a rotation angle or boost velocity. A general transformation, which is some combination of rotations and boosts, can be written in the same way. Usually, it is meaningless to write symbols like δ and ω with two indices belonging to different coordinate systems. Here it does make sense, because the two sets of coordinates differ only by an infinitesimal amount. The matrix S must be a function of $\omega_{\mu\nu}$, and we write its infinitesimal form as

$$S(\Lambda) = I - \frac{\mathrm{i}}{4}\,\omega_{\mu\nu}\sigma^{\mu\nu} + \dots. \tag{7.32}$$

In this expression, I is the unit 4×4 matrix and $\sigma^{\mu\nu}$ denotes a set of 4×4 matrices, to be constructed in terms of the γ. Since $\omega_{\mu\nu}$ is antisymmetric, we can assume that $\sigma^{\mu\nu}$ is also antisymmetric in μ and ν, because a symmetric part would give zero when the implied sums have been carried out. (This antisymmetry means, for example, that $\sigma^{12} = -\sigma^{21}$, but the matrix σ^{12} is not necessarily antisymmetric.) The inverse matrix is $S^{-1} = I + \frac{1}{4}\mathrm{i}\omega_{\mu\nu}\sigma^{\mu\nu} + \dots$, and if we substitute this together with (7.31) and (7.32) into the condition (7.30), it becomes

$$[\gamma^\lambda, \sigma^{\mu\nu}] = 2\mathrm{i}(g^{\lambda\mu}\gamma^\nu - g^{\lambda\nu}\gamma^\mu). \tag{7.33}$$

Readers may verify using the Clifford algebra that this is satisfied if we identify $\sigma^{\mu\nu}$ as

$$\sigma^{\mu\nu} = \frac{\mathrm{i}}{2}\,[\gamma^\mu, \gamma^\nu]. \tag{7.34}$$

The physical significance of the matrix nature of the Dirac wavefunction can be found by the same method that we used to identify the energy and momentum operators. The momentum operator (5.6) is the generator of space translations, in the sense that it generates a Taylor series like (5.43) when we express a function of the new coordinates $(\boldsymbol{x} + \boldsymbol{a})$ in terms of the old ones. We can carry this idea over to Lorentz transformations, but a slight change of notation will be necessary to distinguish the two sets of coordinates. Just for the purposes of this discussion, I will replace the notation $x^{\mu'}$ for the new coordinates by $\bar{x}^\mu$. Consider first a scalar function. Using (7.31), we can write

$$x^\mu = \bar{x}^\mu - g^{\mu\nu}\omega_{\nu\sigma}\bar{x}^\sigma + \dots \tag{7.35}$$

and

$$\phi'(\bar{x}) = \phi(x) = \left(1 - g^{\mu\nu}\omega_{\nu\sigma}\bar{x}^\sigma \frac{\partial}{\partial \bar{x}^\mu} + \dots\right)\phi(\bar{x}). \tag{7.36}$$

If we take into account the antisymmetry of $\omega_{\mu\nu}$, and use p^μ to stand for the momentum operator $\mathrm{i}g^{\mu\nu}\partial_\nu$, this can be rewritten as

$$\phi'(x) = \left(1 - \frac{\mathrm{i}}{2}\,\omega_{\mu\nu}(x^\mu p^\nu - x^\nu p^\mu) + \dots\right)\phi(x). \tag{7.37}$$

This describes the relation between the functional forms of the old and new wavefunctions, and we can drop the bars over the coordinates, which are dummy variables. For Dirac spinors, we must use the transformation law (7.27), and we get an extra term from the matrix S. The result is

$$\psi'(x) = \left(I - \frac{i}{2}\,\omega_{\mu\nu}M^{\mu\nu} + \ldots\right)\psi(x) \tag{7.38}$$

with the generators of Lorentz transformations given by

$$M^{\mu\nu} = \tfrac{1}{2}\sigma^{\mu\nu} + (x^\mu p^\nu - x^\nu p^\mu). \tag{7.39}$$

Since these generators are antisymmetric in μ and ν, only six of them are independent. It is useful to divide them into two groups of three, defined by

$$K^i = M^{0i} \qquad J^i = \tfrac{1}{2}\varepsilon^{ijk}M^{jk} \tag{7.40}$$

where ε^{ijk} is the three-dimensional Levi–Civita tensor, equal to 1 if (i, j, k) is an even permutation of $(1, 2, 3)$, -1 for an odd permutation and zero if two indices are equal. Thus $J^1 = M^{23}$, $J^2 = M^{31}$ and $J^3 = M^{12}$. K^i is the generator of boosts along the ith spatial axis, and J^i is the generator of rotations about the ith axis. It is worth noting that a rotation 'about the z axis', say, is more properly described as a rotation in the x–y plane. That is, it rearranges the x and y coordinates leaving z and t unchanged. The totally antisymmetric tensor ε^{ijk} exists only in three spatial dimensions, so if we want to consider other numbers of dimensions, the J_i could not be defined.

The J_i can be written in three-dimensional notation as

$$J^i = \tfrac{1}{2}\begin{pmatrix}\sigma^i & 0\\ 0 & \sigma^i\end{pmatrix} + (\boldsymbol{r} \times \boldsymbol{p})^i I \tag{7.41}$$

where I is the unit 4×4 matrix, as long as the representation (7.25) is used for the γ matrices. The second term is, of course, the operator representing the 'orbital' angular momentum associated with the motion of the particle, and we therefore interpret the first term as representing an intrinsic angular momentum or *spin*, which is independent of the orbital motion. Although the Dirac equation is a relativistic one, the existence of particles with spin need not be thought of as a relativistic effect. The generators (7.41) are concerned only with rotations and can be used perfectly well in a non-relativistic theory, as is reviewed in appendix 2. In the non-relativistic setting, the independent spin polarization states are specified by the eigenvalue of one spin component, conventionally σ^3, which implies choosing a particular direction in space as the 'spin quantization axis'. In a relativistic theory, this has no Lorentz covariant meaning, because a Lorentz boost mixes spatial and temporal directions.

A covariant description of spin polarization can be given in terms of the *Pauli–Lubanski 4-vector*, defined by

$$W_\mu = \tfrac{1}{2}\varepsilon_{\mu\nu\lambda\sigma}M^{\nu\lambda}p^\sigma \tag{7.42}$$

where $\varepsilon_{\mu\nu\lambda\sigma}$ is the four-dimensional Levi–Civita tensor (see appendix 1). Since ε is totally antisymmetric, the $(x^\nu p^\lambda - x^\lambda p^\nu)$ part of $M^{\nu\lambda}$ makes no contribution to W_μ. In terms of the 3-vectors $\boldsymbol{p}$, $\boldsymbol{\Sigma}$ and $\boldsymbol{K}$, where $\boldsymbol{\Sigma}$ is the spin part of $\boldsymbol{J}$, the components of W^μ are

$$W^0 = \boldsymbol{\Sigma}\cdot\boldsymbol{p} \tag{7.43}$$

$$W^i = \Sigma^i p^0 + (\boldsymbol{K}\times\boldsymbol{p})^i. \tag{7.44}$$

The Lorentz invariant quantity $W^2 = W_\mu W^\mu$ can be evaluated by choosing any convenient reference frame. If we imagine W^2 to act on a momentum eigenfunction, we can replace p^μ with the corresponding eigenvalue. By choosing the rest frame of the particle, where $p^\mu = (m, \boldsymbol{0})$, we find

$$W^2 = -m^2\boldsymbol{\Sigma}^2 = -m^2 s(s+1) \tag{7.45}$$

for a particle of spin s. Thus, a scalar wavefunction with $\boldsymbol{\Sigma} = 0$ represents a spin-0 particle. For a Dirac spinor, $\boldsymbol{\Sigma}$ is the matrix in (7.41), and Σ^2 is $\frac{3}{4}$ times the unit matrix, so the spinor represents spin-$\frac{1}{2}$ particles.

7.3.3 Some properties of the γ matrices

A number of useful properties of the γ matrices follow from the Dirac equation and the Clifford algebra. I will list a number of them, leaving details of their proofs to readers. First, it follows from (7.24) that

$$(\gamma^0)^2 = 1 \quad \text{and} \quad (\gamma^i)^2 = -1 \tag{7.46}$$

for $i = 1, 2$ or 3. If we multiply the Dirac equation (7.21) by γ^0, we get a relativistic Schrödinger equation

$$\mathrm{i}\frac{\partial\psi}{\partial t} = H\psi = (-\mathrm{i}\gamma^0\gamma^i\partial_i + m\gamma^0)\psi. \tag{7.47}$$

The Hamiltonian H must be Hermitian, and from this it follows that γ^0 is Hermitian and the γ^i are anti-Hermitian:

$$\gamma^{0\dagger} = \gamma^0 \quad \text{and} \quad \gamma^{i\dagger} = -\gamma^i. \tag{7.48}$$

According to the Clifford algebra, γ^0 anticommutes with each γ^i, so for $\mu = 0, \ldots, 3$, we can write

$$\gamma^{\mu\dagger} = \gamma^0\gamma^\mu\gamma^0. \tag{7.49}$$

The matrix γ^5 is defined by

$$\gamma^5 = \mathrm{i}\gamma^0\gamma^1\gamma^2\gamma^3 = \frac{\mathrm{i}}{4!}\,\varepsilon_{\mu\nu\lambda\sigma}\gamma^\mu\gamma^\nu\gamma^\lambda\gamma^\sigma. \tag{7.50}$$

It has the properties

$$(\gamma^5)^2 = I \tag{7.51}$$

$$\gamma^\mu\gamma^5 = -\gamma^5\gamma^\mu \quad \text{for any } \mu. \tag{7.52}$$

Although the four matrices γ^μ do not form a 4-vector, it is often necessary to form contractions as if they did. A useful abbreviation is the 'slash' notation

$$\not{a} \equiv \gamma^\mu a_\mu \tag{7.53}$$

where a_μ is any 4-vector. In this notation, the Dirac equation (7.21) takes the form

$$(\mathrm{i}\not{\partial} - m)\psi(x) = 0. \tag{7.54}$$

The Pauli–Lubanski vector (7.42) can be written, for Dirac spinors, as

$$W_\mu = -\tfrac{1}{4}[\gamma_\mu, \not{p}]\gamma^5 \tag{7.55}$$

as readers are invited to prove in exercise 7.6.

7.3.4 Conjugate wavefunction and the Dirac action

The adjoint of the Dirac equation (7.21) is

$$\psi^\dagger(\mathrm{i}\gamma^{\mu\dagger}\psi\,\overleftarrow{\partial}_\mu + m) = 0 \tag{7.56}$$

where $\overleftarrow{\partial}_\mu$ indicates differentiation of the function on its left. This notation is useful in conjunction with the multiplication of the row matrix $\psi^\dagger$ by a γ matrix on the right. If we multiply this equation from the right by γ^0 and use (7.49), we get

$$\bar{\psi}(\mathrm{i}\overleftarrow{\not{\partial}}_\mu + m) = 0 \tag{7.57}$$

where the conjugate wavefunction is defined by

$$\bar{\psi}(x) = \psi^\dagger(x)\gamma^0. \tag{7.58}$$

It is simple to verify that the two equations (7.54) and (7.57) can be derived as Euler–Lagrange equations from the action

$$S = \int \mathrm{d}^4x\,\bar{\psi}(\mathrm{i}\not{\partial} - m)\psi \tag{7.59}$$

by treating ψ and $\bar{\psi}$ as independent variables.

7.3.5 Probability current and bilinear covariants

As in the case of scalar wavefunctions, we would like to identify a 4-vector current density which is conserved; that is it satisfies the equation of continuity. The quantity

$$j^\mu(x) = \bar{\psi}(x)\gamma^\mu\psi(x) \tag{7.60}$$

is easily shown, using the Dirac equation and its adjoint, to be conserved. The component $j^0 = |\psi^\dagger\psi|$, which we would like to identify as the probability density, is positive definite. This would appear to be an advantage, compared with the negative probabilities encountered for the scalar wavefunction, but it will turn out that this is, in a sense, illusory. Since the γ^μ are not themselves the components of a 4-vector, we must show that (7.60) *is* a 4-vector. To do this, we need a property of the transformation matrix $S(\Lambda)$ which, on exponentiating (7.32), is seen to be of the form $S(\Lambda) = \exp(-\mathrm{i}\omega_{\mu\nu}\sigma^{\mu\nu}/4)$. Because of the relation (7.49), we have

$$S^\dagger(\Lambda) = \gamma^0 S^{-1}(\Lambda)\gamma^0. \tag{7.61}$$

Using this and the defining property (7.30), we can write the current density in a new frame as

$$\begin{aligned} j^{\mu'}(x') &= \psi'^\dagger(x')\gamma^0\gamma^{\mu'}\psi'(x') \\ &= \psi^\dagger(x)S^\dagger(\Lambda)\gamma^0\gamma^{\mu'}S(\Lambda)\psi(x) \\ &= \bar{\psi}(x)S^{-1}(\Lambda)\gamma^{\mu'}S(\Lambda)\psi(x) \\ &= \Lambda^{\mu'}{}_{\mu}j^\mu(x) \end{aligned}$$

so j^μ does indeed transform as a 4-vector. Note that the presence of $\bar{\psi}$ rather than $\psi^\dagger$ is essential to this proof.

A number of other tensors can be constructed in the same way. To understand the classification of these, it is necessary to consider a wider class of Lorentz transformations than we have so far. The representative transformation matrices (3.26) and (3.27) each have $\Lambda^0{}_0 \geqslant 1$ and $\det|\Lambda| = +1$. Such transformations are called *proper* Lorentz transformations. Examples of improper transformations are *time reversal* $t' = -t$ and *parity* or spatial reflection $\boldsymbol{x}' = -\boldsymbol{x}$. Each of these has $\det|\Lambda| = -1$. A number of important tensor-like quantities have transformation laws similar to (2.19), except that the right-hand side is multiplied by $\det|\Lambda|$. These are called *pseudotensors*. Three-dimensional examples are provided by the products $\boldsymbol{a} \times \boldsymbol{b}$ of any two vectors, which are called *axial vectors*. Each vector changes sign under parity, but the product does not change sign. (More generally, a quantity whose transformation law contains a factor $(\det|\Lambda|)^n$ is a *tensor density* of weight n.)

The so-called *bilinear covariants* are products of the form $\bar{\psi}\Gamma\psi$, where Γ is a 4×4 matrix. Any 4×4 matrix can, of course, be written in terms of 16 linearly independent ones. Such a set is provided by the matrices I, γ^5, γ^μ, $\gamma^\mu\gamma^5$ and $\sigma^{\mu\nu}$, which have the advantage of giving rise to tensors or pseudotensors. The names and transformation properties of the bilinear covariants are

scalar: $S(x) = \bar{\psi}(x)\psi(x)$ $\qquad S'(x') = S(x)$

pseudoscalar: $P(x) = \bar{\psi}(x)\gamma^5\psi(x)$ $\qquad P'(x') = \det|\Lambda|P(x)$

vector: $V^\mu(x) = \bar{\psi}(x)\gamma^\mu\psi(x)$ $\qquad V^{\mu'}(x') = \Lambda^{\mu'}{}_\mu V^\mu(x)$

axial vector: $A^\mu(x) = \bar{\psi}(x)\gamma^\mu\gamma^5\psi(x)$ $\qquad A^{\mu'}(x') = \det|\Lambda|\Lambda^{\mu'}{}_\mu A^\mu(x)$

tensor: $T^{\mu\nu}(x) = \bar{\psi}(x)\sigma^{\mu\nu}\psi(x)$ $\qquad T^{\mu'\nu'}(x') = \Lambda^{\mu'}{}_\mu\Lambda^{\nu'}{}_\nu T^{\mu\nu}(x).$

The vector covariant is of course the same as (7.60), and the proofs of all the transformation properties are similar to that given above.

7.3.6 Plane-wave solutions

As in the non-relativistic theory, a complete set of plane-wave solutions to the Dirac equation is labelled by the momentum p^μ and a spin component $s = \pm\frac{1}{2}$ along a chosen quantization axis. A covariant description of the spin polarization can be given as follows. In the rest frame, where $p^\mu = (m, \mathbf{0})$, choose a unit 3-vector $\boldsymbol{n}$ as the quantization axis. In a frame in which the momentum is $p^\mu = (p^0, \boldsymbol{p})$, the object

$$n^\mu = \left(\frac{\boldsymbol{p}\cdot\boldsymbol{n}}{m}, \boldsymbol{n} + \frac{(\boldsymbol{p}\cdot\boldsymbol{n})}{m(m + p^0)}\boldsymbol{p}\right) \tag{7.62}$$

is a 4-vector, with $n_\mu n^\mu = -1$ and $p_\mu n^\mu = 0$. The quantity $W\cdot n = W_\mu n^\mu$ is Lorentz invariant. Its value is most easily calculated in the rest frame and is

$$W\cdot n = -m\boldsymbol{\Sigma}\cdot\boldsymbol{n} \tag{7.63}$$

which is the component of spin along $\boldsymbol{n}$ as measured in the rest frame. A complete set of plane-wave solutions is now given by the simultaneous eigenfunctions of $W\cdot n$ and the momentum operator $\mathrm{i}\partial_\mu$. There are both positive and negative energy solutions. Let $p^0 = +(\boldsymbol{p}^2 + m^2)^{1/2}$. The positive energy solutions have the form

$$\psi_{p,s}(x) = \mathrm{e}^{-\mathrm{i}p\cdot x}u(p, s) \tag{7.64}$$

where $u(p, s)$ is a column matrix. To satisfy the Dirac equation (7.54), we must have

$$(\not{p} - m)u(p, s) = 0 \tag{7.65}$$

and, according to the above definition of spin polarization, $(W\cdot n)u(p, s) = -msu(p, s)$. If we use the standard representation (7.25) for the γ matrices and choose $\boldsymbol{n} = (0, 0, 1)$, then in the rest frame

$$u(p,\tfrac{1}{2}) = \begin{pmatrix}1\\0\\0\\0\end{pmatrix} \quad \text{and} \quad u(p,-\tfrac{1}{2}) = \begin{pmatrix}0\\1\\0\\0\end{pmatrix}. \tag{7.66}$$

Corresponding to each positive energy solution there is a negative energy solution

$$\psi^{c}_{p,s}(x) = e^{ip\cdot x}v(p,s) \tag{7.67}$$

where the negative energy spinor $v(p, s)$ satisfies

$$(\not{p} + m)v(p,s) = 0. \tag{7.68}$$

As in the scalar theory, it will be necessary to reinterpret these negative energy solutions in terms of antiparticles. In the scalar case, the negative energy solution is the complex conjugate of a positive energy antiparticle wavefunction. Here, (7.67) is the *charge conjugate* of a positive energy antiparticle wavefunction. The operation of charge conjugation, denoted by the superscript c in (7.67), relates particle and antiparticle states. It involves both complex conjugation and a rearrangement of spinor components. To find the positive energy solution of which (7.67) is the conjugate we define

$$\psi^{c}_{p,s}(x) = \mathscr{C}\psi^{*}_{p,s}(x) \tag{7.69}$$

where $\mathscr{C}$ is a matrix to be found. The spinor $v(p, s) = \mathscr{C}u^*(p, s)$ must satisfy (7.68), given that $u(p, s)$ satisfies (7.65). Taking the complex conjugate of (7.65) and multiplying by $\mathscr{C}$, we find that this will be so provided that

$$\mathscr{C}\gamma^{\mu *}\mathscr{C}^{-1} = -\gamma^{\mu}. \tag{7.70}$$

This is usually expressed differently, by observing that $\gamma^{\mu *}$ is the transpose (denoted by ~) of $\gamma^{\mu\dagger}$. Then, by using (7.46) and (7.49), we can express $\mathscr{C}$ as $C\tilde{\gamma}^0$, where the charge conjugation matrix C has the property

$$C\tilde{\gamma}^{\mu}C^{-1} = -\gamma^{\mu}. \tag{7.71}$$

In the standard representation of the γ matrices, it is found that $C = i\gamma^2\gamma^0$. The charge conjugate spinors corresponding to (7.66) are

$$v(p,\tfrac{1}{2}) = \begin{pmatrix}0\\0\\0\\1\end{pmatrix} \quad \text{and} \quad v(p,-\tfrac{1}{2}) = -\begin{pmatrix}0\\0\\1\\0\end{pmatrix}. \tag{7.72}$$

Some further properties of charge conjugation are explored in the exercises, as is the construction of plane-wave solutions in frames other than the rest frame.

7.3.7 *Massless spin-$\frac{1}{2}$ particles*

Massless spin-$\frac{1}{2}$ particles, such as neutrinos, satisfy the Dirac equation $\mathrm{i}\not{\partial}\psi = 0$. Since they travel with the speed of light, and therefore have no rest frame, the spin polarization vector (7.62) cannot be defined. Instead, spin states can be classified according to *helicity*, which is the component of spin parallel to the 3-vector momentum $\boldsymbol{p}$:

$$h = \boldsymbol{\Sigma}\cdot\boldsymbol{p}/|\boldsymbol{p}|. \tag{7.73}$$

The Pauli–Lubanski vector can be written as $W^\mu = -\frac{1}{2}\gamma^5[\gamma^\mu\not{p} - p^\mu]$ and, for a wavefunction satisfying the massless Dirac equation, we have $W^\mu\psi = \frac{1}{2}\gamma^5 p^\mu\psi$. Thus, plane-wave solutions can be classified as simultaneous eigenfunctions of p^μ and W^μ if they are eigenvectors of γ^5, which in this context is the *chirality* or 'handedness' operator. Since $(\gamma^5)^2 = 1$, any wavefunction can be decomposed as $\psi = \psi_{\mathrm{R}} + \psi_{\mathrm{L}}$, where

$$\psi_{\mathrm{R}} = \tfrac{1}{2}(1 + \gamma^5)\psi \qquad \psi_{\mathrm{L}} = \tfrac{1}{2}(1 - \gamma^5)\psi. \tag{7.74}$$

For a plane wave, these right- and left-handed components are eigenvectors of γ^5 with eigenvalues $+1$ and -1 respectively, and eigenfunctions of W^μ with eigenvalues $\pm\frac{1}{2}p^\mu$. Since $W^0 = \boldsymbol{\Sigma}\cdot\boldsymbol{p}$ and $p^0 = |\boldsymbol{p}|$ for a massless particle, we find that these components have helicity $\pm\frac{1}{2}$. Note that, while the *chiral projections* (7.74) are eigenvectors of γ^5 for any wavefunction ψ, they have definite helicities only when ψ describes *massless* particles.

7.4 Spinor Field Theory

Although the Dirac equation appears to lead to a positive definite probability density, it nevertheless has, as we have just seen, negative energy solutions. In order to interpret these in terms of antiparticles, we must again resort to second quantization. If we write out matrix multiplications explicitly, the action (7.59) is

$$S = \int \mathrm{d}^4x\,\bar{\psi}_i(\mathrm{i}\gamma^\mu{}_{ij}\partial_\mu - m\delta_{ij})\psi_j. \tag{7.75}$$

The momentum conjugate to ψ_i is

$$\Pi_i = \frac{\delta S}{\delta(\partial_0\psi_i)} = \mathrm{i}\bar{\psi}_j\gamma^0{}_{ji} = \mathrm{i}\psi_i^\dagger \tag{7.76}$$

which is the same as (6.26) for the non-relativistic Schrödinger theory. When ψ satisfies the Dirac equation, the action is zero, and in that case the Hamiltonian is

$$H = \int \mathrm{d}^3x\,\Pi_i(\boldsymbol{x}, t)\dot{\psi}_i(\boldsymbol{x}, t) = \int \mathrm{d}^3x\,\bar{\psi}(\boldsymbol{x}, t)\mathrm{i}\gamma^0\partial_0\psi(\boldsymbol{x}, t). \tag{7.77}$$

In accordance with our earlier procedure, we replace the wavefunction with a field operator. This may be expanded in terms of plane-wave solutions as

$$\hat{\psi}(x) = \int \frac{\mathrm{d}^3k}{(2\pi)^3 2\omega(\boldsymbol{k})} \sum_s [\hat{b}(\boldsymbol{k}, s)\mathrm{e}^{-\mathrm{i}k\cdot x}u(k, s) + \hat{d}^\dagger(\boldsymbol{k}, s)\mathrm{e}^{\mathrm{i}k\cdot x}v(k, s)], \tag{7.78}$$

in which $k^0 = (\boldsymbol{k}^2 + m^2)^{1/2}$. The operator $\hat{b}(\boldsymbol{k}, s)$ is to be interpreted as the annihilation operator for a particle of 3-momentum $\boldsymbol{k}$ and spin polarization s, and $\hat{d}^\dagger(\boldsymbol{k}, s)$ as the creation operator for an antiparticle. It is possible to normalize $u(k, s)$ and $v(k, s)$ in such a way that

$$\begin{aligned} \bar{u}(k, s)\gamma^\mu u(k, s') &= \bar{v}(k, s)\gamma^\mu v(k, s') = 2k^\mu \delta_{ss'} \\ \bar{u}(k, s)\gamma^0 v(-k, s') &= 0 \end{aligned} \tag{7.79}$$

(see exercise 7.4) and this ensures a covariant normalization for the particle states.

In terms of the creation and annihilation operators, the Hamiltonian reads

$$\hat{H} = \int \frac{\mathrm{d}^3k}{(2\pi)^3 2\omega(\boldsymbol{k})} \sum_s \omega(\boldsymbol{k})[\hat{b}^\dagger(\boldsymbol{k}, s)\hat{b}(\boldsymbol{k}, s) - \hat{d}(\boldsymbol{k}, s)\hat{d}^\dagger(\boldsymbol{k}, s)]. \tag{7.80}$$

If we were to assume commutation relations similar to (7.17), it may be seen that the antiparticles would contribute negative energies. Other undesirable consequences would also follow. For example, causality would be violated, in the sense that operators representing observable quantities in regions of spacetime at space-like separations would fail to commute. Thus, events in these regions, which cannot communicate via signals travelling at velocities less than or equal to that of light, would not be independent as they ought to be. It is these inconsistencies which give rise to the *spin–statistics theorem* (see p 118). They can be removed if we assume the *anticommutation relations*

$$\{b(\boldsymbol{k}, s), b^\dagger(\boldsymbol{k}', s')\} = \{d(\boldsymbol{k}, s), d^\dagger(\boldsymbol{k}', s')\} = (2\pi)^3 2\omega(\boldsymbol{k})\delta_{ss'}\delta(\boldsymbol{k} - \boldsymbol{k}') \tag{7.81}$$

with all other anticommutators equal to zero. The antiparticle term in (7.80) then changes sign when we reverse the order of the operators, and we also get a constant, as before. Removing the constant is again equivalent to normal ordering, provided that the definition of normal ordering is amended to include a change of sign whenever two fermionic operators are interchanged. The relations (7.81) imply equal-time anticommutation relations for the field components which are

$$\{\hat{\psi}_i(\boldsymbol{x}, t), \hat{\Pi}_j(\boldsymbol{x}', t)\} = \mathrm{i}\delta_{ij}\delta(\boldsymbol{x} - \boldsymbol{x}') \tag{7.82}$$

the anticommutator of two field components or two momentum components being zero.

When anticommutation is taken into account, the number operator for spin-$\frac{1}{2}$ fermions is found to be

$$\hat{N} = \int d^3x\, \hat{\bar{\psi}}(\boldsymbol{x}, t)\gamma^0\hat{\psi}(\boldsymbol{x}, t)$$
$$= \int \frac{d^3k}{(2\pi)^3 2\omega(\boldsymbol{k})} \sum_s [\hat{b}^\dagger(\boldsymbol{k}, s)\hat{b}(\boldsymbol{k}, s) - \hat{d}^\dagger(\boldsymbol{k}, s)\hat{d}(\boldsymbol{k}, s)]. \qquad (7.83)$$

This counts the (number of particles − number of antiparticles), which is the desired result. It can, of course, take both positive and negative values, which is ironic, since the positive definite probability density appeared at first to be a success of the Dirac equation. We see, indeed, that at the level of first quantization the Dirac theory cannot be quite correct. To allow for the antiparticle interpretation, it ought to be possible for $j^0(x)$ to have negative values. There is, in fact, a modification which will do this. Let us consider the plane-wave expansion (7.78) to apply to a wavefunction, the coefficients $b(\boldsymbol{k}, s)$ and $d^*(\boldsymbol{k}, s)$ being numbers rather than operators. For consistency with the anticommutation of the corresponding operators, these should be regarded as *anticommuting numbers*. This means that $b(\boldsymbol{k}, s)b(\boldsymbol{k}', s) = -b(\boldsymbol{k}', s)b(\boldsymbol{k}, s)$ and similarly for any product of b's, d's and their complex conjugates. In particular, the product of an anticommuting number with itself is zero. However, any anticommuting number still commutes with an ordinary commuting (or c-) number. Such anticommuting numbers are said to form a *Grassmann algebra*. The Dirac wavefunction itself is therefore also an anticommuting Grassmann number. For many purposes, we deal only with equations which, like the Dirac equation itself, are linear in the wavefunction, so the anticommutation has no effect. None of the results derived in previous sections are changed. However, certain properties of the bilinear covariants do depend on whether the wavefunction is taken to be commuting or anticommuting, and these will be consistent with corresponding properties of the second-quantized operators only if anticommuting wavefunctions are used. The Hamiltonian and current densities are cases in point.

7.5 Wave Equations in Curved Spacetime

It ought, of course, to be possible to study wave equations and field theories in curved spacetimes. When this is done, it turns out that there are difficulties of interpretation over and above those we have already encountered in Minkowski spacetime, and these difficulties have not, to my mind, been completely resolved. In this section, I shall try to convey the flavour of what is involved. A detailed discussion may be found in Birrell and Davies (1982).

The first requirement, obviously, is that wave equations should be covariant, and therefore the action should be invariant, under general coordinate transformations. Starting from the theories we have already considered, two steps are necessary to construct suitable actions: we must use the covariant spacetime volume element (4.12) and replace partial derivatives with covariant derivatives. It is also possible to add further terms involving the Riemann curvature tensor, which will vanish if the spacetime happens to be flat. In the case of a scalar field, these steps can be carried out straightforwardly. The covariant derivative of a scalar quantity is the same as the partial derivative, and we arrive at an action of the form

$$S = \int \mathrm{d}^4x(-g(x))^{1/2}[g^{\mu\nu}(x)\partial_\mu\phi^*\partial_\nu\phi - m^2\phi^*\phi - \xi R(x)\phi^*\phi] \quad (7.84)$$

where $R(x)$ is the Ricci curvature scalar defined in (2.50) and ξ is a dimensionless number. This additional term is the only possible one which does not involve dimensionful coefficients. The corresponding Euler–Lagrange equation is

$$g^{\mu\nu}\nabla_\mu\nabla_\nu\phi + (m^2 + \xi R)\phi = 0 \quad (7.85)$$

where ∇_μ is the covariant derivative. (Recall that, although $\nabla_\mu\phi = \partial_\mu\phi$, this quantity is a vector, which must be acted on with a covariant derivative. To derive (7.85), we use an integration by parts, and the covariant derivative enters through the covariant version of Gauss' theorem exhibited as equation (A1.18) of appendix 1.) The value of ξ is not determined by any known physical principle and, since the effects of spacetime curvature are too small to measure in the laboratory, it cannot be determined by experiment either. The case $\xi = 0$ is called *minimal coupling*, for obvious reasons. An interesting case is $\xi = \frac{1}{6}$. If $\xi = \frac{1}{6}$ and $m = 0$, the theory possesses a symmetry known as *conformal invariance*. A conformal transformation means replacing the metric $g_{\mu\nu}(x)$ by $\Omega(x)^2 g_{\mu\nu}(x)$, where $\Omega(x)$ is an arbitrary function. If at the same time we replace $\phi(x)$ by $\Omega^{-1}(x)\phi(x)$, then it can be shown that (7.85) is unchanged. Whether we should expect this symmetry to be respected by nature is not clear. At any rate, the case $\xi = \frac{1}{6}$ is known as *conformal coupling*.

To construct a generally covariant version of the Dirac equation requires rather more thought. We have seen that spinor wavefunctions do not have the same transformation properties as any of the tensors considered in chapter 2, so we do not yet know how to form their covariant derivatives. In order to do this, we first recall that it is always possible to set up a system of locally inertial Cartesian coordinates, valid in a sufficiently small region of spacetime. Strictly speaking, this must be an infinitesimal region surrounding, say, the point X with coordinates $x^\mu = X^\mu$. I shall denote these local coordinates by y^a, using indices $a, b, c, \ldots$ to distinguish them from the large-scale coordinates x^μ. In terms of these coordinates, the metric tensor is given at X by the Minkowski form η_{ab}. I shall denote the transformation matrix Λ (equation (2.13)) which relates the two sets of coordinates by the special symbol e:

$$e^\mu{}_a(X) = \left.\frac{\partial x^\mu}{\partial y^a}\right|_{x^\mu = X^\mu} \quad \text{and} \quad e^a{}_\mu(X) = \left.\frac{\partial y^a}{\partial x^\mu}\right|_{x^\mu = X^\mu}. \tag{7.86}$$

If we set up a locally inertial frame of reference at each point of spacetime, in such a way that the directions of their axes vary smoothly from one point to another, then we obtain a set of four vector fields $e^\mu{}_0(x), \ldots, e^\mu{}_3(x)$ which specify, at each point, the directions of these axes. This set of vector fields is known variously as a *vierbein* (a German expression meaning 'four legs'), a *tetrad*, or a *frame field*. In theories which consider numbers of spacetime dimensions other than four, it is called a *vielbein* (a German expression meaning 'many legs'). At a given point X, the vierbein constitutes a set of four 4-vectors $e^a{}_0(X), \ldots, e^a{}_3(X)$ which specify the directions and scales of the large-scale coordinates relative to the inertial coordinates at X. Considered as a whole, the 16 components of the vierbein constitute a kind of rank 2 tensor field whose μ indices transform as a vector under general coordinate transformations and can be raised and lowered using g, while its a indices transform as a 4-vector under Lorentz transformations in the local coordinates and can be raised and lowered using η. By construction, the vierbein satisfies the relations

$$e^\mu{}_a(x)e^a{}_\nu(x) = \delta^\mu_\nu \qquad e^a{}_\mu(x)e^\mu{}_b(x) = \delta^a_b \tag{7.87}$$

and

$$e^\mu{}_a(x)e^\nu{}_b(x)g_{\mu\nu}(x) = \eta_{ab} \qquad e^a{}_\mu(x)e^b{}_\nu(x)\eta_{ab} = g_{\mu\nu}(x). \tag{7.88}$$

Its 16 independent components evidently carry two kinds of information. First, as we see from (7.88), they contain all the information needed to construct the 10 independent components of the metric tensor field. Second, each local inertial frame can be redefined by Lorentz transformations, involving three independent rotations and three boosts, and the remaining six degrees of freedom in the vierbein specify the choices we have actually made.

It is now possible to describe any vector quantity $\boldsymbol{V}$ either in terms of its components $V^\mu(x)$ relative to the large-scale coordinate system, which I shall refer to for brevity as a *coordinate vector*, or in terms of its components $V^a(x)$ relative to the local coordinates at x, which I shall call a *Lorentz vector*. The two sets of components are obviously related by

$$V^\mu(x) = e^\mu{}_a(x) V^a(x) \quad \text{and} \quad V^a(x) = e^a{}_\mu(x) V^\mu(x). \tag{7.89}$$

In fact, any tensor can be expressed in terms of components with any combination of a-type and μ-type indices we happen to find convenient. The advantage of this is clear: we know how to deal with spinors in the local inertial coordinates, and the vierbein permits us to embed these in the curved spacetime. In order to work out the covariant derivative of a spinor, we need a suitable rule for parallel transport. We shall first work out the rule for a Lorentz vector, which will apply, for example, to the current $\bar{\psi}(x)\gamma^a\psi(x)$, and then deduce the corresponding rule for the spinor itself.

To transport $V^a(x)$ to the point $x + \mathrm{d}x$, we need only to translate (2.23) into the language of locally inertial coordinates. The transported vector will be given by

$$V^a(x \to x + \mathrm{d}x) = V^a(x) - \omega^a{}_{b\nu}(x) V^b(x)\, \mathrm{d}x^\nu \tag{7.90}$$

where the coefficients $\omega^a{}_{b\nu}$ are the components of what is called the *spin connection*. They involve both the affine connection, which defines parallel transport of the vector itself, and the vierbein, which relates the locally inertial coordinates at x to those at $x + \mathrm{d}x$. We use the relations $V^\mu(x) = e^\mu{}_a(x) V^a(x)$ and $V^\mu(x \to x + \mathrm{d}x) = e^\mu{}_a(x + \mathrm{d}x) V^a(x \to x + \mathrm{d}x)$ and the expansion $e^\mu{}_a(x + \mathrm{d}x) = e^\mu{}_a(x) + e^\mu{}_{a,\nu}(x)\, \mathrm{d}x^\nu$ to convert (7.90) into a transport equation for V^μ and compare this with the original equation (2.23). We find that the spin connection is given by

$$\omega^a{}_{b\nu} = e^a{}_\mu e^\mu{}_{b,\nu} + e^a{}_\mu e^\sigma{}_b \Gamma^\mu{}_{\sigma\nu}. \tag{7.91}$$

With the spin connection in hand, we can generalize (2.28) to obtain the covariant derivative of a tensor field with both a- and μ-type indices, including a Γ term for each coordinate index and an ω term for each local Lorentz index. The vierbein itself is such a tensor, and by rearranging (7.91) we see that its covariant derivative vanishes:

$$\nabla_\nu e^\mu{}_a = e^\mu{}_{a,\nu} + \Gamma^\mu{}_{\sigma\nu} e^\sigma{}_a - \omega^b{}_{a\nu} e^\mu{}_b = 0. \tag{7.92}$$

This result should give alert readers pause for thought. We saw in §2.4.5 that, in order to make the notion of parallel transport as defined by the affine connection consistent with that defined by the metric, the covariant derivative of the metric tensor field should vanish, and it does so only when the affine connection is the metric connection (2.49). Although we shall usually want Γ to be this metric connection, we have not actually assumed this in order to derive (7.92). To resolve this point to their own satisfaction, readers may like to consider the conditions under which any two of the three notions of parallelism defined by the affine connection, the metric and the vierbein become equivalent. In particular, they should consider the covariant derivatives of $g_{\mu\nu}$, of η_{ab} and of equations (7.88). Let us impose the consistency condition that the magnitude of a transported Lorentz vector should be preserved, so $\eta_{ab}V^a(x \to x + dx)V^a(x \to x + dx) = \eta_{ab}V^a(x)V^b(x)$. It is easy to see that the spin connection must be antisymmetric, in the sense that

$$\omega_{ab\nu}(x) \equiv \eta_{ac}\omega^c{}_{b\nu}(x) = -\omega_{ba\nu}(x). \tag{7.93}$$

By using this condition, readers should be able to show that (2.47) is satisfied, and so Γ must be the metric connection.

We can now turn our attention to spinors, which should satisfy a rule for parallel transport of the form

$$\psi(x \to x + dx) = \psi(x) - \Omega_\nu(x)\psi(x)\,dx^\nu \tag{7.94}$$

where $\Omega_\nu(x)$ is a suitable connection coefficient. This coefficient, like the previous ones, has three indices; the first two are those it possesses by virtue of being a 4×4 spin matrix. To discover what this coefficient is, we demand that, in particular, the scalar quantity $S(x) = \bar{\psi}(x)\psi(x)$ should be invariant under parallel transport, while the Lorentz vector $V^a(x) = \bar{\psi}(x)\gamma^a\psi(x)$ should be transported according to (7.90). From (7.94), we find

$$S(x \to x + dx) = S(x) - \bar{\psi}(x)[\gamma^0\Omega_\nu^\dagger(x)\gamma^0 + \Omega_\nu(x)]\psi(x)dx^\nu \tag{7.95}$$

so our first condition gives

$$\gamma^0\Omega_\nu^\dagger(x)\gamma^0 = -\Omega_\nu(x). \tag{7.96}$$

Using this, we find similarly that V^a is correctly transported provided that

$$[\gamma^a, \Omega_\nu(x)] = \omega^a{}_{b\nu}(x)\gamma^b. \tag{7.97}$$

Taking into account the antisymmetry property (7.93), we can use (7.33) and (7.49) to identify the matrix satisfying these two conditions as

$$\Omega_\nu(x) = -\frac{i}{4}\omega_{ab\nu}(x)\sigma^{ab} = \tfrac{1}{8}\omega_{ab\nu}(x)[\gamma^a, \gamma^b]. \tag{7.98}$$

Thus, the covariant derivative of the spinor is

$$\nabla_\nu\psi(x) = [\partial_\nu + \Omega_\nu(x)]\psi(x). \tag{7.99}$$

It is now a straightforward matter to write down the covariant version of the Dirac equation (7.21). The γ matrices are valid only within the local inertial frame and must be contracted with the covariant derivative by using the vierbein:

$$[\mathrm{i}e^{\mu}{}_{a}(x)\gamma^{a}\nabla_{\mu} - m]\psi(x) = 0. \tag{7.100}$$

We can tidy this up by defining a set of covariant γ matrices

$$\gamma^{\mu}(x) = \mathrm{e}^{\mu}{}_{a}(x)\gamma^{a} \tag{7.101}$$

and it may easily be verified that these satisfy the generally covariant version of the Clifford algebra (7.24):

$$\{\gamma^{\mu}(x), \gamma^{\nu}(x)\} = 2g^{\mu\nu}(x). \tag{7.102}$$

The generally covariant action is clearly

$$S = \int \mathrm{d}^4x(-g(x))^{1/2}\bar{\psi}(x)[\mathrm{i}\gamma^{\mu}(x)\nabla_{\mu} - m]\psi(x) \tag{7.103}$$

and in this case no curvature term can be added with a dimensionless coefficient. If we wish, we can express $(-g)^{1/2}$ as $\det|e^{a}{}_{\mu}|$.

Clearly, wave equations such as (7.85) and (7.100) do not, in general, have simple plane-wave solutions. Only in very special cases, indeed, can their solutions be found in closed form. When we try to reinterpret these equations in terms of quantum fields, we encounter a difficulty of principle, which I shall describe only in qualitative fashion. In previous sections, we found it necessary to expand the field operators using both positive and negative energy solutions. The coefficients of positive energy solutions were interpreted as annihilation operators for particles, while those of the negative energy solutions were creation operators for antiparticles. If we recall the role of the energy operator as the generator of translations in time, it is clear that what we mean by 'energy' depends on what we identify as a 'time' coordinate. If we choose a new coordinate system, it will often be appropriate to expand the field operator using a new set of solutions to the field equations which will, in particular, embody a new distinction between positive and negative energies reflecting our new identification of 'time'. We thus obtain a new set of creation and annihilation operators. Of course, we are merely finding a new expression for the same field operator, so the new creation and annihilation operators can be expressed as linear combinations of the old ones. (The relation between them is known as a *Bogoliubov transformation*.)

Suppose, for simplicity, that both sets of solutions to the wave equation can be labelled by a discrete index and that the new annihilation operators $\hat{b}_i$ are given in terms of the old operators $\hat{a}_i$ and $\hat{a}_i^{\dagger}$ by

$$\hat{b}_i = \sum_j (\alpha_{ij}\hat{a}_j + \beta_{ij}\hat{a}_j^{\dagger}). \tag{7.104}$$

Using the old operators, we identify the vacuum state as that which gives zero when acted on by any of the $\hat{a}$. If we act on this state with the $\hat{b}$, then we shall get zero only if all the coefficients β_{ij} are zero. In general, these coefficients are not zero, so the vacuum state of our old description is not the vacuum state in the new description. It seems, therefore, that there is no uniquely defined vacuum state. Correspondingly, since the vacuum is a state with no particles, there is no unique meaning to the notion of a particle either. Put another way, this means that two observers in different states of motion will disagree as to whether there are any particles present. This curious situation can be made somewhat more palatable if we think only about how particles might be detected by a small piece of apparatus. Simple theoretical models of particle detectors can be devised and imagined to follow some definite trajectory through spacetime. For a given state of the quantum field, the expectation value of the number of particles registered by the detector over some portion of its path does have a well defined value. To be sure, this number of particles will depend on the trajectory of the detector, but that in itself is not too surprising.

It is worth pointing out that this difficulty arises even in Minkowski spacetime. As long as we restrict ourselves to Lorentz transformations between inertial frames of reference, all the β_{ij} in (7.104) are zero, so we can normally take the vacuum state as being uniquely defined. This ceases to be true, however, if we consider non-inertial coordinate systems. For example, it can be shown that a particle detector which is uniformly accelerated relative to its own instantaneous rest frame will detect the presence of particles in the Minkowski vacuum. Remarkably, the detector behaves as if it were surrounded by a thermal bath of particles at a temperature given by $k_B T = \text{acceleration}/2\pi$.

In a curved spacetime, there is in general no preferred coordinate system to correspond to the inertial systems in Minkowski spacetime, and thus no preferred definition of the vacuum state. In view of the equivalence principle, we might expect particles to be created by sufficiently intense gravitational fields and this, essentially, accounts for the *Hawking effect* (Hawking 1974) which permits a black hole in effect to radiate particles and perhaps eventually to evaporate altogether. The black hole actually remains black: the radiated particles can be envisaged as being created in the gravitational field just outside the event horizon and they emerge as black-body radiation at a temperature given by the previous expression, where the acceleration is that due to gravity at the event horizon.

The mathematical details which support these qualitative remarks are somewhat lengthy and technical, and readers are referred to Birrell and Davies (1982) for a lucid presentation of them.

Exercises

7.1. In the Lagrangian density (7.7), let $\phi = 2^{-1/2}(\phi_1 + i\phi_2)$, where ϕ_1 and ϕ_2 are real, and show that $\mathscr{L}$ becomes the sum of independent terms for ϕ_1 and ϕ_2. Identify the two conjugate momenta and carry out the canonical quantization procedure. Show that ϕ_1 and ϕ_2 are the field operators for two particle species, each of which is its own antiparticle. Verify that your commutation relations agree with (7.13) when ϕ is expressed in terms of ϕ_1 and ϕ_2. How are the type 1 and type 2 particle states related to the particle and antiparticle states of §7.2? How does the factor of $2^{-1/2}$ affect the definition of the conjugate momenta, the commutation relations, the definition of creation and annihilation operators and the normalization of particle states?

7.2. Let γ^μ be a set of matrices satisfying (7.24), (7.46) and (7.48) and let U be any constant unitary matrix. Show that the four matrices $U\gamma^\mu U^{-1}$ also have these properties and can therefore be used in the Dirac equation.

7.3. For any 4-vector a^μ, show that $\not{a}\not{a} = a_\mu a^\mu$.

7.4. The spinors (7.66) and (7.72) give plane-wave solutions of the Dirac equation in the rest frame, when the γ matrices (7.25) are used. Denote them by $u(m, s)$ and $v(m, s)$. Show that, in a frame where the momentum is p^μ, the spinors $u(p, s) = (p^0 + m)^{-1/2}(\not{p} + m)u(m, s)$ and $v(p, s) = (p^0 + m)^{-1/2}(-\not{p} + m)v(m, s)$ give plane-wave solutions which satisfy the orthonormality conditions (7.79).

7.5. The idea of charge conjugation requires that $(\psi^c)^c = \eta\psi$, where η is a constant phase factor ($|\eta| = 1$). Why is this? Assuming that $\eta = 1$, show that $\mathscr{C}\mathscr{C}^* = 1$ and $CC^* = -1$ where $\mathscr{C}$ and C are the charge conjugation matrices defined in §7.3.6. Do not assume that the γ matrices are those given in (7.25).

7.6. Show that $\gamma_\mu\gamma^\mu = 4$. Show that $[\gamma_\mu, \gamma_\tau]\gamma^5$ is proportional to $[\gamma_\nu, \gamma_\sigma]$, where (μ, ν, σ, τ) is some permutation of $(0, 1, 2, 3)$. Hence show that $[\gamma_\mu, \gamma_\tau]\gamma^5 = -\mathrm{i}\varepsilon_{\mu\nu\sigma\tau}\gamma^\nu\gamma^\mathrm{T}$.

Show that the Pauli–Lubanski vector (7.42) can be expressed in the form (7.55).

7.7. If $S(\Lambda)$ is a Lorentz transformation matrix which satisfies (7.30), show that $S^{-1}(\Lambda)\gamma^5 S(\Lambda) = \det|\Lambda|\gamma^5$. (It may be helpful to read about the Levi–Civita symbol in appendix 1.)

7.8. If the chiral projection operators are defined as $P_\mathrm{R} = (1 + \gamma^5)/2$ and $P_\mathrm{L} = (1 - \gamma^5)/2$, show that $P_\mathrm{R}^2 = P_\mathrm{R}$, $P_\mathrm{L}^2 = P_\mathrm{L}$ and $P_\mathrm{R}P_\mathrm{L} = P_\mathrm{L}P_\mathrm{R} = 0$. If $\psi_\mathrm{L} = P_\mathrm{L}\psi$, show that $\bar\psi_\mathrm{L} = \bar\psi P_\mathrm{R}$. Show that the charge conjugate of a left-handed spinor is right handed and vice versa.

7.9. If $\psi = \psi_\mathrm{L} + \psi_\mathrm{R}$, show that $\bar\psi\psi = \bar\psi_\mathrm{L}\psi_\mathrm{R} + \bar\psi_\mathrm{R}\psi_\mathrm{L}$ and that $\bar\psi\not{\partial}\psi = \bar\psi_\mathrm{L}\not{\partial}\psi_\mathrm{L} + \bar\psi_\mathrm{R}\not{\partial}\psi_\mathrm{R}$.

7.10. In the standard representation of the γ matrices (7.25) show that the transpose of the charge conjugation matrix C is $\widetilde{C} = -C$. Now define the charge conjugate of the vector current $V^\mu = \bar{\psi}\gamma^\mu\psi$ to be $V^{c\mu} = \bar{\psi}^c\gamma^\mu\psi^c$. Show that $V^{c\mu} = +V^\mu$ if the components of ψ are treated as ordinary numbers and $V^{c\mu} = -V^\mu$ if they are regarded as anticommuting Grassmann numbers. Which treatment is more appropriate in view of the antiparticle interpretation?

8

Forces, Connections and Gauge Fields

One of the central problems faced by theoretical physics is to explain the nature and origin of the forces which act between fundamental particles. In the case of gravity, this is elegantly achieved (at the non-quantum-mechanical level) by general relativity. With hindsight, we may say that an explanation of gravitational forces arises naturally, indeed almost inevitably, from a systematic and explicit account of the geometrical structure of spacetime. The origin of gravitational forces, as described in chapter 4, may be summarized as follows:

(i) To relate physical quantities (represented by tensors) at different points of spacetime, we must introduce a specific geometrical structure, the affine connection, which defines parallel transport.

(ii) The simplest situation is that the connection coefficients are zero everywhere (or can be made so by a suitable choice of coordinates). Departures from this situation are what we recognize as gravitational forces.

(iii) It appears that the particular kinds of departure countenanced by nature can be embodied in a principle of least action.

In essence, *gravitational forces arise from communication between different points of spacetime*. At least at the level of description which accounts for all current experimental observations, it appears that all forces can be considered to arise in essentially this way. I shall first describe how this comes about in the case of electromagnetism, and then discuss how the idea can be generalized to encompass forces of other kinds.

8.1 Electromagnetism

Consider a particle described by a complex wavefunction

$$\phi(x) = \phi_1(x) + i\phi_2(x). \tag{8.1}$$

The absolute phase of ϕ is not an observable quantity. If, for example, each wavefunction in (5.44) is multiplied by the same constant phase factor $\exp(i\theta)$, this factor cancels out of the final result. On the other hand, variations of the phase through spacetime do have a physical significance, because a varying phase angle $\theta(x)$ is differentiated by the momentum operator. This may be expressed differently, if we think of the value of ϕ at the spacetime point x as a point in an 'internal space', namely the (ϕ_1, ϕ_2) plane. The fact that the phase of ϕ is unobservable implies that no particular direction in this plane has any special physical significance. To represent the whole function $\phi(x)$, we must erect a (ϕ_1, ϕ_2) plane at each point of spacetime. The geometrical structure which results is a *fibre bundle*. It is analogous to the Galilean spacetime fibre bundle, in which a three-dimensional Euclidean space is erected at each point in time. Since there is no preferred direction in the (ϕ_1, ϕ_2) plane, variations in the phase of ϕ from one spacetime point to another can be meaningful only if a rule exists for parallel transport through the fibre bundle. In order to attach a meaning to the relative directions of $\phi(x_1)$ and $\phi(x_2)$ in the internal spaces at x_1 and x_2, we need a rule for constructing the wavefunction $\phi(x_1 \rightarrow x_2)$ which exists at x_2 and is to count as 'parallel' to $\phi(x_1)$. The physically meaningful change in ϕ between the points x_1 and x_2 is then given by

$$\delta\phi = \phi(x_2) - \phi(x_1 \rightarrow x_2). \tag{8.2}$$

Evidently, this is quite similar to the way we defined changes in a vector field in (2.22).

An obvious possibility for such a rule is that the phase angles $\tan^{-1}(\phi_2/\phi_1)$ should be equal for $\phi(x_1 \rightarrow x_2)$ and $\phi(x_1)$. It will become apparent that this is the special case corresponding to the absence of electromagnetic fields. Indeed, this rule is equivalent to saying that the ϕ_1 axes at any two points are to count as parallel, and likewise the ϕ_2 axes. In spacetime geometry, the analogous rule, that a single set of self-parallel Cartesian axes can be used to cover the whole manifold, implies that the manifold is flat and that there are no gravitational fields.

A less restrictive rule for parallel transport may be expressed in terms of *connection coefficients* $\Gamma_{ij\mu}$:

$$\phi_i(x \rightarrow x + \Delta x) = \phi_i(x) - \Gamma_{ij\mu}(x)\phi_j(x)\Delta x^\mu. \tag{8.3}$$

This rule has the same form as that for parallel transport of spacetime tensors between infinitesimally separated points via the affine connection (2.23), except that the indices i and j refer to directions in the internal space. Unlike the absolute phase, the magnitude

$|\phi| = (\phi^*\phi)^{1/2} = (\phi_1^2 + \phi_2^2)^{1/2}$ of the wavefunction has a definite physical meaning in terms of probability amplitudes. We therefore include in the definition of parallel transport the requirement that this magnitude remain unchanged. For this to be so, $\Gamma_{ij\mu}$ must be antisymmetric in i and j (see exercise 8.1) and therefore proportional to the two-dimensional Levi–Civita symbol ε_{ij}:

$$\Gamma_{ij\mu}(x) = -\varepsilon_{ij}\lambda A_\mu(x). \tag{8.4}$$

The vector $A_\mu(x)$ will turn out to be essentially the electromagnetic 4-vector potential. The constant λ is intended to allow for different species of particle with different electric charges (proportional to λ). In our present geometrical language, we may say that the wavefunctions representing different particle species exist in independent internal spaces and there is no reason why parallel transport in all these spaces should involve the same connection. If, as we know to be the case, particles of different types respond to the same electromagnetic fields, then we conclude that, as a matter of empirical fact, their connections are determined by the same function $A_\mu(x)$, though possibly with different coefficients.

Because we are concerned only with the phases and not the magnitudes of wavefunctions, it is convenient to deal with a fibre bundle whose fibres consist just of these phases. Each fibre can be envisaged as a copy of the unit circle in the complex plane, whose points are labelled by the phase angle θ, with values between 0 and 2π. The angle $\theta(x)$ in the fibre at the spacetime point x can be thought of as specifying a transformation. Thus, if we write $\phi(x)$ as $\exp(i\theta(x))|\phi(x)|$, then the phase factor transforms $|\phi(x)|$ into $\phi(x)$. More generally, a *phase transformation* changes the phase of any complex wavefunction by an angle between 0 and 2π. If the same transformation is made at each spacetime point, then the wave equation satisfied by ϕ is unchanged, and so are all the matrix elements. In this sense, the transformation is a symmetry of the quantum theory. The set of all these transformations, which are also labelled by θ, constitutes a *symmetry group*. It is possible to consider symmetry groups which are much more general than phase transformations, and the transformations which constitute the group may be labelled by several parameters analogous to θ. The set of all possible values of these parameters is called the *group manifold*. At the most fundamental level, each fibre in a bundle of the kind we are considering is to be thought of as a copy of the group manifold of a symmetry group. In our present case, the symmetry group is called $U(1)$, and its manifold is the unit circle.

In spacetime geometry, we defined objects called tensors by their behaviour under general coordinate transformations. In our fibre bundle, the analogue of a general coordinate transformation is a phase transformation through an angle $\theta(x)$, where $\theta(x)$ is a differentiable function of position. The tensors associated with these transformations are products of ϕ's and ϕ^*'s. For a product $\Phi_{mn} = \phi^{*m}\phi^n$, the transformation law is

$$\Phi'_{mn}(x) = \exp[\mathrm{i}(n - m)\lambda\theta(x)]\Phi_{mn}(x). \tag{8.5}$$

The definition (8.3) of parallel transport leads to a covariant derivative D_μ analogous to (2.24). In terms of the real and imaginary parts of the wavefunction, it is defined by

$$\phi_i(x + \Delta x) - \phi_i(x \to x + \Delta x) = \mathrm{D}_\mu\phi_i(x)\Delta x^\mu + \mathrm{O}(\Delta x^2) \tag{8.6}$$

which, on account of (8.3), gives

$$\mathrm{D}_\mu\phi_i(x) = \partial_\mu\phi_i(x) + \Gamma_{ij\mu}(x)\phi_j(x). \tag{8.7}$$

In terms of the complex wavefunction, this may be rewritten as

$$\mathrm{D}_\mu\phi(x) = [\partial_\mu + \mathrm{i}\lambda A_\mu(x)]\phi(x). \tag{8.8}$$

An essential property of a covariant derivative is that it acts on tensors to produce new tensors. In the present context, this means that $\mathrm{D}_\mu\phi$ must have the same phase transformation property as ϕ itself. As for the affine connection in chapter 2, this requirement leads to a transformation law for the connection A_μ which is different from the law (8.5) for 'tensors'. If $\phi' = \exp(\mathrm{i}\lambda\theta)\phi$, then we must have

$$\mathrm{D}'_\mu\phi' = (\partial_\mu + \mathrm{i}\lambda A'_\mu)(\mathrm{e}^{\mathrm{i}\lambda\theta}\phi) = \mathrm{e}^{\mathrm{i}\lambda\theta}\mathrm{D}_\mu\phi = \mathrm{e}^{\mathrm{i}\lambda\theta}(\partial_\mu + \mathrm{i}\lambda A_\mu)\phi \tag{8.9}$$

and so the transformation law for A_μ is

$$A'_\mu(x) = A_\mu(x) - \partial_\mu\theta(x). \tag{8.10}$$

The action of the covariant derivative on a tensor which transforms according to (8.5) must therefore be

$$\mathrm{D}_\mu\Phi_{mn} = [\partial_\mu + \mathrm{i}(n - m)A_\mu]\Phi_{mn}. \tag{8.11}$$

The set of transformation rules given by (8.5) and (8.10) is usually called a *local gauge transformation* and the connection coefficient A_μ is called a *gauge field*. The derivative D_μ may be called a *gauge-covariant derivative* to distinguish it from the generally covariant derivative ∇_μ.

The intrinsic geometrical structure of spacetime is determined, as we saw in chapter 2, by the metric and by the affine connection. Once the presence of this structure in the dynamical equations of physics has been made explicit through their components $g_{\mu\nu}$ and $\Gamma^\mu{}_{\nu\sigma}$, we expect that these equations should be generally covariant: that is their forms should be independent of our choice of a coordinate system. If the dynamical equations are derived from a principle of least action, general covariance is guaranteed by choosing the action to be a scalar.

In the same way, the geometry of the U(1) fibre bundle of electromagnetism (i.e. the relationships between phases of wavefunctions at different points in spacetime) is determined by the gauge field $A_\mu(x)$. Once the gauge field has been incorporated into the equations of motion, we expect these equations to be *gauge covariant*. That is, their forms should be preserved by gauge transformations. This will automatically be so if they are derived from a gauge-invariant action. Since we are working in Minkowski space, we shall also require the action to be a Lorentz scalar.

Let us first construct the wave equation for a spin-$\frac{1}{2}$ particle in a prescribed electromagnetic field. In the case of spacetime geometry, an action which is invariant under general coordinate transformations could be built from tensors by contracting all their indices, so that the transformation matrices cancel. Correspondingly, to make a gauge-invariant action, we can use products of gauge tensors, with transformation laws of the form (8.5). Clearly, the product of one such tensor with the complex conjugate of another of the same type will be invariant. Consider the Dirac action (7.59). It should be clear that this will become gauge invariant if we replace the ordinary derivative ∂_μ with the gauge-covariant derivative D_μ:

$$S_{\text{Dirac}} = \int \mathrm{d}^4x\,\bar{\psi}(x)(\mathrm{i}\not{\partial} - \lambda \not{A}(x) - m)\psi(x). \tag{8.12}$$

The equation which follows from varying $\bar{\psi}$ is

$$(\mathrm{i}\not{\partial} - \lambda \not{A}(x) - m)\psi(x) = 0. \tag{8.13}$$

This is known as the *minimal coupling prescription*. It is the simplest modification of the original Dirac equation which makes it gauge covariant and reduces to the original one when $A_\mu = 0$. A variety of other equations could be invented by introducing further gauge-covariant terms which vanish when $A_\mu = 0$, but there appears to be no good physical reason for doing so.

Some physical consequences of this modified Dirac equation will be explored in the next chapter. A mathematical consequence is that the symmetry of the theory under *global* phase transformations—those which change the phase of the wavefunction by the same amount at each spacetime point—has been promoted to a *local* symmetry, since the phase may be changed by a position-dependent amount $\theta(x)$, provided that a compensating change (8.10) is made in the gauge field. This is precisely analogous to relativistic geometry. In special relativity, Lorentz transformations with a position-independent matrix Λ are global symmetries, and the affine connection coefficients are zero in Cartesian coordinate systems. When the affine connection is explicitly included, general coordinate transformations with position-dependent Λ are symmetries, in the sense of general covariance. In the absence of gravitational fields, coordinate systems may be found in which the affine connection coefficients are everywhere zero. We shall shortly see that, in the absence of electromagnetic fields, the gauge field can be expressed as the gradient of a scalar function, $A_\mu(x) = \partial_\mu \omega(x)$. Therefore, by choosing $\theta(x) = \omega(x) + \text{constant}$ in (8.10), A_μ can be set to zero everywhere. This amounts to choosing a special set of coordinate systems in the fibre bundle which are analogous to the inertial frames of special relativity.

In addition to (8.13), which describes the response of a charged particle to electromagnetic fields, we need an equation (the analogue of the Einstein field equations) which determines how electromagnetic fields are generated by a distribution of charged particles. The way to find these is again to add to the action a part involving the connection. This must be gauge invariant and therefore constructed from quantities which are tensors under gauge transformations. The only such quantity which can be built from the gauge field alone is the curvature of the bundle defined, like the Riemann tensor, as the commutator of two covariant derivatives

$$F_{\mu\nu} = -\frac{\mathrm{i}}{\lambda}[\mathrm{D}_\mu, \mathrm{D}_\nu] = \partial_\mu A_\nu - \partial_\nu A_\mu. \tag{8.14}$$

This is in fact gauge invariant, and the simplest Lorentz scalar that can be constructed from it is $F_{\mu\nu}F^{\mu\nu}$. We see that $F_{\mu\nu}$ is none other than the Maxwell field strength tensor given in (3.50) and (3.51). It turns out, as with gravity, that an extremely successful theory is obtained by including in the action only a term proportional to this quantity. This is, essentially, the first term of (3.54). If we identify j^μ_e in (3.54) as proportional to the current density (7.60), then the second term of (3.54) is reproduced by the $A\!\!\!/$ term in (8.12). There is at present no definitive understanding in either theory of why the simplest allowed form of the action should be the one actually selected by nature, although some properties of interacting quantum field theories which we touch on briefly in the next chapter suggest a possible explanation.

To make the correspondence with Maxwell's theory exact, we must examine more closely the role of the electric charge. So far, we have established only that the simplest action contains a term *proportional* to $F_{\mu\nu}F^{\mu\nu}$. Allowing for n species of spin-$\frac{1}{2}$ particles, the total action may be written as

$$S = \int \mathrm{d}^4x \left(-\frac{1}{4e^2} F_{\mu\nu}F^{\mu\nu} + \sum_{i=1}^{n} \bar{\psi}_i(x)(\mathrm{i}\partial\!\!\!/ - \lambda_i A\!\!\!/(x) - m_i)\psi_i(x) \right) \tag{8.15}$$

where e is a constant whose value is not known *a priori*. This constant, which may be identified as a fundamental electric charge, is clearly somewhat analogous to the constant G which appears in the theory of gravity. Note, however, that the curvature term in the Einstein–Hilbert gravitational action is linear in the Riemann tensor $R_{\mu\nu\sigma\tau}$, which can be contracted to give a non-trivial scalar curvature R. In electromagnetism, the contraction $g^{\mu\nu}F_{\mu\nu}$ is identically zero, because $g^{\mu\nu}$ is symmetric and $F_{\mu\nu}$ antisymmetric, and the simplest non-trivial Lorentz scalar is quadratic in $F_{\mu\nu}$. This is symptomatic of some important differences between the two theories. The standard form of electromagnetism is obtained by rescaling the gauge field:

$$A_\mu(x) \to eA_\mu(x) \qquad F_{\mu\nu}(x) \to eF_{\mu\nu}(x). \tag{8.16}$$

after which the action becomes

$$S = \int \mathrm{d}^4x\left(-\frac{1}{4}F_{\mu\nu}F^{\mu\nu} + \sum_{i=1}^{n}\bar{\psi}_i(x)(\mathrm{i}\not{\partial} - \lambda_i e\not{A}(x) - m_i)\psi_i(x)\right). \tag{8.17}$$

The equations derived from this action describe the electromagnetic interactions of n species of particle with masses m_i and charges $\lambda_i e$. Evidently, the $(n + 1)$ constants $(\lambda_1, \ldots, \lambda_n, e)$ are not all independent. We can choose e to be the magnitude of the electronic charge by setting $\lambda_{\text{electron}} = -1$. Then the charges of the remaining particles are multiples, λ_i, of this fundamental charge. There is no reason, however, why the λ_i should be integers or even rational numbers. The fact that the electric charges of all observed particles are integral multiples of a fundamental charge has no explanation within the theory of electromagnetism alone. A possible explanation is offered by the *grand unified theories* of strong, weak and electromagnetic interactions which will be outlined in chapter 12. Notice also that had $F_{\mu\nu}$ contained terms quadratic in A_μ which, as we shall see shortly, is the case in non-Abelian gauge theories, the rescaling of the gauge field in (8.13) would not have removed the charge e entirely from the curvature term, and e would have been a genuine independent parameter.

8.2 Non-Abelian Gauge Theories

The internal spaces in which wavefunctions exist may be more complicated than the complex plane. Consider, for example, the nucleons—the proton and neutron. In processes involving the strong interaction (of which more in chaper 12), they appear on an equal footing: the strong interaction is said to be *charge independent*. This observation, together with the fact that their masses are very similar, leads to the idea that the proton and neutron can be regarded as different states of the same particle—the *nucleon*. The nucleon wavefunction is then a two-component matrix

$$\psi_N(x) = \begin{pmatrix} \psi_p(x) \\ \psi_n(x) \end{pmatrix}. \tag{8.18}$$

Actually, since the nucleons are spin-$\frac{1}{2}$ particles, each of the two components is itself a four-component spinor, but this does not at present concern us. A state with $\psi_n = 0$ is a pure proton state and vice versa, while a state in which both components are non-zero is a superposition of the two. This is quite analogous to the non-relativistic description of spin-$\frac{1}{2}$ polarization states (see appendix 2). In particular, any unitary transformation (i.e. a rearrangement of the components which leaves the magnitude $(\bar{\psi}_N \psi_N)^{1/2}$ unchanged) can be expressed as

$$\psi'_N = \exp[i(\theta I + \tfrac{1}{2}\boldsymbol{\alpha}\cdot\boldsymbol{\tau})]\psi_N \equiv U(\theta, \boldsymbol{\alpha})\psi_N \tag{8.19}$$

where I is the unit 2×2 matrix, τ^1, τ^2, τ^3 are the Pauli matrices and $\theta, \alpha^1, \alpha^2, \alpha^3$ are real angles. Such transformations are involved, for example, in reactions which change the state of a nucleon but not the total number of nucleons, such as beta decay ($n \to p + e^- + \bar{\nu}_e$) or pion–nucleon scattering ($\pi^- + p \to n + \pi^0$). The matrices τ^i have the same numerical values as the spin matrices (7.26), but the symbol τ emphasizes that they refer to a different internal property of the particles. This property is called *isotopic spin*, or more commonly *isospin*, denoted by $\boldsymbol{T}$. The transformations parametrized by θ are phase transformations which will not concern us for the moment. The others, of the form

$$U(\boldsymbol{\alpha}) = \exp(\tfrac{1}{2}i\boldsymbol{\alpha}\cdot\boldsymbol{\tau}) \tag{8.20}$$

can be regarded as rotations in an internal three-dimensional *isospin space*. The proton and neutron states correspond to 'isospin up' and 'isospin down' with respect to a chosen quantization axis in this space.

There now arises a question similar to that which led to electromagnetism. The two-component wavefunction at the spacetime point x is to be thought of as existing in a copy of isospin space erected at x, and we would like to know how the directions of the (T_1, T_2, T_3) axes at different points are related. In contrast to the complex phase, these directions have definite physical meanings, because the proton and neutron are physically identifiable states. Parallel transport of a wavefunction may be defined by introducing a connection as in (8.3), except that i and j now label the components in (8.18) rather than real and imaginary parts. If Γ is zero, then a parallelly transported wavefunction which represents, say, a neutron at x also represents a neutron at $x + \Delta x$. If Γ is not zero, then the wavefunction may, after being transported to $x + \Delta x$, represent a superposition of proton and neutron states. Since the connection in (8.3) turned out to be related to the electromagnetic field, we may anticipate that the isospin connection is similarly related to some kind of force field. Evidently, one effect of this force is to turn neutrons into protons, so it might provide a means of describing beta decay.

The right-hand side of (8.3) now corresponds to an infinitesimal rotation of the kind (8.20), so the connection coefficient has the form

$$\Gamma_{ij\mu}(x) = -\tfrac{1}{2}\mathrm{i}A^a_\mu(x)(\tau^a)_{ij}. \tag{8.21}$$

There are three independent gauge fields A^a_μ, corresponding to the three independent isospin rotations. This connection acts in the fibre bundle whose typical fibre is the set of all transformations of the form (8.20) or, equivalently, the set of all values of the α^a which lead to distinct transformations. This can be taken as the set of all positive and negative values such that $\boldsymbol{\alpha}\cdot\boldsymbol{\alpha} \leqslant 4\pi^2$, with the proviso that all values for which the equality holds correspond to the same transformation and therefore to a single point of the fibre (see exercise 8.2). This set of transformations constitutes the group SU(2). It is a *non-Abelian* group, which means that two rotation matrices $U(\boldsymbol{\alpha})$ and $U(\boldsymbol{\beta})$ do not commute unless $\boldsymbol{\alpha}$ and $\boldsymbol{\beta}$ point in the same direction. The group U(1) of electromagnetism is an *Abelian* group, because any two phase transformations commute with each other. One consequence of the non-Abelian nature of isospin rotations is that no arbitrary constant λ appears in the connection (8.21) to distinguish different particle species. This is because, as we shall see in more detail below, the gauge field A^a_μ has an intrinsic scale. For example, a rotation through an angle of π about the T_1 or T_2 axis changes a proton state with $T_3 = \frac{1}{2}$ into a neutron state with $T_3 = -\frac{1}{2}$. The same rotation must produce the same reversal of T_3 when acting on any set of particle wavefunctions which form an isospin multiplet. Therefore, the size of the rotation angles in (8.20) has a definite meaning, common to all particle species, and we have no freedom to introduce an arbitrary parameter as in the Abelian case (8.5). On the other hand, different particle species may fall into isospin multiplets of different sizes. Just as with angular momentum, an isospin-T multiplet has $(2T + 1)$ members. For the moment, the three pions (π^+, π^0, π^-) may serve as an example of an isospin-1 triplet. At present, however, in order to describe the mathematics of non-Abelian theories in its simplest terms, I am not taking proper account of the observed properties of elementary particles. When we come to study the application of these theories to real physical particles, it will be necessary to revise the way in which the particles are assigned to isospin multiplets. The wavefunction for an isospin-T multiplet undergoes parallel transport with a connection similar to (8.21) except that the Pauli matrices are replaced with a suitable set of three $(2T + 1) \times (2T + 1)$ matrices, called the isospin-T *representation* of the group SU(2). The same gauge field appears in each case, however.

Given the gauge connection, we have a *gauge-covariant derivative*

$$\mathrm{D}_\mu = \partial_\mu + \mathrm{i}A_\mu(x) \tag{8.22}$$

where A_μ is a matrix defined by

$$A_\mu(x) = A_\mu^a(x)T^a \tag{8.23}$$

and T^a are the isospin matrices appropriate to the particular multiplet of wavefunctions on which the derivative acts. Under a gauge transformation, each multiplet transforms as

$$\psi'(x) = U(\boldsymbol{\alpha})\psi(x) = \exp(\mathrm{i}\boldsymbol{\alpha}(x)\cdot\boldsymbol{T})\psi(x). \tag{8.24}$$

To find the transformation law for the gauge fields, consider

$$\mathrm{D}'_\mu\psi' = (\partial_\mu + \mathrm{i}A'_\mu)U\psi = (U\partial_\mu + \partial_\mu U + \mathrm{i}A'_\mu U)\psi. \tag{8.25}$$

The requirement is that this should equal $U\mathrm{D}_\mu\psi$, so A_μ must transform as

$$A'_\mu = UA_\mu U^{-1} + \mathrm{i}(\partial_\mu U)U^{-1}. \tag{8.26}$$

If U were just the phase factor $\exp(\mathrm{i}\theta(x))$, this would be the same as (8.10).

The non-Abelian analogue of the electromagnetic field strength tensor is, naturally, the curvature tensor $-\mathrm{i}[\mathrm{D}_\mu, \mathrm{D}_\nu]$. This is more closely analogous to the Riemann tensor, in the sense that it involves the non-commuting properties of both the derivative ∂_μ and the matrices T^a. It is given by

$$F_{\mu\nu} = \partial_\mu A_\nu - \partial_\nu A_\mu + \mathrm{i}[A_\mu, A_\nu]. \tag{8.27}$$

Of course, the matrix form of this expression depends on the particular representation of the gauge group (here SU(2)) to which the matrices T^a belong. However, in every representation, these matrices satisfy the commutation relations of the *Lie algebra*

$$[T^a, T^b] = \mathrm{i}C^{abc}T^c \tag{8.28}$$

where the set of *structure constants* C^{abc} is totally antisymmetric in a, b and c. For SU(2), they are given by $C^{abc} = \varepsilon^{abc}$ (see appendix 2). The T^a are the generators of the symmetry transformations (in our case, isospin rotations) and in group theory language are called the generators of the symmetry group. Using the definition (8.23) of the matrices A_μ, we find that

$$F_{\mu\nu} = F_{\mu\nu}^a T^a \tag{8.29}$$

where the field strengths

$$F_{\mu\nu}^a = \partial_\mu A_\nu^a - \partial_\nu A_\mu^a - C^{abc}A_\mu^b A_\nu^c \tag{8.30}$$

are the same for any representation.

Unlike the electromagnetic field strength tensor, (8.27) is not a gauge-invariant object. In fact, its transformation law is

$$F'_{\mu\nu} = UF_{\mu\nu}U^{-1} \tag{8.31}$$

as readers are invited to verify in exercise 8.3. From this it follows that the three field strengths (8.30) ($a = 1, 2, 3$) belong to an isospin-1 multiplet. To understand this, notice first that (8.31) implies the transformation

$$F^{a\prime}_{\mu\nu} = \mathcal{U}^{ab}(\boldsymbol{\alpha})F^{b}_{\mu\nu} \tag{8.32}$$

where the coefficients $\mathcal{U}^{ab}$ are defined by

$$U(\boldsymbol{\alpha})T^{b}U^{-1}(\boldsymbol{\alpha}) = \mathcal{U}^{ab}(\boldsymbol{\alpha})T^{a}. \tag{8.33}$$

It is a group-theoretical fact (which I shall not prove) that, if we regard these coefficients as elements of a 3×3 matrix, it can be written as

$$\mathcal{U}(\boldsymbol{\alpha}) = \exp(\mathrm{i}\boldsymbol{\alpha}\cdot\mathcal{T}) \tag{8.34}$$

where the three 3×3 matrices $\mathcal{T}^{a}$ form a special representation of SU(2) called the *adjoint representation*. Every Lie group possesses such a representation, in which the number of members of the multiplet is equal to the number of independent generators. The matrices of the adjoint representation can be expressed in terms of the structure constants as

$$(\mathcal{T}^{a})_{bc} = -\mathrm{i}C^{abc}. \tag{8.35}$$

The proof that these matrices satisfy the algebra (8.28) is the subject of exercise 8.4.

Once again, we need to construct a gauge-invariant action for the gauge fields. The simplest possibility is

$$S = -\frac{1}{4g^{2}}\int \mathrm{d}^{4}x F^{a}_{\mu\nu}F^{a\mu\nu} \tag{8.36}$$

where g is a *coupling constant* analogous to the electric charge. As in (8.16), we now rescale the gauge field by a factor of g, so the field strength tensor becomes

$$F^{a}_{\mu\nu} = \partial_{\mu}A^{a}_{\nu} - \partial_{\nu}A^{a}_{\mu} - gC^{abc}A^{b}_{\mu}A^{c}_{\nu} \tag{8.37}$$

after dividing by an overall factor of g. In the quantum theory, the gauge field becomes a field operator for 'intermediate vector bosons' which mediate the corresponding force. In the case of electromagnetism, these are photons, which are neutral particles. When the action (8.36) is expressed in terms of the rescaled field strength (8.37), the overall factors of g^{2} cancelling out, it contains products of three A's multiplied by g and products of four A's multiplied by g^{2}. It will become clear in the next chapter that such terms represent interactions between the vector bosons of the non-Abelian theory, whose strength is measured by g. Indeed, it is already obvious that the actions for free particles considered in chapters 6 and 7 are only quadratic in the field operators. Thus, these particles carry the 'charge' g of the force which they themselves mediate, and this fact has important physical consequences. The situation is similar in the case of gravity. If the metric tensor is expanded as $g_{\mu\nu} = \eta_{\mu\nu} + h_{\mu\nu}$, we can attempt to develop a quantum theory of gravity, treating $h_{\mu\nu}$ as a field operator for *gravitons*, which are 'intermediate tensor bosons' mediating the gravitational force. The gravitational analogue of charge is energy density which is, of course, possessed by the gravitons themselves. Thus, gravitons interact amongst themselves through gravitational forces, and this comes about through the fact that the last term in the gravitational action (4.14) contains terms which are higher than quadratic in $h_{\mu\nu}$. I should point out, however, that the quantum theory of gravity constructed in this way appears to be mathematically unsound, for reasons I shall touch on later. No experiment has yet succeeded in detecting a graviton, but it is generally believed that this is because of the extreme weakness of their interactions rather than because they do not exist.

If we include spin-$\frac{1}{2}$ fermions upon which the gauge field acts, the total action is

$$S = \int d^4x\left(-\frac{1}{4}F^a_{\mu\nu}F^{a\mu\nu} + \sum_{i=1}^{n}\bar{\psi}_i(x)(i\partial\!\!\!/ - gA\!\!\!/(x) - m_i)\psi_i(x)\right). \quad (8.38)$$

This is now expressed in a rather compact notation. The sum is over multiplets of wavefunctions ψ_i, each having $(2T^{(i)} + 1)$ members in the case of SU(2) isospin. Each member is itself a Dirac spinor, so ψ_i may be represented schematically in the form

$$\psi_i = \begin{pmatrix} (\vdots) \\ (\vdots) \\ \vdots \\ (\vdots) \end{pmatrix} \begin{matrix} \}4 \\ \\ \end{matrix} \qquad 2T^{(i)} + 1.$$

The matrix $A\!\!\!/$ is

$$A\!\!\!/ = A^a_\mu \gamma^\mu T^{(i)a} \quad (8.39)$$

where $T^{(i)a}$ is the ath generator matrix in the isospin-$T^{(i)}$ representation. The Dirac matrix γ^μ acts on each four-component spinor independently, while $T^{(i)a}$ treats the spinors as single elements.

From the action (8.38) we derive an Euler–Lagrange equation for the gauge field which is the non-Abelian analogue of Maxwell's equations:

$$D_\mu F^{\mu\nu} = J^\nu \quad \text{or} \quad \partial_\mu F^{a\mu\nu} - gC^{abc}A^b_\mu F^{c\mu\nu} = J^{a\nu}. \quad (8.40)$$

The current is given by

$$J^{a\nu} = g \sum_i \bar{\psi}_i \gamma^\nu T^{(i)a} \psi_i. \tag{8.41}$$

For example, in the case of the nucleon doublet,

$$\begin{aligned} J^{3\nu} &= g(\bar{\psi}_p, \bar{\psi}_n)\gamma^\nu \begin{pmatrix} \frac{1}{2} & 0 \\ 0 & -\frac{1}{2} \end{pmatrix} \begin{pmatrix} \psi_p \\ \psi_n \end{pmatrix} \\ &= g[\tfrac{1}{2}\bar{\psi}_p \gamma^\nu \psi_p - \tfrac{1}{2}\bar{\psi}_n \gamma^\nu \psi_n] \\ &= g \sum_{p,n} T^3 \times (\text{probability current density}). \end{aligned} \tag{8.42}$$

There is also, of course, a Dirac equation of the form (8.13) for each multiplet of wavefunctions, λ being replaced by g and $\not{A}$ by (8.39).

We saw in chapter 3 that, as a consequence of gauge invariance, the electromagnetic current j^μ_e is conserved in the classical theory. As readers may easily check using the Dirac equation (8.13), the quantum-mechanical current $j^\mu_e = \lambda\bar{\psi}\gamma^\mu\psi$ (which becomes $\lambda e\bar{\psi}\gamma^\mu\psi$ after the rescaling (8.16)) is also conserved. The conservation law $\partial_\mu j^\mu_e = 0$ is a gauge-covariant equation because the current is a gauge scalar, with $n = m = 1$ in (8.5), and its gauge-covariant derivative (8.11) is the same as the ordinary derivative. In the non-Abelian theory, however, the current (8.41) is not a gauge scalar. It is a multiplet of currents, whose members are labelled by a, which belongs to the adjoint representation of the gauge group and satisfies the covariant equation

$$D_\mu J^\mu = 0 \quad \text{or} \quad \partial_\mu J^{a\mu} - gC^{abc} A^b_\mu J^{c\mu} = 0. \tag{8.43}$$

The current is said to be *covariantly conserved*, but it clearly is not conserved in the usual sense. This does not, however, imply a breakdown of the general rule that a symmetry implies the existence of a conserved quantity. If we differentiate the non-Abelian Maxwell equation (8.40), and take into account the antisymmetry of $F^{\mu\nu}$, we find that the modified current

$$\tilde{J}^{a\nu} = J^{a\nu} + gC^{abc} A^b_\mu F^{c\mu\nu} \tag{8.44}$$

is conserved in the ordinary sense:

$$\partial_\nu \tilde{J}^{a\nu} = \partial_\nu \partial_\mu F^{a\mu\nu} = 0. \tag{8.45}$$

The two terms in (8.44) have a simple physical significance. The current represents the flow of isospin, in the same way that the electromagnetic current represents the flow of charge. The first contribution is that of the fermions, and the second is that of the gauge fields or, in the quantized theory, of the vector bosons which, as we have seen, themselves carry isospin.

The components of the field strength tensor (8.37) can be thought of as 'electric' and 'magnetic' fields $\boldsymbol{E}^a$ and $\boldsymbol{B}^a$. As we saw in chapter 3,

(3.45) implies that there are no magnetic monopoles, except at the expense of singularities in the potential $A_\mu(x)$. In the non-Abelian theory, the corresponding equation is

$$\partial_i B^{ai} = gC^{abc} A_i^b B^{ci}. \tag{8.46}$$

Because the right-hand side is non-zero, the non-Abelian theory allows the possibility of 'magnetic monopoles' without singularities in the gauge field. Of course, the non-Abelian 'magnetic field' is not what we ordinarily recognize as a magnetic field. In unified theories, which are more complicated than the one we have so far discussed, electromagnetism is combined with other forces in a manner which permits the appearance of objects with the properties of genuine magnetic monopoles.

8.3 Non-Abelian Theories and Electromagnetism

It is now necessary to understand how the phase transformations of electromagnetism fit in with the SU(2) isospin rotations we have been considering. The general unitary transformation (8.19) includes a phase transformation which we have so far ignored. Since θ multiplies the unit matrix, any phase transformation commutes with any isospin rotation, so the set of transformations of the form (8.19) constitute a product group, written as SU(2) × U(1). This means that each transformation is the product of two independent transformations, one from each group. In the transformations considered in the last section, only the identity transformation of U(1) was involved. Now, the U(1) component of this product group cannot correspond directly to electromagnetism because it changes the phase of the electrically charged proton and the neutral neutron by the same amount. To represent electromagnetism in this context, we must look for transformations of the form (8.19) in which the angles θ and $\boldsymbol{\alpha}$ are related in such a way that the net transformation changes the phase of ψ_p while leaving the phase of ψ_n unchanged. The relation which achieves this is

$$\theta = \tfrac{1}{2} Y\omega \qquad \alpha_1 = \alpha_2 = 0 \qquad \alpha_3 = \omega \tag{8.47}$$

where ω is an arbitrary angle and Y is a constant, which in this case is $Y = 1$. With this relation, we have

$$\theta I + \tfrac{1}{2}\boldsymbol{\alpha}\cdot\boldsymbol{\tau} = \omega \begin{pmatrix} 1 & 0 \\ 0 & 0 \end{pmatrix} \tag{8.48}$$

and

$$U(\theta, \boldsymbol{\alpha}) = \begin{pmatrix} \mathrm{e}^{\mathrm{i}\omega} & 0 \\ 0 & 1 \end{pmatrix} \tag{8.49}$$

which is the desired transformation matrix. Since any two matrices of the form (8.49) commute with each other, the set of all such transformations is a U(1) subgroup of SU(2) × U(1), and quite suitable for representing electromagnetism.

If this scheme is to work, it must be possible to assign to each isospin multiplet a value of Y, called *hypercharge*, in such a way that (8.49) correctly reflects the charges of all the particles in the multiplet. That is, if we use in (8.48) the isospin matrices appropriate for the particular multiplet, the transformation matrix must turn out to have the form

$$U(\theta, \boldsymbol{\alpha}) = \begin{pmatrix} e^{iQ_1\omega} & & 0 \\ & e^{iQ_2\omega} & \\ 0 & & \ddots \end{pmatrix} \tag{8.50}$$

where the Q_i are the charges of the particles in the multiplet. This will be so if the charges are related to the T^3 quantum numbers of the particles by

$$Q = T^3 + \tfrac{1}{2}Y. \tag{8.51}$$

It so happens that relations of just this kind, the *Gell-Mann–Nishijima relations*, are needed for the phenomenological classification of observed particles. For example, $Y = 1$ and $T^3 = \pm\frac{1}{2}$ for the nucleon doublet and $Y = 0$, $T^3 = (1, 0, -1)$ for the pions.

8.4 Relevance of Non-Abelian Theories to Physics

Had we not already known of the existence of electromagnetic forces, the geometrical considerations of §8.1 might have led us to predict the occurrence of such forces in nature. Can we, then, identify forces in nature which correspond to the extension of these geometrical ideas to non-Abelian symmetry groups? The answer to this is a qualified ‘yes’. The idea of non-Abelian gauge theories was first suggested by Yang and Mills in 1954, and theories of this kind are generally known as *Yang–Mills theories*. At that time, it appeared that observed particles such as protons, neutrons and pions were truly fundamental, and the theory of Yang and Mills was based on the approximate nuclear isospin symmetry which relates these particle states in the way I have described. It is now believed that the nucleons, pions and other strongly interacting particles are themselves composed of more fundamental particles, the *quarks*. The experimental evidence for this, although compelling, is indirect. It appears that quarks are permanently bound inside the observed particles, and no quark has ever been detected in isolation. The nuclear isospin symmetry, part of what is now known as *flavour* symmetry, appears to be more or less accidental and the proton and neutron, for example, are not to be regarded as different states of the same particle in the straightforward way suggested by (8.18). However, it is consistent with our present knowledge to group the quarks, and also the *leptons*, which include the electron, muon and tau particle, together with their associated neutrinos, into multiplets of a different symmetry called *weak isospin*. This is also an SU(2) symmetry and can be combined, as above, with phase transformations to give SU(2) × U(1).

The gauge theory associated with this symmetry can be identified as describing the electromagnetic and weak interactions. As it happens, the proton and neutron can loosely be considered as forming a weak isospin doublet, in the sense that converting a proton into a neutron involves changing one of its constituent quarks, called an 'up' (u) quark, into a 'down' (d) quark, and these two quarks form a weak isospin doublet. Therefore, the picture of beta decay as parallel transport in the presence of a non-trivial gauge connection survives in this version of the theory. Quantum-mechanically, what happens is that a d quark in a neutron, say, turns into a u quark by emitting a gauge quantum, a particle called W^-, whose field operator is one of the gauge fields, which then decays into an electron and an antineutrino.

To construct a theory of such processes, which I shall describe more thoroughly in chapter 12, an important obstacle must be overcome. Unlike electromagnetic forces, the weak interaction which is responsible for beta decay has a very short range. As will become clear in the next chapter, this implies that the gauge quanta must have rather large masses. In fact, the W^- is observed to be about 100 times as massive as the proton. By analogy with (7.7), we see that the gauge field action should contain a term something like $m^2 A^a_\mu A^{a\mu}$. No such term appears in (8.38), for the very good reason that it is not gauge invariant. In order to interpret the SU(2) × U(1) theory in terms of electroweak interactions, therefore, we have to understand how massive gauge quanta can emerge from a gauge-invariant theory. This requires the idea of *spontaneous symmetry breaking*, which will be introduced in chapter 9.

8.5 The Theory of Kaluza and Klein

Now that we have seen how theories of electromagnetism and other forces arise from much the same sort of geometrical considerations as the relativistic theory of gravity, it is natural to wonder whether the analogy can be made any more concrete. In other words, are the origins of gravity and other forces not merely similar but identical? Kaluza (1921) and Klein (1926) put forward a theory in which gravity and electromagnetism appear as two different aspects of exactly the same phenomenon. According to this theory, the vector potential A_μ is part of the metric tensor of a five-dimensional spacetime.

Setting aside, temporarily, the fact that we perceive only four dimensions, let us call the five-dimensional metric tensor $\tilde{g}_{AB}$. To emphasize the extra dimension, I shall let the indices A and B take the values 0, 1, 2, 3, 5. We redefine the components of $\tilde{g}_{AB}$ as follows:

$$\tilde{g}_{5\mu} = \tilde{g}_{\mu 5} = \tilde{g}_{55} A_\mu \qquad \tilde{g}_{\mu\nu} = g_{\mu\nu} + \tilde{g}_{55} A_\mu A_\nu \tag{8.52}$$

where the indices μ and ν run from 0 to 3 as usual. The action for five-dimensional gravity is

$$S = \frac{1}{16\pi\tilde{G}} \int \mathrm{d}^5 x (-\tilde{g})^{1/2} \tilde{R} \tag{8.53}$$

where the gravitational constant $\tilde{G}$, the metric determinant $\tilde{g}$ and the curvature scalar $\tilde{R}$ are the five-dimensional ones. We now make two assumptions:

(i) $g_{\mu\nu}$ and A_μ are independent of x^5 and g_{55} is just a constant;

(ii) the five-dimensional spacetime manifold has the structure illustrated in figure 8.1. In the fifth dimension it is of finite extent and closes to form a cylinder of radius r_5.

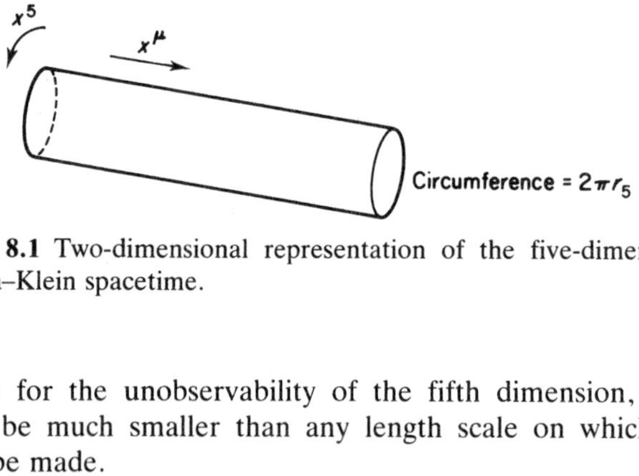

Figure 8.1 Two-dimensional representation of the five-dimensional Kaluza–Klein spacetime.

To account for the unobservability of the fifth dimension, we simply take r_5 to be much smaller than any length scale on which measurements can be made.

If, using these assumptions, (8.52) is substituted into (8.53), the result is

$$S = \int \mathrm{d}^4 x (-g)^{1/2} \left(\frac{1}{16\pi G} R - \frac{1}{4e^2} F_{\mu\nu} F^{\mu\nu} \right) \tag{8.54}$$

where g and R are the four-dimensional quantities, and $F_{\mu\nu}$ is the Maxwell field strength tensor with ordinary derivatives $\partial_\mu A_\nu$ replaced by covariant ones, $\nabla_\mu A_\nu$. This is precisely the action we need to describe a spacetime in which both gravitational and electromagnetic fields are present. The four-dimensional gravitational constant G and the charge e are given in terms of the original parameters by

$$G = \widetilde{G}/2\pi r_5 \quad \text{and} \quad e^2 = 8\widetilde{G}/\widetilde{g}_{55} r_5. \tag{8.55}$$

Readers may like to be warned that this simple and natural-looking result is quite complicated to verify. Thus, we use (2.49) to work out the five-dimensional affine connection coefficients, separating out those which have only μ indices from those which have one or more indices equal to 5. We substitute the result into (2.35) to get the five-dimensional Riemann tensor and contract this with the five-dimensional metric tensor to get $\widetilde{R}$. That the result of all this boils down to (8.54) strikes me as a minor miracle!

Appealing though this theory is, little attention was paid to it for a long time. Partly, no doubt, this was because it leads to no new observable effects. An unsatisfactory feature is that the two assumptions needed to obtain the final result have no particular justification. The theory would be greatly improved if some dynamical explanation could be found: that is if it could be shown that a more general five-dimensional spacetime would naturally evolve into one approximately described by (8.54). Unfortunately, no such mechanism is known. It is worth mentioning that assumption (i) can be relaxed by expanding $g_{\mu\nu}$ and A_μ as Fourier series in x_5. For the reason indicated in exercise 8.5, the additional terms correspond to wavefunctions or field operators for particles with very large masses, which we would not expect to have observed. In this sense, assumption (i) can be regarded as a natural consequence of assumption (ii).

More complicated non-Abelian gauge theories can be obtained in much the same way, by starting with more dimensions and *compactifying* them in various ways. In recent years, the Kaluza–Klein idea has been much studied because a number of theories, the *supergravity* and *superstring* theories, either can be more simply formulated in more than four dimensions or are mathematically consistent only in some number of dimensions greater than four. Opinion is divided as to whether the extra dimensions need actually exist in compactified form or are simply a mathematical device. At the time of writing, all such theories are largely speculative and no wholly satisfactory theory has, at least to my knowledge, been completely worked out.

Exercises

8.1. If the real and imaginary parts of ϕ are changed to $\phi_i + \delta\phi_i$, what is the first-order change in the magnitude of ϕ? Show that parallel transport using the connection coefficient (8.4) leaves the magnitude of ϕ unchanged.

8.2. In the transformation matrix (8.20), let $\boldsymbol{\alpha} = \alpha \boldsymbol{n}$, where $\boldsymbol{n}$ is a unit vector. Show that $(\boldsymbol{\tau}\cdot\boldsymbol{n})^2 = 1$ and that $\exp(\mathrm{i}\alpha\boldsymbol{\tau}\cdot\boldsymbol{n}/2) = \cos(\alpha/2) + \mathrm{i}\sin(\alpha/2)(\boldsymbol{\tau}\cdot\boldsymbol{n})$. Show that an angle $\alpha + 4\pi$ leads to the same transformation as α and that all distinct transformations are included if α is restricted to the range $-2\pi \leqslant \alpha \leqslant 2\pi$. Hence show that the range of values of $\boldsymbol{\alpha}$ which all correspond to distinct transformations is $\boldsymbol{\alpha}\cdot\boldsymbol{\alpha} \leqslant 4\pi^2$, except that all values of $\boldsymbol{\alpha}$ for which the equality holds correspond to $U = -1$.

8.3. A matrix U and its inverse U^{-1} are related by $UU^{-1} = I$. Bearing in mind that U and U^{-1} do not necessarily commute, show that, if U depends on x, $\partial_\mu U^{-1} = -U^{-1}(\partial_\mu U)U^{-1}$. For the gauge-transformed field (8.26), show that

$$\partial_\mu A'_\nu = U\{\partial_\mu A_\nu + [U^{-1}\partial_\mu U, A_\nu] + \mathrm{i}U^{-1}(\partial_\mu\partial_\nu U) + \mathrm{i}[U^{-1}\partial_\mu U, U^{-1}\partial_\nu U]\}U^{-1}.$$

Hence verify (8.31).

8.4. For any three matrices T^a, T^b and T^c, verify the *Jacobi identity*

$$[[T^a, T^b], T^c] + [[T^b, T^c], T^a] + [[T^c, T^a], T^b] = 0.$$

Taking these matrices to obey the Lie algebra (8.28), show that the structure constants C^{abc} satisfy the relation

$$C^{abd}C^{dce} + C^{bcd}C^{dae} + C^{cad}C^{dbe} = 0.$$

Hence show that the matrices defined by (8.35) satisfy the Lie algebra.

8.5. Show that the five-dimensional Kaluza–Klein metric $\tilde{g}_{AB}$ can be written in the matrix form

$$\tilde{g}_{AB} = \begin{pmatrix} I & (\tilde{g}_{55})^{1/2}A_\mu \\ 0 & (\tilde{g}_{55})^{1/2} \end{pmatrix} \begin{pmatrix} g_{\mu\nu} & 0 \\ 0 & 1 \end{pmatrix} \begin{pmatrix} I & 0 \\ (\tilde{g}_{55})^{1/2}A_\nu & (\tilde{g}_{55})^{1/2} \end{pmatrix}.$$

The elements of each matrix represent, in clockwise order from the top left, a 4×4 matrix, a four-component column, a single element, and a four-component row. $g_{\mu\nu}$ is the four-dimensional metric and I the 4×4 unit matrix. Hence show that the five-dimensional inverse matrix $\tilde{g}^{AB}$ has elements $\tilde{g}^{\mu\nu} = g^{\mu\nu}$, $\tilde{g}^{5\mu} = \tilde{g}^{\mu 5} = -A^\mu$ and $\tilde{g}^{55} = A_\mu A^\mu + (g_{55})^{-1/2}$, and that the five-dimensional metric determinant is $\det|\tilde{g}_{AB}| = \tilde{g}_{55}\det|g_{\mu\nu}|$.

Consider a scalar field with the five-dimensional action

$$S = \int \mathrm{d}^5x(-\tilde{g})^{1/2}\tilde{g}^{AB}\partial_A\tilde{\phi}^*\partial_B\tilde{\phi}.$$

Assume that $\tilde{\phi}(x, x_5) = \exp(\mathrm{i}\lambda x_5)\phi(x)$ where x denotes the four-dimensional coordinates. When the extra dimension is compactified, show that $\phi(x)$ can be interpreted as the field for scalar particles with charge λe and a mass given by $m^2 = -\lambda^2/\tilde{g}_{55}$. What values of λ are permissible?

9

Interacting Relativistic Field Theories

The relativistic wave equations and field theories encountered in chapter 7 described only the properties of free, non-interacting particles. The wave equation for a free particle is always of the form (differential operator)$\phi = 0$, and therefore the corresponding Lagrangians are quadratic in the fields. We have already seen that gauge theories give rise, in a natural way, to Lagrangians which contain terms of higher than quadratic order in the fields, and these terms describe interactions. In (8.38), for example, $\bar{\psi}\not{A}\psi$ describes an interaction between a fermion and a gauge field, while the higher-order terms in $F^a_{\mu\nu}F^{a\mu\nu}$ describe interactions of the gauge fields amongst themselves. It is, of course, only in the presence of interactions that physically interesting events can occur. At the same time, the physical interpretation of interacting quantum field theories is rather difficult. The interpretation of free field theories is based on expansions such as (7.78) in terms of solutions of the appropriate wave equation, the coefficients being interpreted as creation and annihilation operators. When a fermion interacts with a gauge field, the Dirac equation is modified as in (8.13). If the gauge field is itself an operator, this equation cannot be regarded as a wave equation for ψ alone, and the plane-wave solutions of the free theory have no definite significance. It is, of course, possible to write the field as a Fourier transform, but the momentum k^μ no longer satisfies the constraint $k_\mu k^\mu = m^2$. Although field operators still have the canonical commutation relations, such as (7.82) for Dirac spinors, the coefficients in the Fourier expansion no longer have the simple commutation relations required for creation and annihilation operators.

To make sense of interacting theories, it is generally advantageous to have in mind some particular kind of experiment whose outcome we want to predict. More often than not, the experiments to which relativistic field theory is relevant are high-energy scattering experiments. These are, indeed, the most fruitful method of probing the fundamental structure of matter, and it is with a view to interpreting such experiments that much of the mathematics of interacting field theories has been developed. I shall begin, therefore, by discussing the field-theoretic aspects of this interpretation.

9.1 Asymptotic States and the Scattering Operator

The multi-particle states encountered in free field theories are eigenstates of the Hamiltonian, so they can exist unchanged for as long as the system is left undisturbed. In an interacting theory, the eigenstates of the Hamiltonian cannot, in general, be characterized by a definite number of particles with definite energies and momenta. Indeed, it is not often possible to discover exactly what these eigenstates are. It is reasonable to suppose that the ground state is recognizable as the vacuum. Another reasonable assumption is that a single, stable particle can exist in isolation for an indefinite time, so that these single-particle states would also be energy eigenstates. If the second assumption is valid, it might appear that each stable particle would be represented in the theory by a field operator which creates it from the vacuum and, conversely, that each field operator in the theory could act on the vacuum to create a stable single-particle state. This, however, is not so. For example, the standard model of particle physics described in chapter 12 contains, amongst others, field operators for quarks and for muons. While muons are indeed observed experimentally, they eventually decay (with a lifetime of about 2×10^{-6} s) into electrons and neutrinos, so a single muon state cannot be a true energy eigenstate. Quarks, on the other hand, are never observed in isolation, so a single quark state is not even approximately an eigenstate of the Hamiltonian. Within the standard model, the proton is a true eigenstate, but the operator which creates it from the vacuum is a complicated combination of quark and other field operators. (This statement is believed to be true, being consistent with observations and with approximate calculations, but it has not, as far as I know, been rigorously proved.) According to grand unified theories, protons can also decay into lighter particles, and so even the proton is not an eigenstate. At the time of writing, however, proton decay has not been observed.

A second difficulty of interpretation is that, although single particles have, within experimental resolution, well defined energies and momenta, they also follow quite well defined paths (seen, for example, as narrow tracks in cloud chamber photographs) and so cannot, strictly speaking, be described by plane waves. This is not a difficulty of principle, because it is quite possible to represent these particles by localized wave packets, whose spread in momentum is well within the range allowed by experimental resolution. Such wave packets are, however, inconvenient to deal with. The standard formalism of interacting field theories is based on a compromise between the strict mathematics and the need for a straightforward interpretation of actual observations. The arguments I am about to present are not really adequate for problems such as the confinement of quarks, but the necessary modifications can be introduced at a later stage.

The processes in which particles scatter or decay are always observed to occur within a very small spacetime region, called the *interaction region*. Outside the interaction region, particles behave, to an extremely good approximation, as if they were free. The initial and final multi-particle states can therefore be approximated as eigenstates of the Hamiltonian of a non-interacting theory. The real reason for this is that particle wavefunctions outside the interaction regions are localized wave packets which do not overlap appreciably. It is convenient to imagine, however, that the interactions are actually 'switched off' at times well before and well after the scattering or decay event takes place. This should be allowable, since the interactions are having no significant effect at these times. I shall denote all the field operators collectively by ϕ (dropping the $\hat{}$ for simplicity of notation), and the free-particle Hamiltonian by $H_0(\phi)$. Then, taking the event to occur at around $t = 0$, we replace the true Hamiltonian by

$$H(\phi) = H_0(\phi) + e^{-\varepsilon|t|}H_{\text{int}}(\phi) \tag{9.1}$$

where H_{int} is the part of the Hamiltonian which contains interactions and ε is a small parameter which will be set to zero at a suitable stage of the calculation. The modified Hamiltonian reduces to H_0 at $t = \pm\infty$, but if ε is small enough, it is essentially the same as the true Hamiltonian within the interaction region. This mathematical device is known as *adiabatic switching*. At very early or very late times, referred to as the 'in' region and 'out' region respectively, we no longer need localized wave packets to prevent the particles from interacting, and the wavefunctions of the incoming and outgoing particles can be taken as plane waves.

The field operators $\phi(\boldsymbol{x}, t)$ are, of course, Heisenberg-picture operators, whose evolution with time depends on the Hamiltonian. In the 'in' and 'out' regions, they should behave approximately as free fields. We therefore assume that

$$\begin{aligned} \phi(\boldsymbol{x}, t) &\approx Z^{1/2}\phi_{\text{in}}(\boldsymbol{x}, t) \quad \text{for } t \to -\infty \\ &\approx Z^{1/2}\phi_{\text{out}}(\boldsymbol{x}, t) \quad \text{for } t \to +\infty \end{aligned} \tag{9.2}$$

where ϕ_{in} and ϕ_{out} are free field operators and Z is a constant, called the *wavefunction renormalization constant*, which allows the magnitude of the 'in' and 'out' fields to be adjusted in accordance with the correct normalization of the states they create. (Close inspection reveals that some care is needed in interpreting (9.2), but I must refer readers to more specialized textbooks for a discussion of this point.) Unlike the interacting fields, the 'in' and 'out' fields can be expanded in terms of plane-wave solutions of the appropriate wave equations, the coefficients being interpreted as particle creation and annihilation operators. The initial state of particles about to undergo scattering will be of the form

$$|k_1, \ldots, k_N; \text{in}\rangle = a_{\text{in}}^{\dagger}(k_N) \ldots a_{\text{in}}^{\dagger}(k_1)|0\rangle . \tag{9.3}$$

In most cases, of course, N will be 1 for a decaying particle or 2 for a collision process. The creation operators will be those appropriate for the particular particle species involved. Possible final states, or 'out' states, may be constructed in the same way using 'out' operators. The 'in' and 'out' states are known collectively as *asymptotic states.*

The 'in' states are eigenstates of the Hamiltonian $H_0(\phi_{\text{in}})$, but not of the true Hamiltonian $H(\phi)$ which governs the actual time evolution. In the Heisenberg picture, a state vector such as (9.3) stands for the whole history of the system, but its meaning depends on the Hamiltonian. Thus, (9.3) stands for that state which, in the remote past, consisted of N particles with momenta $k_1, \ldots, k_N$, but this does not mean that the state continues to consist of these N particles. The analogously defined 'out' state stands for that state which, in the remote future, will consist of . . .. Thus, the probability amplitude to detect final state particles with momenta $k_1', \ldots, k_{N'}'$ given the initial state (9.3) is

$$\langle k_1', \ldots, k_{N'}'; \text{out}|k_1, \ldots, k_N; \text{in}\rangle \tag{9.4}$$

and one of the primary tasks of field theory is to calculate these amplitudes. I shall not dwell here on the mechanics of converting these amplitudes into directly measurable quantities such as decay rates and scattering cross-sections. Some relevant formulae are given in appendix 4 and further details can be found in the specialized literature. It is reasonable to assume that the same multi-particle states can exist in the 'out' region as in the 'in' region, and so there should be a one-to-one correspondence between 'in' and 'out' states. This correspondence is expressed in terms of the *scattering operator* S:

$$|k_1, \ldots, k_N; \text{in}\rangle = S|k_1, \ldots, k_N; \text{out}\rangle . \tag{9.5}$$

Thus, the amplitude (9.4) can be expressed as a matrix element of S between two 'out' states and is called an *S-matrix element*. To preserve the normalization of the asymptotic states, S must be a unitary operator.

9.2 Reduction Formulae

The S-matrix elements can be expressed in terms of the field operators of the interacting theory by means of the *LSZ reduction formula*, named after its inventors H Lehmann, K Symanzik and W Zimmermann. I shall derive an example of such a formula for the case of a single scalar field. The creation and annihilation operators for particles in the 'in' and 'out' regions can be expressed in terms of the 'in' and 'out' fields by (7.12). We now apply the identity

$$\int_{-\infty}^{\infty} \mathrm{d}t \, \frac{\partial f(t)}{\partial t} = \lim_{t\to\infty} f(t) - \lim_{t\to-\infty} f(t) \tag{9.6}$$

and the relations (9.2) to write

$$\begin{aligned} a_{\text{in}}(\boldsymbol{k}) - a_{\text{out}}(\boldsymbol{k}) &= \left[\lim_{t\to-\infty} - \lim_{t\to\infty}\right] \mathrm{i}Z^{-1/2} \int \mathrm{d}^3x \, \mathrm{e}^{\mathrm{i}k\cdot x} \overleftrightarrow{\partial}_0 \phi(x) \\ &= -\mathrm{i}Z^{-1/2} \int_{-\infty}^{\infty} \mathrm{d}x^0 \partial_0 \left(\int \mathrm{d}^3x \, \mathrm{e}^{\mathrm{i}k\cdot x} \overleftrightarrow{\partial}_0 \phi(x) \right). \end{aligned} \tag{9.7}$$

If we use the fact that $k^2 = m^2$ and integrate by parts, ignoring any surface term, we can rewrite this as

$$a_{\text{in}}(\boldsymbol{k}) - a_{\text{out}}(\boldsymbol{k}) = -\mathrm{i}Z^{-1/2} \int \mathrm{d}^4x \, \mathrm{e}^{\mathrm{i}k\cdot x} (\Box + m^2) \phi(x). \tag{9.8}$$

Let us use this result to find and expression for the probability amplitude $\langle k'; \text{out} | k; \text{in} \rangle$ for a particle of momentum k' to be found in the distant future, given a single-particle state of momentum k in the distant past. The first step is to write $\langle k'; \text{out}|$ as $\langle 0 | a_{\text{out}}(k')$ and re-express a_{out} using (9.8). The action of $a_{\text{in}}(k')$ on $|k; \text{in}\rangle$ is given by (6.10), but with a relativistic normalization factor as in (7.17) and (7.18), so we get

$$\begin{aligned} &\langle k'; \text{out} | k; \text{in} \rangle \\ &\quad = (2\pi)^3 2\omega_k \delta(\boldsymbol{k} - \boldsymbol{k}') + \mathrm{i}Z^{-1/2} \int \mathrm{d}^4x \mathrm{e}^{\mathrm{i}k'\cdot x} (\Box + m^2) \langle 0 | \phi(x) | k; \text{in} \rangle. \end{aligned} \tag{9.9}$$

Now, we want to use the same method to create $|k; \text{in}\rangle$ from the vacuum. Obviously, we have

$$\langle 0 | \phi(x) | k; \text{in} \rangle = \langle 0 | \phi(x) a^{\dagger}_{\text{in}}(k) | 0 \rangle \tag{9.10}$$

but if we use (9.8) directly we will get an unwanted term $\langle 0 | \phi a^{\dagger}_{\text{out}} | 0 \rangle$. If, instead, we could arrange to get $\langle 0 | a^{\dagger}_{\text{out}} \phi | 0 \rangle$, this term could be eliminated, because $\langle 0 | a^{\dagger}_{\text{out}} = (a_{\text{out}} | 0 \rangle)^{\dagger} = 0$. To this end, remember that a_{in} and a_{out} arise from the limits $t \to -\infty$ and $t \to \infty$ respectively in the time integral in (9.8). Therefore, we can arrange the desired ordering of operators by defining the *time-ordered product*

$$T[\phi(x)\phi^\dagger(y)] = \phi(x)\phi^\dagger(y) \quad \text{if } x^0 > y^0$$
$$= \phi^\dagger(y)\phi(x) \quad \text{if } y^0 > x^0 \tag{9.11}$$

in which the operator referring to the latest time stands on the left. Then, using the adjoint of (9.8), we find

$$\mathrm{i}Z^{-1/2}\int \mathrm{d}^4y\mathrm{e}^{-\mathrm{i}k\cdot y}(\Box_y + m^2)\langle 0|T[\phi(x)\phi^\dagger(y)]|0\rangle$$
$$= \langle 0|\phi(x)a^\dagger_{\mathrm{in}}(k)|0\rangle - \langle 0|a^\dagger_{\mathrm{out}}(k)\phi(x)|0\rangle \tag{9.12}$$

and the second term vanishes. Finally, we substitute this into (9.9) to obtain the reduction formula

$$\langle k'; \mathrm{out}|k; \mathrm{in}\rangle = (2\pi)^3 2\omega_k\delta(\boldsymbol{k} - \boldsymbol{k}') + (\mathrm{i}Z^{-1/2})^2$$
$$\times \int \mathrm{d}^4x\,\mathrm{d}^4y\mathrm{e}^{\mathrm{i}(k'\cdot x - k\cdot y)}(\Box_x + m^2)(\Box_y + m^2)\langle 0|T[\phi(x)\phi^\dagger(y)]|0\rangle. \tag{9.13}$$

The S-matrix element has now been expressed entirely in terms of the original interacting field, so at this point we can take $\varepsilon = 0$ in (9.1) and forget about the 'in' and 'out' fields.

The quantity $\mathrm{i}\langle 0|T[\phi(x)\phi^\dagger(y)]|0\rangle$ is called the *Feynman propagator* for the field ϕ. If translational invariance holds, in both space and time, then it depends only on $(x - y)$ and may be written as a Fourier transform

$$G_{\mathrm{F}}(x - y) = \mathrm{i}\langle 0|T[\phi(x)\phi^\dagger(y)|0\rangle = \int \frac{\mathrm{d}^4k}{(2\pi)^4}\mathrm{e}^{-\mathrm{i}k\cdot(x-y)}\widetilde{G}_{\mathrm{F}}(k). \tag{9.14}$$

If we use this Fourier transform in the reduction formula and integrate by parts to let the derivatives act on the exponential, we get

$$\langle k'; \mathrm{out}|k; \mathrm{in}\rangle = (2\pi)^3 2\omega_k\delta^3(\boldsymbol{k} - \boldsymbol{k}')$$
$$- \mathrm{i}(\mathrm{i}Z^{-1/2})^2(2\pi)^4\delta^4(k - k')(k^2 - m^2)^2\widetilde{G}_{\mathrm{F}}(k). \tag{9.15}$$

Since k and k' are the 4-momenta of free particles, they satisfy $(k^2 - m^2) = (k'^2 - m^2) = 0$. Therefore, the second term is zero unless $\widetilde{G}_{\mathrm{F}}(k)$ has a singularity at $k^2 = m^2$. The form of the propagator depends on the nature of the interactions. If they are such that the particles created by ϕ are stable, then $\widetilde{G}_{\mathrm{F}}(k)$ will turn out to behave roughly as $(k^2 - m^2)^{-1}$. The second term in (9.15) is then zero. In that case, the *single-particle* 'in' and 'out' states satisfy the same orthogonality relation (7.18) as in a free field theory. This means that $|k; \mathrm{in}\rangle$ and

$|k; \text{out}\rangle$ are the same state, as we would expect for a single stable particle. If, on the other hand, the ϕ particles can decay into lighter ones, it will turn out that G_F is of the form $(k^2 - m^2)^{-2}\Gamma(k)$, where $\Gamma(k)$ is related to the probability per unit time for the decay process to occur. In that case, (9.15) can roughly be interpreted as the statement (probability of survival) = 1 − (probability of decay). The set of 4-momenta which satisfy $k^2 = m^2$ is called the *mass shell*. Quantities like the propagator, known generically as *Green functions*, are well defined for more general 4-momenta, but S-matrix elements such as (9.15) involve only the *residues* of poles of these Green functions at $k^2 = m^2$, the *on-shell* residues.

It should be clear that the operations which led to the reduction formula (9.13) can be repeated for initial and final states which contain more than one particle. Thus, all S-matrix elements can be expressed in terms of vacuum expectation values of time-ordered products of field operators, $\langle 0|T[\phi(x_1) \ldots \phi^\dagger(x_N)]|0\rangle$, where N is the total number of incoming and outgoing particles. The T product orders all the operators according to their time arguments with the latest on the left and the earliest on the right. Readers who wish to carry out detailed calculations in field theory will find the technicalities explained in specialized textbooks, but I shall briefly mention a few of them. When spin-$\frac{1}{2}$ particles are involved, the exponentials in (9.13) are replaced by plane-wave solutions of the Dirac equation, and $(\Box + m^2)$ by the Dirac operator $(\mathrm{i}\partial\!\!\!/ - m)$. Thus, for single particles, (9.13) becomes

$$\begin{aligned}\langle k', s'; \text{out}|k, s; \text{in}\rangle = {} & (2\pi)^3 2\omega_k \delta_{ss'}\delta(\boldsymbol{k} - \boldsymbol{k}') \\ & - Z^{-1}\int \mathrm{d}^4x\,\mathrm{d}^4y\,\mathrm{e}^{\mathrm{i}(k'\cdot x - k\cdot y)}\bar{u}(k', s')(\mathrm{i}\partial\!\!\!/_x - m) \\ & \times \langle 0|T[\psi(x)\bar{\psi}(y)]|0\rangle(-\mathrm{i}\overleftarrow{\partial\!\!\!/}_y - m)u(k, s).\end{aligned} \tag{9.16}$$

Included in the definition of the T product is a change of sign for each interchange of fermion fields needed to bring the initial product into the correct time order. The quantization of gauge fields will be discussed later in this chapter from a different point of view from the one we have adopted up to now. There are technical complications in defining time-ordered products in a Lorentz covariant manner, which interested readers will find discussed in the more specialized literature.

By means of reduction formulae, all probability amplitudes for collision and decay processes can be expressed in terms of vacuum expectation values of time-ordered products of field operators. Except in very special cases, these expectation values can be calculated only approximately. Suitable methods of approximation can be developed by continuing to work with field operators, but a much more convenient framework for calculation is available, namely the *path integral* formalism, which I shall now describe.

9.3 Path Integrals

9.3.1 Path integrals in non-relativistic quantum mechanics

To reduce things to their simplest terms, consider first the non-relativistic theory of a single particle, moving in one dimension in a potential $V(x)$. To make the analogy with field theory as close as possible, I will take the mass of the particle to be $m = 1$. A quantity somewhat analogous to the Green functions of quantum field theory is the matrix element

$$G_{\mathrm{fi}}(t_1, t_2) = \langle x_{\mathrm{f}}, t_{\mathrm{f}}|T[\hat{x}(t_1)\hat{x}(t_2)]|x_{\mathrm{i}}, t_{\mathrm{i}}\rangle. \tag{9.17}$$

The ket $|x_{\mathrm{i}}, t_{\mathrm{i}}\rangle$ is a Heisenberg-picture vector representing that state in which the particle is at the point x_{i} at the initial time t_{i}, so it is an eigenvector of the Heisenberg-picture operator $\hat{x}(t)$ at the instant $t = t_{\mathrm{i}}$ only. The bra $\langle x_{\mathrm{f}}, t_{\mathrm{f}}|$ is defined similarly, and t_1 and t_2 lie between t_{i} and t_{f}. The idea of a path integral, due to P A M Dirac and R P Feynman, is that this matrix element can be expressed as an integral over all paths $x(t)$ which the particle might follow between x_{i} at time t_{i} and x_{f} at time t_{f}. An integral over paths can be defined by splitting the time interval $t_{\mathrm{f}} - t_{\mathrm{i}}$ into N segments, of lengths Δt, doing an ordinary multiple integral over the $N - 1$ points $x(t_{\mathrm{i}} + n\Delta t)$ and taking the limit $N \to \infty$, as illustrated in figure 9.1. Symbolically, this may be written as

$$\int \mathcal{D}x(t)(\ldots) = \lim_{N\to\infty} \int_{-\infty}^{\infty} \prod_{n=1}^{N-1} \mathrm{d}x_n(\ldots). \tag{9.18}$$

(A somewhat more rigorous treatment can be given in terms of probability measures over suitable classes of functions.)

To show how (9.17) can be expressed in terms of such an integral, we first translate it into the Schrödinger picture, taking the Schrödinger and Heisenberg pictures to coincide at t_{i}. For the case $t_2 > t_1$, we have

$$G_{\mathrm{fi}}(t_1, t_2) = \langle x_{\mathrm{f}}|\exp[-\mathrm{i}(t_{\mathrm{f}} - t_2)\hat{H}]\hat{x}\exp[-\mathrm{i}(t_2 - t_1)\hat{H}]\hat{x}\exp[-\mathrm{i}(t_1 - t_{\mathrm{i}})\hat{H}]|x_{\mathrm{i}}\rangle. \tag{9.19}$$

Now, the expression

$$\int_{-\infty}^{\infty} \mathrm{d}x|x\rangle\langle x| \tag{9.20}$$

is equal to the identity operator since, for any $\langle\Phi|$ and $|\Psi\rangle$

$$\langle\Phi|\left\{\int \mathrm{d}x|x\rangle\langle x|\right\}|\Psi\rangle = \int \mathrm{d}x\phi^*(x)\psi(x) = \langle\Phi|\Psi\rangle. \tag{9.21}$$

Since $\hat{x}|x\rangle = x|x\rangle$, we can express (9.19) as

$$\int_{-\infty}^{\infty} \mathrm{d}x_1\, \mathrm{d}x_2 \langle x_\mathrm{f}|\exp[-\mathrm{i}(t_\mathrm{f} - t_2)\hat{H}]|x_2\rangle \times x_2\langle x_2|\exp[-\mathrm{i}(t_2 - t_1)\hat{H}]|x_1\rangle x_1\langle x_1|\exp[-\mathrm{i}(t_1 - t_\mathrm{i})\hat{H}]|x_\mathrm{i}\rangle. \tag{9.22}$$

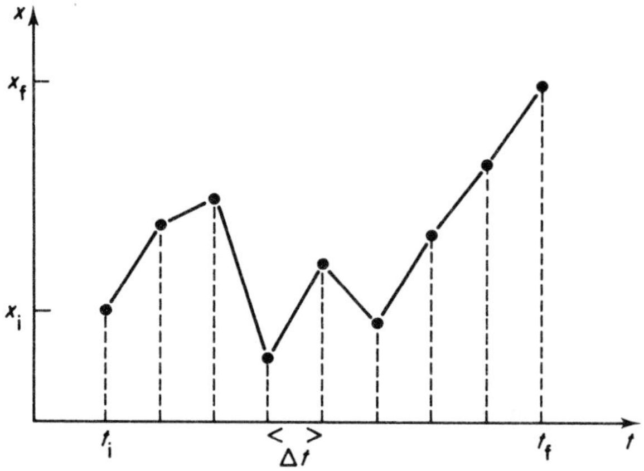

Figure 9.1 Construction of a Feynman path integral over all trajectories leading from x_i at time t_i to x_f at time t_f.

In the same way, we can split up each of the remaining matrix elements into a large number of short time intervals:

$$\langle x_\mathrm{f}|\exp[-\mathrm{i}(t_\mathrm{f} - t_\mathrm{i})\hat{H}]|x_\mathrm{i}\rangle = \int_{-\infty}^{\infty} \prod_{n=1}^{N-1} \mathrm{d}x_n \langle x_\mathrm{f}|\mathrm{e}^{-\mathrm{i}\Delta t\hat{H}}|x_{N-1}\rangle \ldots \langle x_1|\mathrm{e}^{-\mathrm{i}\Delta t\hat{H}}|x_\mathrm{i}\rangle. \tag{9.23}$$

If $\Delta t = (t_\mathrm{f} - t_\mathrm{i})/N$ is small enough, the exponential in each matrix element can be expanded as

$$\langle x_2|\mathrm{e}^{-\mathrm{i}\Delta t\hat{H}}|x_1\rangle \approx \langle x_2|[\hat{I} - \tfrac{1}{2}\mathrm{i}\Delta t\hat{p}^2 - \mathrm{i}\Delta t V(\hat{x})]|x_1\rangle. \tag{9.24}$$

Taking each operator in turn, we can evaluate the matrix elements as

$$\langle x_2|\hat{I}|x_1\rangle = \delta(x_2 - x_1) = (2\pi)^{-1}\int \mathrm{d}k \exp[\mathrm{i}k(x_1 - x_2)]$$

$$\langle x_2|V(\hat{x})|x_1\rangle = (2\pi)^{-1}\int \mathrm{d}k \exp[\mathrm{i}k(x_1 - x_2)]V(x_2)$$

$$\langle x_2|\hat{p}^2|x_1\rangle = (2\pi)^{-1}\int \mathrm{d}k\,\mathrm{d}k' \exp[\mathrm{i}(kx_1 - k'x_2)]\langle k'|\hat{p}^2|k\rangle$$

$$= (2\pi)^{-1}\int \mathrm{d}k \exp[\mathrm{i}k(x_1 - x_2)]k^2.$$

On re-exponentiating, we find

$$\langle x_2|\mathrm{e}^{-\mathrm{i}\Delta t\hat{H}}|x_1\rangle = (2\pi)^{-1}\int \mathrm{d}k \exp[\mathrm{i}k(x_1 - x_2) - \tfrac{1}{2}\mathrm{i}\Delta t k^2 - \mathrm{i}\Delta t V(x_2)] \tag{9.25}$$

up to terms of order $(\Delta t)^2$. We now shift the integration variable by $k \to k + (x_1 - x_2)/\Delta t$, after which the k integral produces just a constant:

$$\langle x_2|\mathrm{e}^{-\mathrm{i}\Delta t\hat{H}}|x_1\rangle = \text{constant} \times \exp\left\{\mathrm{i}\Delta t\left[\frac{1}{2}\left(\frac{x_1 - x_2}{\Delta t}\right)^2 - V(x_2)\right]\right\}. \tag{9.26}$$

In the limit $\Delta t \to 0$, this becomes exact. For a longer time interval, we can therefore use (9.23) to write

$$\langle x_\mathrm{f}|\exp[-\mathrm{i}(t_\mathrm{f} - t_\mathrm{i})\hat{H}]\,|x_\mathrm{i}\rangle$$

$$= \text{constant} \times \lim_{N\to\infty} \int_{-\infty}^{\infty} \prod_{n=1}^{N-1} \mathrm{d}x_n \exp\left\{\mathrm{i}\Delta t \sum_{n=1}^{N}\left[\frac{1}{2}\left(\frac{x_n - x_{n-1}}{\Delta t}\right)^2 - V(x_n)\right]\right\} \tag{9.27}$$

where $x_0 = x_\mathrm{i}$ and $x_N = x_\mathrm{f}$. If we now consider the points x_n to belong to a path $x(t)$, with $x_n = x(t_\mathrm{i} + n\Delta t)$, then $(x_n - x_{n-1})/\Delta t = \dot{x}(t)$, and the expression in square brackets in (9.27) is the classical Lagrangian

$$L = \tfrac{1}{2}\dot{x}^2 - V(x). \tag{9.28}$$

When we apply this result to (9.22), x_1 and x_2 become $x(t_1)$ and $x(t_2)$ respectively. Obviously, we could extend the original matrix element (9.17) to include any number of x's, with the general result

$$\langle x_\mathrm{f}, t_\mathrm{f}|T[\hat{x}(t_1)\ldots\hat{x}(t_n)]|x_\mathrm{i}, t_\mathrm{i}\rangle$$

$$= \int \mathcal{D}x(t)x(t_1)\ldots x(t_n)\exp\left(\mathrm{i}\int_{t_\mathrm{i}}^{t_\mathrm{f}} L(\dot{x}, x)\,\mathrm{d}t\right). \tag{9.29}$$

The path integral is over all paths for which $x(t_\mathrm{i}) = x_\mathrm{i}$ and $x(t_\mathrm{f}) = x_\mathrm{f}$, and all the constants from momentum integrations have been absorbed into the definition of the symbol $\mathcal{D}x(t)$. The position variables $x(t_j)$

inside the path integral are, of course, commuting numbers, so their order does not matter. It should, however, be obvious from the derivation of this result that the path integral does represent the matrix element of a *time-ordered* product of position operators, rather than a product ordered in some other way.

9.3.2 Functional integrals in quantum field theory

Despite some slight technical complications which I shall no go into, the vacuum expectation values of time-ordered products of field operators which appear in the reduction formulae for S-matrix elements can be represented by integrals similar to (9.29). For a scalar field, we have

$$\langle 0|T[\hat{\phi}(x_1)\ldots\hat{\phi}^\dagger(x_n)]|0\rangle = \int \mathcal{D}\phi(x)\phi(x_1)\ldots\phi^*(x_n)\mathrm{e}^{\mathrm{i}S(\phi)} \qquad (9.30)$$

where $S(\phi)$ is the action. The integral is over complex functions $\phi(x)$ and is often called a *functional integral* rather than a path integral. The adjoint field operator is represented in the integral by the complex conjugate function $\phi^*(x)$, and if the field is Hermitian the integral is only over real functions.

In the case of fermions, the fields in the functional integral must be taken as Grassmann variables, to take into account the anticommuting properties of the original field operators. I shall not enter into details of the properties of Grassmann integrals, which are given in specialized field theory textbooks, but I will mention their special features as the need arises.

It might seem that functional integrals would be extremely difficult to evaluate and so, more often than not, they are. In practice, however, it is usually possible to extract the results we require by means of manipulations which avoid our having to compute a functional integral directly. As a first example, let us evaluate the Feynman propagator (9.14) for a free scalar field. It is convenient to introduce a *generating functional* for the Green functions (9.30), defined by

$$Z_0(J, J^*) = \int \mathcal{D}\phi \exp\left\{\mathrm{i}\int \mathrm{d}^4x[\mathcal{L}_0 + J^*(x)\phi(x) + J(x)\phi^*(x)]\right\} \qquad (9.31)$$

where $\mathcal{L}_0$ is the free field Lagrangian density (7.7) and the normalization of the integral is such that $Z_0(0, 0) = 1$. The propagator is given by

$$G_\mathrm{F}(x-y) = -\mathrm{i}\frac{\delta}{\delta J^*(x)}\frac{\delta}{\delta J(y)}Z_0(J, J^*)\bigg|_{J=J^*=0} \qquad (9.32)$$

and other Green functions can obviously be generated by further differentiations. The meaning of the functional derivative $\delta/\delta J(x)$ is explained in detail in appendix 1.

In this and other calculations, it is necessary to re-express spacetime integrals using integration by parts. For simplicity, I shall always assume that boundary conditions can be applied which ensure that surface terms vanish. Readers should, however, be aware that this cannot always be done. In particular, the non-linear field equations, which are the Euler–Lagrange equations of interacting field theories, frequently have topologically non-trivial solutions, described in the literature as solitons, monopoles, instantons, vortex strings and the like, and, when these are important, the boundary conditions must be considered more carefully. With this proviso, the exponent in (9.31) can be written in the form

$$\int d^4x \left\{ -\left[\phi^*(x) - \int d^4y g(x-y) J^*(y)\right] \times (\Box_x + m^2)\left[\phi(x) - \int d^4z g(x-z) J(z)\right]\right\} + \int d^4x \int d^4y J^*(x) g(x-y) J(y) \tag{9.33}$$

where $g(x-y)$ is a Green function which satisfies the equation

$$(\Box + m^2) g(x-y) = \delta(x-y). \tag{9.34}$$

Since the functional integral is the limit of a product of ordinary integrals over the range $-\infty$ to ∞, it is possible to shift the integration variable by

$$\phi(x) \to \phi(x) + \int d^4z g(x-z) J(z) \tag{9.35}$$

whereupon we find that

$$Z_0(J, J^*) = Z_0(0,0) \exp\left[i \int d^4x \int d^4y J^*(x) g(x-y) J(y)\right]. \tag{9.36}$$

In view of the normalization $Z_0(0,0) = 1$, we have succeeded in evaluating the generating functional without actually carrying out a functional integral, as long as we can find the function $g(x-y)$. For most purposes, these manipulations are all we shall ever need. It is easy to verify that $g(x-y)$ can be expressed as a Fourier transform

$$g(x-y) = -\int \frac{d^4k}{(2\pi)^4} \frac{e^{-ik\cdot(x-y)}}{k^2 - m^2}. \tag{9.37}$$

This is not well defined as it stands, however, because the integrand has poles at $k_0 = \pm(\boldsymbol{k}^2 + m^2)^{1/2}$. In fact, if the k_0 integral is carried out as a contour integral, then several different solutions to (9.34) can be found by routing the contour round the poles in different ways. Equivalently, the poles can be shifted into the complex plane by a small amount ε, which is taken to zero after the integration. Now, according to (9.32), the Feynman propagator is equal to $g(x-y)$, so we must choose that solution which agrees with the original definition (9.14). This can be calculated directly using the properties of the field operators, and the correct definition is found to be (see exercise 9.3)

$$G_F(x-y) = -\lim_{\varepsilon\to 0}\int \frac{d^4k}{(2\pi)^4}\frac{e^{-ik\cdot(x-y)}}{k^2-m^2+i\varepsilon}. \tag{9.38}$$

Our final result for the generating functional is therefore

$$Z_0(J, J^*) = \exp\left[i\int d^4x\int d^4y\, J^*(x)G_F(x-y)J(y)\right]. \tag{9.39}$$

The appearance of this prescription of replacing m^2 by $m^2 - i\varepsilon$ may be understood as follows. The functional integral (9.31) is not really well defined, because the integrand is, for any value of ϕ, a complex number of unit magnitude. In effect, the $m^2 - i\varepsilon$ prescription adds to the exponent a term $-\varepsilon\int d^4x\phi^2(x)$. This provides a convergence factor which makes the integrand decay to zero at large values of ϕ.

For spin-$\frac{1}{2}$ particles, the Feynman propagator is a 4×4 matrix defined by

$$S_{Fij}(x-y) = -i\langle 0|T[\psi_i(x)\bar{\psi}_j(y)]|0\rangle. \tag{9.40}$$

It satisfies the spinor version of (9.34), namely

$$(i\partial\!\!\!/ - m)S_F(x-y) = \delta(x-y) \tag{9.41}$$

and is given by

$$\begin{aligned} S_F(x-y) &= -(i\partial\!\!\!/ + m)G_F(x-y) \\ &= \lim_{\varepsilon\to 0}\int \frac{d^4k}{(2\pi)^4}e^{-ik\cdot(x-y)}\frac{(k\!\!\!/ + m)}{k^2-m^2+i\varepsilon}. \end{aligned} \tag{9.42}$$

9.4 Perturbation Theory

The simplest example of an interacting field theory is a scalar field theory with Lagrangian density $\mathcal{L}_0 - V(\phi, \phi^*)$. The generating functional for its Green functions can be written as

$$Z(J, J^*) = \int \mathcal{D}\phi \exp\left[-i\int d^4x V(\phi, \phi^*)\right]\exp\left\{i\int d^4x[\mathcal{L}_0 + J^*\phi + J\phi^*]\right\}. \tag{9.43}$$

Since differentiation of the second exponential by $J(x)$ or $J^*(x)$ multiplies it by $i\phi^*$ or $i\phi$, we can express this as

$$Z(J, J^*) = N\exp\left[-i\int d^4x V\left(-i\frac{\delta}{\delta J^*}, -i\frac{\delta}{\delta J}\right)\right]Z_0(J, J^*) \tag{9.44}$$

where N is a normalizing factor determined by the condition $Z(0, 0) = 1$. If ϕ is complex then, since $\mathscr{L}$ must be real, the potential V can depend only on $\phi^*\phi$. For reasons I shall discuss shortly, the only useful theory of this kind is defined by

$$V(\phi, \phi^*) = \tfrac{1}{4}\lambda(\phi^*\phi)^2 \tag{9.45}$$

where λ is a coupling constant. There is no known method of computing this generating functional or any of the individual Green functions exactly. A commonly used method of approximation is *perturbation theory*, which means an expansion in powers of λ. To see how this expansion works, let us first calculate the normalization factor N in (9.44) correct to first order in λ. On expanding the exponential and setting $Z_0(0, 0) = 1$, we obtain

$$Z(0, 0) = N\left[1 - \frac{\mathrm{i}}{4}\lambda \int \mathrm{d}^4x \left(\frac{\delta}{\delta J(x)}\right)^2 \left(\frac{\delta}{\delta J^*(x)}\right)^2 Z_0(J, J^*)\bigg|_{J=J^*=0} + \mathrm{O}(\lambda^2)\right]. \tag{9.46}$$

When the expression (9.39) for Z_0 is expanded, we see that after differentiating and setting $J = J^* = 0$, only the term containing $(\int J^* G_{\mathrm{F}} J)^2$ survives. By carrying out the functional differentiation, we find that the normalizing constant is

$$N = 1 - \tfrac{1}{2}\mathrm{i}\lambda \int \mathrm{d}^4x [G_{\mathrm{F}}(0)]^2 + \mathrm{O}(\lambda^2). \tag{9.47}$$

Taking this result into account, we can find a similar approximation to the propagator of the interacting theory, conveniently defined by

$$G(x - y) = \langle 0|T[\phi(x)\phi^\dagger(y)|0\rangle = -\frac{\delta}{\delta J^*(x)}\frac{\delta}{\delta J(y)} Z(J, J^*)\bigg|_{J=J^*=0} \tag{9.48}$$

which is

$$G(x - y) =$$
$$-\mathrm{i}G_{\mathrm{F}}(x - y) - \mathrm{i}\lambda \int \mathrm{d}^4z (-\mathrm{i})^3 G_{\mathrm{F}}(x - z) G_{\mathrm{F}}(z - z) G_{\mathrm{F}}(z - y) + \mathrm{O}(\lambda^2). \tag{9.49}$$

It Fourier transform can be written, using (9.38), as

$$\widetilde{G}(p) = \frac{\mathrm{i}}{p^2 - m^2 + \mathrm{i}\varepsilon} - \frac{\lambda}{(p^2 - m^2 + \mathrm{i}\varepsilon)^2} \int \frac{\mathrm{d}^4k}{(2\pi)^4} \frac{1}{k^2 - m^2 + \mathrm{i}\varepsilon} + \mathrm{O}(\lambda^2). \tag{9.50}$$

This expression, and those arising in the perturbation series for all other Green functions, are conveniently represented by *Feynman diagrams*. The diagrams corresponding to (9.50) are shown in figure 9.2, and are constructed according to the following rules:

(i) 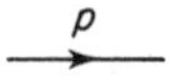

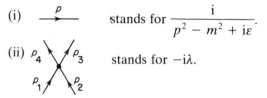

stands for $\dfrac{\mathrm{i}}{p^2 - m^2 + \mathrm{i}\varepsilon}$.

(ii) 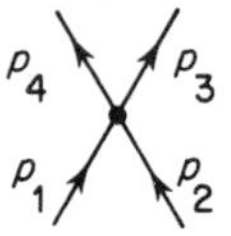

stands for $-\mathrm{i}\lambda$.

together with the condition $p_1 + p_2 = p_3 + p_4$ for momentum conservation,

(iii) All internal momenta, such as k in figure 9.2, whose values are not fixed by momentum conservation are integrated over.

(iv) Each diagram has a combinatorial factor arising from the expansions of exponentials and the chain rule for differentiation. Many field theory textbooks supply rules for calculating this factor, but in my experience it is best obtained from first principles.

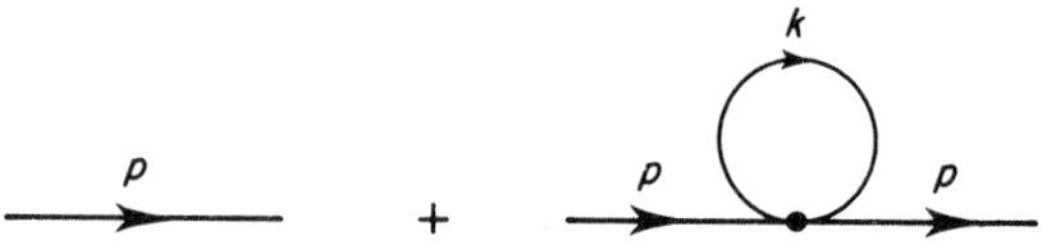

Figure 9.2 Diagrammatic representation of equation (9.50).

At a given order in λ, there are fixed numbers of vertices and unperturbed propagators available, and there is a contribution to the Green function from each diagram which can be formed from these elements. For example, figure 9.3 shows some of the diagrams which contribute to the S-matrix element for two-particle elastic scattering. Each diagram has four external propagators, one for each of the two incoming and two outgoing particles. The S-matrix element itself is similar to (9.15), but with a factor $(k_i^2 - m^2)$ for each particle multiplying the Green function. Evidently, these are just cancelled by the external propagators in figure 9.3, leaving a non-zero result.

Figure 9.3 Examples of Feynman diagrams which contribute to the elastic scattering amplitude for two spin-0 particles.

The Feynman rules for theories containing fermions differ in two respects from those given above. One is that each propagator line represents the matrix (9.42). Most often, only two fermion lines meet at any given vertex. For example, a term of the form $e\bar{\psi}\not{A}\psi$ in the action (8.17) gives rise to a vertex of the form

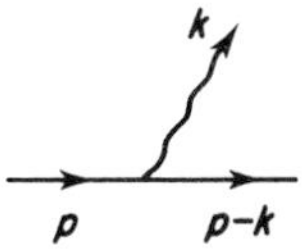

where the wavy line denotes the photon propagator, to be discussed in the next section. As far as the fermion is concerned, this vertex, together with the propagators, corresponds to the matrix product

$$-\mathrm{i}eS_{\mathrm{F}}(p)\gamma^{\mu}S_{\mathrm{F}}(p-k) \qquad (9.51)$$

whose index μ will be contracted with a corresponding index belonging to the photon propagator. Each internal fermion propagator will be multiplied by a matrix on either side. An external propagator will be multiplied by a matrix on one side (where it meets a vertex), leaving one free Dirac index. This free index is the one belonging to a field operator in the original matrix element, such as (9.40), and will eventually be contracted with a Dirac operator and a wavefunction, as in the reduction formula (9.16). The second difference is the appearance of some power of -1 in the combinatorial factor. These signs arise from the anticommutation properties of the Grassmann variables in the functional integral. Every closed loop of fermion propagators gives a factor of -1 and extra minus signs come from the ordering of field operators in a time-ordered product. Once again, I must ask readers who wish to become proficient in these calculations to consult a specialized text for detailed explanations of the technicalities.

Feynman diagrams such as those in figures 9.2 and 9.3 are often thought of as representing actual physical processes. For example, the first diagram of figure 9.3 is thought of as an immediate transformation of the initial two-particle state into the final two-particle state, while the higher-order diagrams represent indirect transformations via intermediate states in which particles corresponding to the internal propagators are created and subsequently annihilated. The net effect of each of these processes is the same, in the sense that they each involve the same initial and final states. The overall probability amplitude is the sum of the amplitudes for all possible ways in which this net transformation can occur. A particle whose transitory existence is represented by an internal propagator differs from a real, observable particle, because its 4-momentum does not have to satisfy the mass shell constraint $k^2 = m^2$. For this reason, the intermediate particles are called *virtual particles*.

This provides a pictorial language which is often useful for discussing the mathematics of perturbation theory. Clearly, however, this language is closely tied to our use of an expansion in powers of λ or some other coupling constant. It can therefore be expected to provide a reliable picture of actual physical events only when this expansion gives an accurate approximation to the quantities we are attempting to calculate.

9.5 Quantization of Gauge Fields

I have mentioned several times that gauge fields, such as the electromagnetic 4-vector potential A_μ introduced in chapter 8, can be treated as field operators whose associated particles are vector bosons, such as photons, but I have so far avoided giving details. The reason for this reticence is that there are problems in the quantum-mechanical treatment which do not arise for scalar or spinor fields and which are most conveniently overcome by the use of path integrals. Recall that a spinor field has four components. These correspond to four distinct states of the corresponding particles, namely particles and antiparticles, each having two independent spin polarizations. The electromagnetic field A_μ also has four components, but photons are found to exist only in two independent states, corresponding to right- and left-circularly polarized radiation, and are their own antiparticles. Therefore, two of the four field degrees of freedom are in some way redundant.

Mathematically, this is seen as follows. In the absence of charged particles, the action of electromagnetism is the first term of (8.17). With $F_{\mu\nu}$ given by the antisymmetric expression (8.14), this action is independent of $\partial_0 A_0$ and therefore the canonical momentum Π^0 conjugate to A_0 is identically zero. Therefore, there are at most three independent momenta, which are found, by analogy with (6.26), to be

$$\Pi^i = \frac{\delta S}{\delta(\partial_0 A_i)} = F^{i0} = E^i. \tag{9.52}$$

Since there are at most three independent momenta, there can also be at most three independent field variables. To reduce the matter to its simplest terms, let us regard A_0 as the redundant component. The four Euler–Lagrange equations are Maxwell's equations (3.52) with the current set to zero. The one obtained by varying A_0 cannot be regarded as an equation of motion on the same footing as the others, since A_0 is not a bona fide dynamical variable, but must rather be regarded as a further constraint on the remaining field components. (Readers familiar with such matters will realize that A_0 is playing the role of a Lagrange multiplier.) The offending Maxwell equation is Gauss' law (3.44) which, given (9.52), may be written as

$$\nabla\cdot\boldsymbol{E} = \partial_i\Pi^i = 0. \tag{9.53}$$

This is a relation between the three momenta, which implies that only two of them are really independent. We conclude that there are really only two genuine field variables and two conjugate momenta, corresponding to the two observed polarization states of the photon.

This conclusion is closely related to the property of gauge invariance. The two redundant degrees of freedom correspond to the fact that changes in the fields of the form (3.60) have no physical effect. Consider the four Maxwell equations, which may be written as

$$\Box A_\mu - \partial_\mu(\partial_\nu A^\nu) = 0. \tag{9.54}$$

As long as we are dealing with classical fields, we can make a gauge transformation (3.60), choosing $\theta(x)$ such that $\Box\theta = \partial_\nu A^\nu$, and the transformed gauge field will then satisfy the constraint

$$\partial_\nu A^\nu = 0. \tag{9.55}$$

By imposing this constaint we implicitly remove one of the redundant degrees of freedom and, for the constrained field, (9.54) becomes

$$\Box A_\mu = 0. \tag{9.56}$$

This is the Klein–Gordon equation for a massless particle, so the photon is indeed massless. The second redundant degree of freedom consists in the fact that we can make further gauge transformations, with $\Box\theta = 0$, without spoiling the constraint (9.55). It is, however, difficult to write down a second constraint on A_μ which removes this degree of freedom explicitly.

It is possible to construct a quantum theory of photons in the same spirit as the theories of spin-0 and spin-$\frac{1}{2}$ particles we discussed in chapter 7 and in earlier sections of this chapter. Because of the redundant degrees of freedom, though, the technical details are rather cumbersome, and I shall simply state the essential results. By constructing the appropriate Pauli–Lubanski vector (7.42), we can show that the photon has spin 1. The two circular polarizations correspond to states of helicity ± 1 and there is no helicity-0 state. In accordance with the spin–statistics theorem, photons are bosons. For scattering processes which involve photons in the initial or final state, reduction formulae similar to (9.13) or (9.16) can be derived in which the contribution from a photon is

$$\mathrm{i}Z^{-1/2}\int \mathrm{d}^4x\,\mathrm{e}^{\pm \mathrm{i}k\cdot x}\langle 0|T[\ldots \varepsilon(k)\cdot j_{\mathrm{e}}(x)\ldots]|0\rangle. \tag{9.57}$$

The current density j_{e} is the current which appears in Maxwell's equations (3.52) and reappears on the right-hand side of (9.54) as an operator when electromagnetic interactions are included. This expression could have been written in a form more similar to (9.13), with the left-hand side of (9.54) replacing the current. The advantage of the form (9.57) is that it avoids ambiguities concerning the definition of time-ordered products of gauge fields, as well as the question of whether the constraint (9.55) is to be imposed and, if so, how. In terms of Feynman diagrams, we have simply cancelled out the external photon propagators before evaluating the Green function rather than afterwards. The current operator in (9.57) inserts into a diagram the vertex to which this propagator would have been attached. The 4-vector $\varepsilon^\mu(k)$ represents the spin polarization state of the incoming or outgoing photon. That is to say, the wavefunction

$$A^\mu_k(x) = \varepsilon^\mu(k)\mathrm{e}^{-\mathrm{i}k\cdot x} \tag{9.58}$$

is the vector potential which, in classical electromagnetism, gives an electromagnetic plane wave of the appropriate polarization.

The reduction formula (9.57) serves to make contact with observable physical processes in a way which temporarily avoids the problem of quantizing the gauge field. This problem can, of course, no longer be avoided when we come to calculate the vacuum expectation value itself, since we expect Feynman diagrams to contain internal photon or other gauge field propagators as well as external ones. In the case of a scalar field, whose quantum-mechanical properties are straightforward, the path integral representation (9.30) could be deduced from the canonical formalism of field operators. It is possible to adopt a different attitude and regard the path integral as *defining* a quantum theory, given that we have an action S which defines the corresponding classical theory. This *path integral quantization scheme* is an alternative to the canonical scheme of §5.4, upon which our theory up to this point has rested. If we adopt this point of view then, at first sight, it appears that we simply have to calculate with an appropriate generating functional

$$Z(\text{sources}) = \int \mathcal{D}(\text{fields})\mathrm{e}^{\mathrm{i}S+\text{source terms}\}}. \tag{9.59}$$

The functional integral is over all the fields in the theory, and the source terms are similar to those in (9.31), namely

$$\int \mathrm{d}^4x\{J^*\phi + J\phi^* + J_\mu A^\mu + \bar{\eta}\psi + \bar{\psi}\eta + \ldots\} \tag{9.60}$$

with one term for each field. The sources J, J^μ, η, etc are the arguments of Z, and the action is an expression such as (8.17) or (8.38), perhaps with the addition of scalar fields, depending on the particular theory considered.

We should, of course, be rather suspicious of this procedure if it enabled us to ignore entirely the problems associated with redundant gauge degrees of freedom. In fact these problems reappear in the following way. Since the action S is gauge invariant, the integrand in

(9.59) is independent of the gauge degrees of freedom when we set the sources to zero, and the functional integrals over these degrees of freedom lead to a meaningless infinity. It is, in fact, impossible to do perturbation theory with (9.59) as it stands, because we cannot find propagators for the gauge fields. In the case of electromagnetism, if we follow the same steps as for the scalar field, we find that the propagator, denoted by $\mathrm{D}_{\mathrm{F}\mu\nu}$ should satisfy an equation similar to (9.34), but with the Klein–Gordon operator replaced by the Maxwell operator in (9.54):

$$\Box \mathrm{D}_{\mathrm{F}\mu\nu}(x-y) - \partial_\mu \partial^\lambda \mathrm{D}_{\mathrm{F}\lambda\nu}(x-y) = -g_{\mu\nu}\delta(x-y). \qquad (9.61)$$

This equation has no solution.

A way round this difficulty was found by L D Fadeev and V N Popov. Their argument is slightly complicated, and again I shall just state the result. It is possible to modify the action by adding two terms to the Lagrangian density of the gauge fields:

$$-\tfrac{1}{4}F^a_{\mu\nu}F^{a\mu\nu} \rightarrow -\tfrac{1}{4}F^a_{\mu\nu}F^{a\mu\nu} - (1/2\xi)f(A) + \bar{c}\Delta(A)c. \qquad (9.62)$$

The function $f(A)$ is a function of the gauge fields, whose purpose is to remove the gauge invariance of the original action, thereby allowing a propagator to be constructed. We are allowed a considerable freedom in choosing this function, although only a limited number of choices are convenient in practice. The new fields c and $\bar{c}$, which are to be integrated over in the generating functional, correspond to fictitious particles, usually called *ghosts*. Although these are spin-0 particles, the mathematics requires their fields to be Grassmann variables, so they are fermions, contradicting the spin–statistics theorem which applies to all physical particles. The quantity Δ is a differential operator, whose exact form depends on our choice of $f(A)$. In the case of electromagnetism (where, of course, the index a does not appear), a convenient choice of $f(A)$ is

$$f(A) = (\partial_\mu A^\mu)^2. \qquad (9.63)$$

With this choice, Δ turns out to be independent of A_μ. In this case, the ghosts do not interact with other particles and can be ignored. By modifying the action in this way, we naturally modify the Green functions as well. In particular, they now depend on the arbitrary parameter ξ. As a consequence of the original gauge invariance, however, it can be shown that S-matrix elements and other physically measurable gauge-invariant quantities are unaffected by the modification and are independent of ξ. I shall give an example of this in due course. The $f(A)$ term in (9.62) is often referred to as a *gauge-fixing* term. This is somewhat misleading, as it suggests that a constraint has been applied to eliminate the redundant gauge degrees of freedom. What really happens is that these degrees of freedom, together, in general, with the ghosts, conspire to have no net effect on physical quantities.

When (9.63) is used for $f(A)$, readers may readily verify that the equation for the propagator becomes

$$\Box \mathrm{D}_{\mathrm{F}\mu\nu}(x-y) - (1-\xi^{-1})\partial_\mu\partial^\lambda \mathrm{D}_{\mathrm{F}\lambda\nu}(x-y) = -g_{\mu\nu}\delta(x-y) \quad (9.64)$$

and that its solution is

$$\mathrm{D}_{\mathrm{F}\mu\nu}(x-y) = \int \frac{\mathrm{d}^4k}{(2\pi)^4}\frac{\mathrm{e}^{-\mathrm{i}k\cdot(x-y)}}{k^2+\mathrm{i}\varepsilon}\left(g_{\mu\nu} + (\xi-1)\frac{k_\mu k_\nu}{k^2}\right). \quad (9.65)$$

If we include a term $m^2(A_\mu^a A^{a\mu})/2$ in the Lagrangian density, we get a theory of massive vector bosons, with the propagator

$$\mathrm{D}_{\mathrm{F}\mu\nu}(x-y) = \int \frac{\mathrm{d}^4k}{(2\pi)^4}\frac{\mathrm{e}^{-\mathrm{i}k\cdot(x-y)}}{k^2-m^2+\mathrm{i}\varepsilon}\left(g_{\mu\nu} + (\xi-1)\frac{k_\mu k_\nu}{k^2-\xi m^2}\right). \quad (9.66)$$

As it stands, such a theory is not gauge invariant, and we are not entitled to use the extra Fadeev–Popov terms. Unlike (9.65), (9.66) has a finite limit when we remove the gauge-fixing term by taking ξ to infinity:

$$\mathrm{D}_{\mathrm{F}\mu\nu}(x-y) = \int \frac{\mathrm{d}^4k}{(2\pi)^4}\frac{\mathrm{e}^{-\mathrm{i}k\cdot(x-y)}}{k^2-m^2+\mathrm{i}\varepsilon}\left(g_{\mu\nu} - \frac{k_\mu k_\nu}{m^2}\right). \quad (9.67)$$

At the level of free particles, this non-gauge-invariant theory makes good sense. The massive spin-1 particles have three spin polarization states and there are no redundant degrees of freedom (see exercise 9.4). In interacting theories, however, massive vector bosons are troublesome, as we shall shortly discover.

9.6 Renormalization

Earlier on, we derived an expression (9.50) for the first-order correction to the scalar propagator which results from interactions in the theory defined by (9.45). This correction and further corrections at higher orders of perturbation theory are properly thought of as a *self-energy*, or as a correction to the mass of the particle brought about by interactions. Thus, the parameter m which appears in the Lagrangian density is not the true mass of the particle. It is usually called the *bare mass*, and I shall denote it henceforth by m_0. The pole of the complete propagator must appear at the true mass $p^2 = m^2$ and, of course, the 'in' and 'out' states should be defined in terms of the true mass which therefore still appears in the reduction formulae. As we shall see below, the integral in (9.50) is purely imaginary and I shall denote it by $-\mathrm{i}\Sigma(m_0)$. Then (9.50) may be written as

$$\widetilde{G}(p) = \frac{i}{p^2 - m_0^2 + i\varepsilon}\left(1 - \lambda\frac{\Sigma}{p^2 - m_0^2 + i\varepsilon}\right)^{-1} + O(\lambda^2)$$
$$= \frac{i}{p^2 - m_0^2 - \lambda\Sigma + i\varepsilon} + O(\lambda^2). \qquad (9.68)$$

This is not merely *ad hoc*. Amongst the whole set of Feynman diagrams which contribute to the propagator, there is the infinite sum of diagrams shown in figure 9.4, which is easily shown to be a geometric series. Thus, the true mass is given by

$$m^2 = m_0^2 + \lambda\Sigma(m_0) + O(\lambda^2). \qquad (9.69)$$

This relation is said to represent *mass renormalization*.

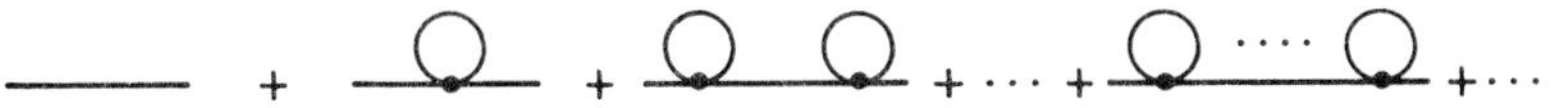

Figure 9.4 The Feynman diagrams whose sum forms the geometric series (9.68).

At this lowest order of perturbation theory, the residue of this propagator at $p^2 = m^2$ is still i. At higher orders, this residue is no longer equal to i. This means that, when acting on the vacuum state, the field operators of the interacting theory create single-particle states whose normalization is different from those of the non-interacting theory. In order to have a clear physical interpretation of our calculated scattering amplitudes, we demand that the 'in' and 'out' states should have the standard normalization of the non-interacting theory. To this end, we define the *wavefunction renormalization constant* Z which appears in the reduction formulae by the requirement

$$\lim_{p^2 \to m^2} Z^{-1}(p^2 - m^2)\widetilde{G}(p) = i. \qquad (9.70)$$

For reasons which will shortly become apparent, it is convenient to define a renormalized field

$$\phi_R(x) = Z^{-1/2}\phi(x) \qquad (9.71)$$

and renormalized Green functions

$$G_R^{(n)}(x_1, \ldots, x_n) = \langle 0|T[\phi_R(x_1) \ldots \phi_R(x_n)]|0\rangle_c \qquad (9.72)$$

which take into account the adjusted normalization. The subscript c in (9.72) denotes the *connected* Green functions. These are obtained by ignoring all Feynman diagrams which consist of two or more disconnected parts. For example, the complete four-point Green function (the vacuum expectation value involving four fields) contains, amongst many others, the diagrams shown in figure 9.5, but only diagrams (a) and (c)

are connected. The disconnected diagrams are associated with particles which continue from the initial state to the final state without colliding, while the connected diagrams refer to particles which actually collide, and are therefore of greater interest. The complete Green functions, should we ever want them, can be expressed in terms of connected ones.

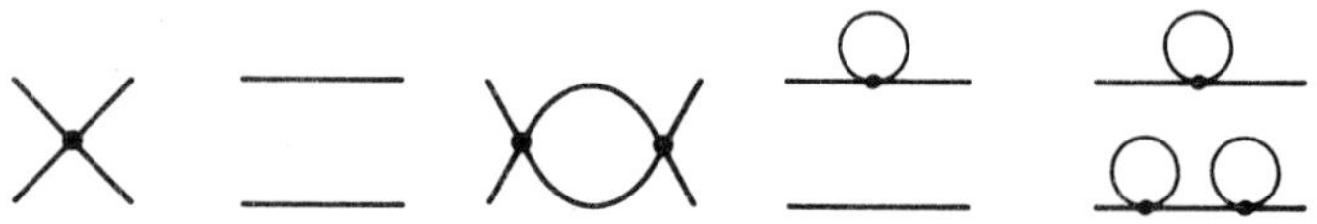

Figure 9.5 Some Feynman diagrams which contribute to the four-point Green function. Only (a) and (c) are connected diagrams.

Finally, it is necessary to realize that the coupling constant which appears in the action, and which I shall now denote by λ_0, is not a physically measurable quantity. If, for example, we measure the cross-section for 2 particle → 2 particle scattering, then the measured quantity includes contributions from every Feynman diagram in $G^{(4)}$ and we cannot single out the contribution from the single vertex λ_0. In order to compare the results of our calculations with experimental data, we must exchange λ_0 for a *renormalized coupling constant* λ which *is* measurable. There is considerable latitude in how we actually do this. A suitable definition might be

$$\lambda = \left[\prod_{i=1}^{4}(p_i^2 - m^2)\right] G_{\mathrm{R}}^{(4)}(p_1, \ldots, p_4)\big|_{p_i = p_i(\mu)}, \tag{9.73}$$

where $p_i(\mu)$ are a set of chosen momentum values, whose magnitude must depend on a parameter μ having the dimensions of mass. At any rate, if we are to continue using perturbation theory, the relation between λ and λ_0 must be of the form

$$\lambda = \lambda_0 + \mathrm{O}(\lambda_0^2) \tag{9.74}$$

so that a power series in λ_0 can be re-expressed as a series in λ. The exact physical significance of λ depends, of course, on the method used to define it and, in particular, on the chosen value of μ.

The preceding remarks show, I hope, that renormalization is a natural and essential part of the physical interpretation of a field theory. There is, however, a more sinister aspect to renormalization which must now be revealed. Let us evaluate the self-energy

$$\Sigma(m_0) = \mathrm{i}\int \frac{\mathrm{d}^4 k}{(2\pi)^4} \frac{1}{k^2 - m_0^2 + \mathrm{i}\varepsilon}. \tag{9.75}$$

If the k_0 integral is done as a contour integral, the poles appear as in figure 9.6. The contour of integration can be rotated, avoiding these poles, to run along the imaginary axis, in effect replacing k_0 by ik_0. This is called a *Wick rotation*. The result is an integral in four-dimensional Euclidean space, whose integrand depends only on the magnitude of k, so that in polar coordinates the angular integrations give just a constant factor:

$$\Sigma(m_0) = \int \frac{\mathrm{d}^4k}{(2\pi)^4} \frac{1}{k^2 + m_0^2} = \frac{1}{8\pi^2} \int_0^\infty \frac{k^3\,\mathrm{d}k}{k^2 + m_0^2}. \tag{9.76}$$

When k is large, the integral behaves as k^2, so it diverges quadratically at its upper limit: it is infinite! In practice, this does not matter. When we express the propagator (9.68) in terms of the true mass, it is equal to $\mathrm{i}(p^2 - m^2 + \mathrm{i}\varepsilon)^{-1}$ plus higher-order corrections, and Σ does not appear in our final answer for any physical quantity. On the other hand, many other infinite integrals can be expected to occur. While these are embarrassing, we can still obtain sensible, finite results for measurable quantities provided that all infinite integrals disappear after renormalization. In quantum electrodynamics, our embarrassment is somewhat alleviated by the fact that the renormalized theory yields predictions which agree with experiment to some 10 significant figures. What we require is that the renormalized Green functions should have well defined finite values when they are expressed in terms of true particle masses and renormalized coupling constants. If this is true for a particular field theory, the theory is said to be *renormalizable*. At the present stage of our knowledge, it appears that only renormalizable theories can be used as models of physical reality. Whether this is really true is not quite clear. We are, after all, only able to do approximate calculations, and it could be that the fault lies in our methods of approximation rather than in the field theory itself.

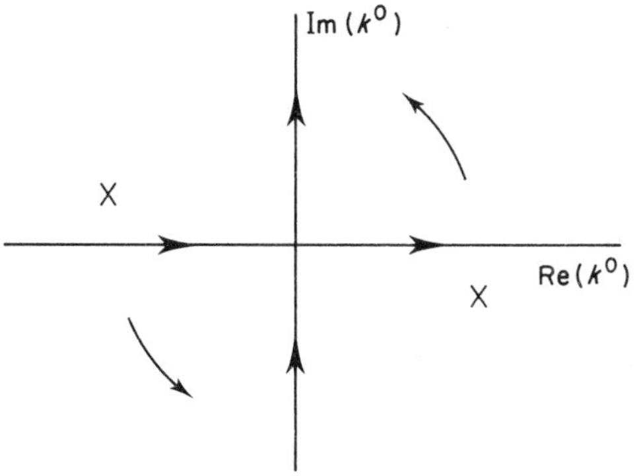

Figure 9.6 Wick rotation of the integration contour in the complex k^0 plane. Crosses mark the poles of the Feynman propagator, which do not impede the anticlockwise rotation of the contour.

The task of finding out whether a given field theory is renormalizable or not is a lengthy and highly technical one, and I shall do no more than state some of the essential results.

(a) A simple, though not infallible, criterion for renormalizability is provided by *dimensional analysis*. Since we are using natural units ($\hbar = c = 1$), there is a single free dimension, which I shall take to be measured in units of mass. Thus, the dimension of any quantity can be expressed as $(\text{mass})^D$. A momentum has $D = 1$. Since the two terms of a differential operator such as $(\Box + m^2)$ must have the same dimensions, ∂_μ has $D = 1$ and, correspondingly, the volume element d^4x has $D = -4$. The action S appears in a functional integral as the argument of an exponential and must therefore be dimensionless ($D = 0$), which means that $D = 4$ for a Lagrangian density. For a scalar field, whose Lagrangian density includes (7.7), this implies that $D = 1$. Similar arguments show that a spinor field has $D = 3/2$ and a gauge field has $D = 1$. Now the power of k with which the integral (9.76) diverges, namely 2, is equal, for fairly obvious reasons, to the dimension D of the integral.

Suppose that a coupling constant λ has dimension D_λ and a Green function G has dimension D_G. We evaluate the Green function as a power series

$$G = G_0 + \lambda G_1 + \lambda^2 G_2 + \dots . \tag{9.77}$$

Each coefficient G_n is a multiple momentum integral of dimension $D_G - nD_\lambda$, which may be expected to diverge with this power. Assume that we have enough freedom, using mass, coupling constant and wavefunction renormalization, to eliminate all infinities at order $n = 1$. If D_λ is negative, the infinities become more severe at higher orders, and we might expect to reach a point where we no longer have enough freedom to eliminate them. On the other hand, if D_λ is zero or positive, then things get no worse at higher orders. A more detailed argument along these lines shows that, indeed, the theory is likely to be renormalizable if $D_\lambda \geqslant 0$. Indeed, if D_λ is positive, then the infinities may cease altogether after some order, and the theory is called *super-renormalizable*. Consideration of (9.45), (8.17) and (8.38) reveals that λ, e and g are all dimensionless and, other things being equal, these theories should be renormalizable. One reason for restricting the actions to contain only the terms we have considered is that other possible terms would involve coupling constants of negative dimension and destroy renormalizability.

(b) When a theory possesses *symmetries* such as gauge invariance, these restrict the terms which may appear in the action and therefore also restrict the number of independent parameters and the number of renormalizations which can be used to eliminate divergences. However, the same symmetries also restrict the ways in which infinite integrals can appear. In general, to construct a renormalizable theory, it is necessary to include in the action all possible terms which are allowed by symmetries and do not involve coupling constants of negative dimension.

(c) The dimensional criterion works for scalar field theory because the propagator $i(k^2 - m^2 + i\varepsilon)^{-1}$ behaves for large k like k^D, where D, equal to -2, is the dimension of the propagator. The same is true of the momentum–space propagator obtained from (9.42) for spin-$\frac{1}{2}$ fermions and that obtained from (9.65) for photons. For massive spin-1 particles, however, the term $k_\mu k_\nu/m^2$ in (9.67) leads to more severe divergences than are allowed for by dimensional analysis. As a result, interacting theories of massive spin-1 particles are found to be non-renormalizable, even when the dimensional criterion is satisfied. The propagator (9.66) does not lead to this problem, because of the extra power of k^2 in the denominator of the expression $k_\mu k_\nu/(k^2 - \xi m^2)$. However, the gauge-fixing term which allows us to use a propagator of this kind can be introduced only in a gauge-invariant theory. Therefore, a renormalizable theory of massive spin-1 particles must be gauge invariant and, as we saw in chapter 8, special measures are necessary to achieve this.

(d) In some gauge theories which have dimensionless couplings and might be expected to be renormalizable, there occur certain 'anomalous' Feynman integrals whose divergences cannot be renormalized away. How and why these *anomalies* occur is the subject of a large and technical literature, and the details cannot be pursued here. The root cause is a subtle breakdown of gauge invariance in functional integrals. Even when a fully gauge-invariant action is used, the integration measure $\mathcal{D}$(fields) in (9.59) may fail to be gauge invariant. For this reason, a field theory which is gauge invariant at the classical level may cease to be so upon quantization. Several different kinds of anomalies have been identified. Those which afflict gauge theories arise in Feynman diagrams from closed fermion loops and can be traced to gauge non-invariance of the fermionic path integral. The only way to remove these anomalies is to arrange for anomalies from several different fermion species to cancel amongst themselves. Indeed, the standard theory of weak and electromagnetic interactions (to be discussed in chapter 12) is potentially anomalous, and the sets of particle species, called *families* or *generations* of quarks and leptons, which are required for the cancellation of anomalies, are exactly those whose existence is inferred from experiment.

9.7 Quantum Electrodynamics

Quantum electrodynamics, or QED for short, is the field theory which describes the behaviour of charged particles with only electromagnetic interactions. It is, of course, most useful when the effects of other interactions are negligible, and this is most nearly true when we study the properties of the charged leptons—electrons and muons. (There are also the tau particles, but these are produced only in high-energy collisions and their properties cannot be determined with the same accuracy.) The electrodynamics of electrons and muons is the most accurate theory in existence, if accuracy is measured by the agreement between theoretical and experimental data. I shall illustrate the applications of interacting field theories by discussing some well known consequences of QED, namely the Coulomb potential, the Lamb shift of spectral lines in simple atoms, and the magnetic dipole moments of charged particles. Although the formalism has been developed with a view to interpreting scattering experiments, none of the quantities of interest here is conveniently described in these terms. Moreover, the detailed calculations involve much complicated algebra, although they are quite straightforward in principle. I shall therefore use somewhat qualitative arguments to identify the quantities which need to be calculated and omit detailed algebra when it does not illuminate questions of principle.

9.7.1 The Coulomb potential

From the point of view of perturbation theory, the interactions between charged particles come about through the exchange of virtual photons. A few of the Feynman diagrams which contribute to the scattering of two particles are shown in figure 9.7. To see how this description is related to the more elementary idea of a potential energy, let us first go back to chapter 6, where we wrote down in equation (6.21) the potential energy operator for particles interacting through a potential. I am going to show that all reference to photons can be eliminated from QED, leaving a theory of charged particles alone. In this version of the theory, we can, under suitable circumstances, obtain a potential energy operator of the form (6.21) which involves the familiar Coulomb potential.

For a single species of charged particle, and with the gauge-fixing function introduced in (9.63), the Lagrangian density for QED may be written as

$$\mathcal{L}_{\text{QED}} = A_\mu[g^{\mu\nu}\Box - (\xi^{-1} - 1)\partial^\mu\partial^\nu]A_\nu - j_\text{e}^\mu A_\mu + \bar{\psi}(\mathrm{i}\not{\partial} - m)\psi \quad (9.78)$$

where, for particles of charge q, the electromagnetic current is

$$j_\text{e}^\mu = q\bar{\psi}\gamma^\mu\psi. \quad (9.79)$$

The idea now is to carry out the functional integral over A_μ, leaving an effective action for ψ alone:

$$\exp[\mathrm{i}S_{\text{eff}}(\psi)] = \int \mathcal{D}A \exp[\mathrm{i}S(\psi, A)]. \quad (9.80)$$

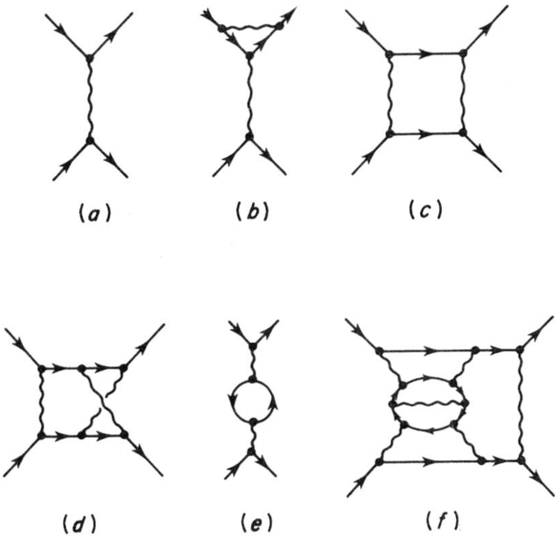

Figure 9.7 Some diagrams which contribute to the elastic scattering amplitude for two electrons. Diagrams (*a*) and (*e*) are the first two of a geometric series analogous to figure 9.4.

This is easy to do, since, as far as the A integral is concerned, the current density can be considered as a source, similar to that in (9.31) or (9.60). In the same way that we derived (9.36), but using the photon propagator, we obtain

$$S_{\mathrm{eff}} = \int \mathrm{d}^4x\bar{\psi}(x)(\mathrm{i}\not{\partial} - m)\psi(x) + \tfrac{1}{2}\int \mathrm{d}^4x\,\mathrm{d}^4y j_{\mathrm{e}}^{\mu}(x)\mathrm{D}_{\mathrm{F}\mu\nu}(x-y)j_{\mathrm{e}}^{\nu}(y). \tag{9.81}$$

Obviously, we would like to identify the last term as

$$-\tfrac{1}{2}\int \mathrm{d}t \int \mathrm{d}^3x\,\mathrm{d}^3y\rho(\boldsymbol{x}, t)V(\boldsymbol{x}-\boldsymbol{y})\rho(\boldsymbol{y}, t) \tag{9.82}$$

where $\rho = \psi^\dagger\psi$ is the particle density.

The idea of a potential energy $V(\boldsymbol{x}, \boldsymbol{y})$ between two particles located at $\boldsymbol{x}$ and $\boldsymbol{y}$ is really a classical one. To extract a comparable notion from the quantum-mechancial action (9.81), I shall imagine the current density (9.79) to represent an actual distribution of real charged particles although, in reality, it stands for a quantum-mechanical operator and appears only in intermediate stages of a calculation of, say, a scattering cross-section. Readers may like to consider for themselves how this step might be justified more rigorously. We can verify immediately that the potential is independent of the arbitrary gauge-fixing parameter ξ. This follows from the conservation of electric charge, expressed by the equation of continuity $\partial_\mu j_{\mathrm{e}}^{\mu} = 0$. The ξ-dependent term in (9.81) has no effect because

$$\int d^4y\, e^{ik\cdot y} k_\nu j_e^\nu(y) = \int d^4y(-i\partial_\nu e^{ik\cdot y}) j_e^\nu(y) = i\int d^4y e^{ik\cdot y}\partial_\nu j_e^\nu(y) = 0. \tag{9.83}$$

We now obtain the standard Coulomb potential by considering a *static* distribution of charged particles, with $j_e^\mu = q(\rho, \mathbf{0})$ and ρ independent of time. Then the integrals over x^0 and y^0 in (9.81) act only on the photon propagator. One of these, integrating the exponential in (9.65), gives $2\pi\delta(k^0)$, while the other corresponds to the time integral in (9.82). We now have a complete correspondence between (9.81) and (9.82) with

$$V(\mathbf{r}) = q^2 \int \frac{d^3k}{(2\pi)^3} \frac{e^{i\mathbf{k}\cdot\mathbf{r}}}{\mathbf{k}^2} = \frac{q^2}{4\pi|\mathbf{r}|}. \tag{9.84}$$

If the charge distribution is not static, then the interaction cannot be described just by an electric potential. There will, for example, be corrections for the magnetic force between particles in relative motion. In the case of a force mediated by exchange of massive particles, say of mass M, we should expect the potential to be of the *Yukawa* form

$$V(\mathbf{r}) = q^2 \int \frac{d^3k}{(2\pi)^3} \frac{e^{i\mathbf{k}\cdot\mathbf{r}}}{\mathbf{k}^2 + M^2} = \frac{q^2 e^{-M|\mathbf{r}|}}{4\pi|\mathbf{r}|}. \tag{9.85}$$

The *range* of such a force, as measured by the exponential decay, is a distance equal to $1/M$ in natural units or to $\hbar/Mc$ in laboratory units.

9.7.2 Vacuum polarization

Evidently, the Coulomb potential is associated with the transfer of a single virtual photon. The very simplest approximation to QED, which considers only single-photon exchange between real particles, is, roughly speaking, a classical approximation. If, for example, we calculate the scattering cross-section for two electrons using only the single-photon diagram of figure 9.7(a), the result, in the non-relativistic limit, is

$$\frac{d\sigma}{d\Omega} = \frac{\alpha^2 m^2}{16p^4}\left(\frac{1}{\sin^4(\theta/2)} + \frac{1}{\cos^4(\theta/2)} - \frac{1}{\sin^2(\theta/2)\cos^2(\theta/2)}\right) \tag{9.86}$$

where $\alpha = e^2/4\pi \approx 1/137$ is the fine structure constant and, in the centre of mass frame, θ is the scattering angle and p the magnitude of the 3-momentum of each particle. This is a modified version of the classical Rutherford formula, corrections arising from the electrons' being identical spin-$\frac{1}{2}$ particles. Quantum-mechanical corrections, which are all the diagrams containing closed loops, are small in QED, because each additional photon brings with it a pair of vertices and therefore a factor of α. Under some circumstances, however, they can be measured by accurate experiments.

Some, though of course not all, of these corrections can be regarded as modifications of the photon propagator. For example, figure 9.7(e) is obtained from 9.7(a) by inserting a closed loop of virtual charged particles, and the same modification can be made to any photon appearing in any diagram. The total effect of such modifications can be represented by replacing each unperturbed photon propagator by the complete propagator, whose first few terms are shown in figure 9.8. After making this replacement, of course, individual diagrams like figure 9.7(e) do not appear. By using the complete photon propagator in (9.84), we should obtain a modified Coulomb potential, which describes some of the quantum corrections to classical electrodynamics. This modified potential is said to result from *vacuum polarization*. Picturesquely, the idea is that the electric field of a charged particle polarizes the surrounding vacuum, in the sense that the original particle becomes surrounded by a distribution of virtual charged particle–antiparticle pairs, and the net potential is that due to this modified charge distribution.

Figure 9.8 Some diagrams which contribute to the complete photon propagator.

In momentum space, the contribution to the complete propagator $\mathrm{D}_{\mu\nu}(q)$ of the second diagram of figure 9.8 is

$$\mathrm{i}e^2\mathrm{D}_{\mathrm{F}\mu\sigma}(q)\int\frac{\mathrm{d}^4k}{(2\pi)^4}\,\frac{\mathrm{Tr}[\gamma^\sigma(\not{k}+m)\gamma^\tau(\not{k}+\not{q}+m)]}{(k^2-m^2+\mathrm{i}\varepsilon)[(k+q)^2-m^2+\mathrm{i}\varepsilon]}\mathrm{D}_{\mathrm{F}\tau\nu}(q) \quad (9.87)$$

and the set of all diagrams consisting of strings of these loops is, like (9.68), a geometric series. Because the photon propagator always appears inside Feynman diagrams multiplied by e^2, it is useful to consider the quantity $\alpha\mathrm{D}_{\mu\nu}(q)$. The contribution of the above set of diagrams to the $g_{\mu\nu}$ part of this quantity is

$$\frac{\alpha_0}{[1+\alpha_0 I(q^2)]}\,\frac{g_{\mu\nu}}{(q^2+\mathrm{i}\varepsilon)} \quad (9.88)$$

where $I(q^2)$ is an infinite integral arising from that in (9.87), and I have added the subscript to α_0 to indicate the need for renormalization.

Our hope is that (9.88) will turn into a finite expression when we rewrite it in terms of the true fine structure constant α. We would expect (and this can be verified *a posteriori*) that quantum corrections to the usual Coulomb potential should be appreciable only for charged particles separated by a very short distance. As in (9.84), the static potential corresponds to $q^0 = 0$ or $q^2 = -|\boldsymbol{q}|^2$. We identify the true fine structure constant as the coefficient of $1/|\boldsymbol{r}|$ in the large-distance limit of the potential, which means considering the exchange of a long-wavelength photon, or the limit $|\boldsymbol{q}| \to 0$. Thus, in the approximation where we use only the diagrams which led to (9.88), we have

$$\alpha = \frac{\alpha_0}{1 + \alpha_0 I(0)}. \tag{9.89}$$

Then (9.88) becomes

$$\frac{\alpha}{\{1 + \alpha[I(q^2) - I(0)]\}} \frac{g_{\mu\nu}}{(q^2 + \mathrm{i}\varepsilon)}. \tag{9.90}$$

The difference $I(q^2) - I(0)$ is finite and can be expressed in the form

$$I(q^2) - I(0) =$$

$$-\frac{1}{3\pi}\int_0^1 \mathrm{d}x\left\{\left[1 - 2\left(-\frac{m^2}{q^2}\right)\right]\ln\left[1 + x(1-x)\left(-\frac{q^2}{m^2}\right)\right] + 2x(1-x)\right\}$$

$$= -\frac{1}{15\pi}\left(\frac{-q^2}{m^2}\right) + \mathrm{O}\left(\left(\frac{-q^2}{m^2}\right)^2\right). \tag{9.91}$$

Although we have considered only a single species of charged particle, there will in reality be similar contributions from every species which exists in nature. Clearly, however, the major contribution will be that of the lightest species, namely the electron. The next lightest particle, the muon, is about 200 times heavier and its contribution to large-distance or low-energy vacuum polarization effects is much smaller.

9.7.3 The Lamb shift

The modified Coulomb potential which is the spatial Fourier transform of (9.88) will not be exactly proportional to $1/r$. This has a measurable effect upon the atomic spectrum of hydrogen. Readers will recall that in the elementary non-relativistic theory of the hydrogen atom the energy levels are independent of angular momentum and that this fact depends crucially on the form of the Coulomb potential. In a relativistic treatment based on the Dirac equation, the degeneracy is partly lifted by spin–orbit coupling, which leads to the fine structure splitting, but, for example, the $2S_{1/2}$ and $2P_{1/2}$ levels are still degenerate. If the Coulomb potential is not exactly proportional to $1/r$, then this degeneracy too is lifted. Actually, there are other effects of the loop diagrams of QED which cause a more pronounced 2S − 2P splitting than does the vacuum polarization. The measurements of W E Lamb and R C Retherford in 1947 showed the $2P_{1/2}$ level to lie below the $2S_{1/2}$ by an amount corresponding to a frequency $\Delta E/h$ of some 1000 MHz, while a calculation of the vacuum polarization effect alone suggests a shift of about 27 MHz in the opposite direction. However, detailed calculations, including all QED effects and also some nuclear effects, agree with more recent measurements which give a shift of about 1057.9 MHz, within the experimental accuracy of 0.02 MHz. Since this uncertainty is about a thousand times less than the contribution of vacuum polarization, this agreement may be taken as confirming the modification of the Coulomb law.

9.7.4 The running coupling constant

The modified Coulomb potential can be interpreted as $V(r) = \alpha(r)/r$, where $\alpha(r)$ is an effective distance-dependent coupling constant. Pictorially, if vacuum polarization is interpreted as the screening of the bare charge of a particle by a cloud of virtual electron–positron pairs, then the apparent charge of the particle depends upon how far into this cloud we have penetrated before looking at it. In Fourier-transformed language, the apparent charge depends upon the wavelength, and thus upon the energy and momentum of a real or virtual photon which interacts with the charged particle. Using (9.90), we define a *running coupling constant* $\alpha(-q^2)$ by

$$\alpha(-q^2) = \frac{\alpha}{1 + \alpha[I(q^2) - I(0)]}. \tag{9.92}$$

The argument $-q^2$ is used because we are often interested in negative values of q^2, as in the calculation of the static potential.

There are a number of important theoretical issues associated with this running coupling constant. In the first place, there is a close link with the process of renormalization. Instead of using the true fine structure constant, we could in principle define a renormalized coupling constant in terms of the value of (9.88) at $q^2 = -\mu^2$, μ being an arbitrary parameter as in (9.73). Then, in (9.89) and (9.90), $I(0)$ would be replaced by $I(-\mu^2)$. We easily find that

$$\alpha(\mu^2) = \frac{\alpha}{1 + \alpha[I(-\mu^2) - I(0)]} \tag{9.93}$$

which is the same equation as (9.92).

The existence of the running coupling constant can be taken to mean that the effective strength of electromagnetic interactions varies with energy. The variation is appreciable only when $(-q^2) \gg m^2$, and in that limit we have

$$\alpha(-q^2) \approx \alpha\left[1 - \frac{\alpha}{3\pi}\ln\left(-\frac{q^2}{m^2}\right)\right]^{-1}. \tag{9.94}$$

At energies of a few hundred GeV, which are accessible in modern particle accelerators, $\alpha(-q^2)$ has increased by only about 2% from its zero-energy value α. On the other hand, we see that $\alpha(-q^2)$ becomes infinite when $(-q^2) = m^2 \exp(411\pi)$. This energy is so vast as to be irrelevant to any conceivable experiment, but there is cause for concern on theoretical grounds. The pole in (9.90) at $q^2 = 0$ is, as we know, associated with the existence of real photons of zero mass. An infinite value of the running coupling constant would seem to imply the existence of a particle with imaginary mass M given by $M^2 = -m^2 \exp(411\pi)$, sometimes referred to as the *Landau ghost*. This would be a *tachyonic*, or faster-than-light particle, since $v^2/c^2 = 1 - M^2/E^2$ at energy E. Such particles are generally believed to be impossible and so the Landau ghost seems to indicate some fundamental flaw in QED. A related problem is that the infinite constant $I(0)$ in (9.89) is positive. This appears to mean that there is no positive value of α_0 (and therefore no real value of e_0), even zero or infinity, for which the renormalized α is non-zero. If every permissible value of α_0 leads to α being zero, then the theory is in fact non-interacting (we are not allowed to set $\alpha = 1/137$) and is said to be *trivial*. This question is somewhat confused, because the arguments are based on approximations of one kind or another and the bare coupling α_0 has no direct physical meaning. There is no doubt that perturbative QED is an excellent theory of electromagnetism at experimentally accessible energies, but it is believed by some that it would break down at sufficiently high energies and, indeed, that it ultimately makes sense only when embedded in a more complete theory.

9.7.5. Anomalous magnetic moments

A charged, spinning particle might be expected to possess a magnetic dipole moment, and so it does. An extremely accurate test of QED is provided by measurements of the magnetic moments of the electron and muon. To see how these are calculated, it is helpful first to study the non-relativistic limit of the Dirac equation (8.13) which, for an electron of charge $-e$, reads

$$(\mathrm{i}\not{\partial} + e\not{A} - m)\psi = 0. \tag{9.95}$$

When the kinetic energy is much smaller than the rest energy m, we can, approximately, derive the Schrödinger equation from this. We first multiply on the left by γ^0 to give

$$\mathrm{i}\frac{\partial\psi}{\partial t} = (-\mathrm{i}\gamma^0\gamma^i\partial_i - eA^0 + e\gamma^0\gamma^i A^i + m\gamma^0)\psi \tag{9.96}$$

and note that, in the standard representation of the γ matrices, we have

$$\gamma^0\gamma^i = \begin{pmatrix} 0 & \sigma^i \\ \sigma^i & 0 \end{pmatrix} \quad \text{and} \quad \gamma^0 = \begin{pmatrix} I & 0 \\ 0 & -I \end{pmatrix}. \tag{9.97}$$

When m is large compared with the kinetic energy, the most rapid time dependence of ψ is in a factor $\exp(-\mathrm{i}mt)$. For a free positive energy particle in its rest frame, the solution is $\exp(-\mathrm{i}mt)$ times one of the spinors (7.66). For small kinetic and electromagnetic energies, therefore, we anticipate a solution of the form

$$\psi = \exp(-\mathrm{i}mt)\begin{pmatrix}\chi \\ \theta\end{pmatrix} \tag{9.98}$$

where χ and θ are two-component spinors and θ is small. On substituting this into (9.96), we obtain two coupled equations for χ and θ:

$$\mathrm{i}\frac{\partial\chi}{\partial t} = -\sigma^i(\mathrm{i}\partial_i - eA^i)\theta - eA^0\chi \tag{9.99}$$

$$\mathrm{i}\frac{\partial\theta}{\partial t} = -\sigma^i(\mathrm{i}\partial_i - eA^i)\chi - eA^0\theta - 2m\theta. \tag{9.100}$$

When m is large and θ small, the solution to (9.100) is approximately

$$\theta \approx -\frac{1}{2m}\sigma^i(\mathrm{i}\partial_i - eA^i)\chi \tag{9.101}$$

and by substituting this into (9.100) we find

$$\mathrm{i}\frac{\partial\chi}{\partial t} = -\frac{1}{2m}\sigma^i\sigma^j(\boldsymbol{\nabla} + \mathrm{i}e\boldsymbol{A})^i(\boldsymbol{\nabla} + \mathrm{i}e\boldsymbol{A})^j\chi - e\Phi\chi \tag{9.102}$$

where $\Phi = A^0$ is the electric potential.

Now, the Pauli matrices satisfy the identity

$$\sigma^i\sigma^j = \delta^{ij} + \mathrm{i}\varepsilon^{ijk}\sigma^k \tag{9.103}$$

which leads to the final result

$$\mathrm{i}\frac{\partial\chi}{\partial t} = \left(-\frac{1}{2m}(\boldsymbol{\nabla} + \mathrm{i}e\boldsymbol{A})^2 - e\Phi + \frac{e}{m}\frac{1}{2}\boldsymbol{\sigma}\cdot\boldsymbol{B}\right)\chi. \tag{9.104}$$

The first two terms on the right-hand side give the usual Schrödinger equation for a particle of charge $-e$ in an electric potential Φ and magnetic vector potential $\boldsymbol{A}$. The last term represents the interaction of a magnetic moment $\boldsymbol{\mu} = (-e/m)(\boldsymbol{\sigma}/2)$ with the magnetic field $\boldsymbol{B} = \boldsymbol{\nabla} \times \boldsymbol{A}$. Since the spin angular momentum operator is $\boldsymbol{s} = \boldsymbol{\sigma}/2$, we have

$$\boldsymbol{\mu} = -g_s\mu_\mathrm{B}\boldsymbol{s} \tag{9.105}$$

where $\mu_\mathrm{B} = e/2m$ is the *Bohr magneton* and $g_s = 2$. This is a somewhat surprising prediction of the Dirac equation, since the corresponding g factor for orbital angular momentum is 1.

Experimentally, this prediction is approximately verified for electrons and muons, but there is a correction of about 0.1% arising from higher-order quantum effects in QED. The way this comes about is quite similar to the modification of the Coulomb potential by vacuum polarization. In (9.95), the middle term is $e\gamma^\mu A_\mu \psi$, and the γ^μ is the same as that which appears in the QED vertex (9.51). In any Feynman diagram, we can replace any vertex by one of an infinite set of modified vertices, a few of which are shown in figure 9.9, and still have a valid diagram. Thus, there is an effective vertex Γ^μ, namely the sum of these modified vertex diagrams, which replaces γ^μ in the same sense that the complete photon propagator replaces the unperturbed one. Essentially, the *anomalous magnetic moment* is calculated by making this replacement in the above analysis, but the technical details are a little complicated. The anomaly is defined by $a = (g_s - 2)/2$ and its lowest-order contribution is $\alpha/2\pi$. The best theoretical and experimental values for the electron anomaly are

$$a_{\text{theor}} = (1\,159\,652.4 \pm 0.4) \times 10^{-9}$$
$$a_{\text{exp}} = (1\,159\,652.4 \pm 0.2) \times 10^{-9}.$$

For muons, there is similar agreement between experimental and theoretical results, although the accuracy of each is slightly less. For the proton and neutron, the g factors obtained from the Dirac equation are 2 and 0 respectively, but they are found experimentally to be approximately 5.58 and -3.82. The reason for these large discrepancies is that the Dirac equation applies to point particles. The experimental values for the various magnetic moments may be taken as evidence that, whereas the electron and muon are truly elementary particles, the proton and neutron have an internal structure, being composed of more elementary constituents, the quarks. Although theoretical models of the quark structure of nucleons are by no means as accurate as QED, the magnetic moments can be reasonably well accounted for on this basis.

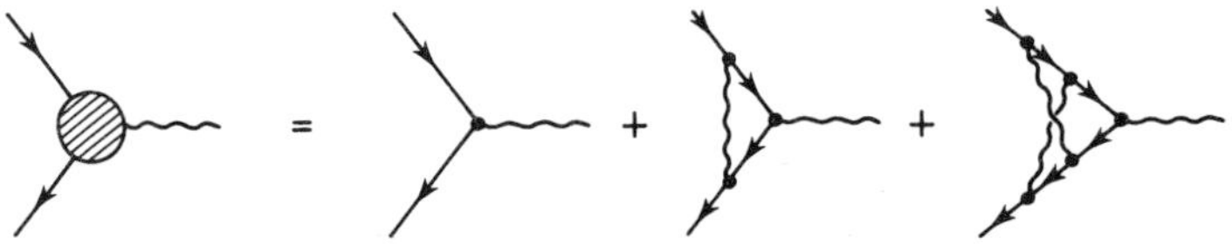

Figure 9.9 The effective electron–photon vertex which gives rise to an anomalous magnetic moment.

Exercises

9.1. In many contexts, Green functions of various kinds are encountered as a means of solving differential equations. If $\phi_0(x)$ is a solution of the Klein–Gordon equation $(\Box + m^2)\phi_0 = 0$, show that a solution of the equation $(\Box + m^2)\phi(x) = j(x)$ is given by

$$\phi(x) = \phi_0(0) + \int d^4y\, G_F(x - y)j(y).$$

9.2. In equation (7.11), denote the positive energy part of $\phi(x)$ by $\phi_a(x)$ and the negative energy part by $\phi_c^*(x)$. Show that

$$\int d^3x\, G_F(\boldsymbol{x}' - \boldsymbol{x}, t' - t)\frac{\overleftrightarrow{\partial}}{\partial t}\phi(\boldsymbol{x}, t)$$

$$= \theta(t' - t)\phi_a(\boldsymbol{x}', t') - \theta(t - t')\phi_c^*(\boldsymbol{x}', t')$$

where θ is the step function (appendix 1). Can you justify Feynman's description of an antiparticle as 'a particle travelling backwards in time'?

9.3. Write down an expression for the time-ordered product of two bosonic or two fermionic field operators, using the step functions $\theta(x^0 - y^0)$ and $\theta(y^0 - x^0)$ to distinguish the two time orderings. Use Cauchy's theorem to show that the step function can be represented as

$$\theta(t - t') = \lim_{\varepsilon \to 0} \frac{1}{2\pi i} \int_{-\infty}^{\infty} d\omega\, \frac{e^{i\omega(t-t')}}{\omega - i\varepsilon}.$$

By expressing the free field operators in terms of creation and annihilation operators, verify the expressions (9.38) and (9.42) for the scalar and spinor propagators.

9.4. The field operator A_μ for a non-interacting spin-1 particle of mass m has the Lagrangian density $\mathscr{L} = -\frac{1}{4}F_{\mu\nu}F^{\mu\nu} + (m^2/2)A_\mu A^\mu$, where $F_{\mu\nu}$ is given by (8.14). Show that the Euler–Lagrange equations are the *Proca equations*

$$\partial_\mu F^{\mu\nu} + m^2 A^\nu = 0.$$

From these equations, show that $\partial_\mu A^\mu = 0$ and that A_μ satisfies the Klein–Gordon equation.

9.5. The symbol $\Box^{-1}$ means that, if $\Box A = B$, then $A = \Box^{-1}B$. For example, $\Box^{-1}\exp(iq\cdot x) = -\exp(iq\cdot x)/q^2$. The transverse and longitudinal projection operators $T_{\mu\nu}$ and $L_{\mu\nu}$ are defined by $T_{\mu\nu} = g_{\mu\nu} - \partial_\mu\partial_\nu\Box^{-1}$ and $L_{\mu\nu} = \partial_\mu\partial_\nu\Box^{-1}$.

Show that (a) $L_{\mu\nu} + T_{\mu\nu} = g_{\mu\nu}$, (b) $L_{\mu\sigma}L^\sigma{}_\nu = L_{\mu\nu}$, (c) $T_{\mu\sigma}T^\sigma{}_\nu = T_{\mu\nu}$, (d) $L_{\mu\sigma}T^\sigma{}_\nu = T_{\mu\sigma}L^\sigma{}_\nu = 0$. Solve (9.64) by expressing the differential operator in terms of these projection operators and by expressing $D_{F\mu\nu}(x - y)$ in terms of projection operators acting on $\delta(x - y)$. (For this purpose, set $\varepsilon = 0$.)

9.6. A charged particle of mass m undergoes an electromagnetic scattering process, emitting a single virtual photon which subsequently interacts with another particle. If p^μ and p'^μ are the initial and final momenta of the particle ($p^2 = p'^2 = m^2$), then the momentum of the virtual photon is $q^\mu = p'^\mu - p^\mu$. Show that $q^2 \leqslant 0$.

10

Equilibrium Statistical Mechanics

When we deal with systems containing many particles, it soon becomes essential to adopt statistical methods of analysis. To a large extent, statistical mechanics has been developed with a view to studying condensed matter systems, such as solids and fluids, upon which laboratory experiments can be performed. In some cases, the quantum-mechanical properties of the constituent particles are crucial. This is true, for example, when we study the properties of electrons in metals or semiconductors, or of superfluid helium. In other cases, it is sufficient to treat the constituent particles according to classical mechanics, although it may still be necessary to determine their properties, such as the forces which act between them, from the underlying quantum theory. The properties of most normal fluids and many magnetic properties of solid materials, for example, can be adequately and conveniently treated by classical methods.

There are, moreover, important connections between statistical mechanics and the relativistic field theories which have been our concern in previous chapters. Indeed, the entire history of quantum mechanics and quantum field theory might be said to have started with Planck's attempts to understand black-body radiation in terms of statistical mechanics. The most obvious connection is that it may be necessary to consider the behaviour of large assemblages of high-energy particles, whose proper description is in terms of quantum field theory. Black-body radiation is a case in point, although it can be understood without the full machinery of field theory. Other examples are the hot dense gases found, it is thought, in the cores of some stars or in the early universe and, perhaps, small amounts of hot matter formed in high-energy collisions of heavy ions. At the mathematical level, there are close formal similarities between the thermal averages of statistical mechanics and the functional integral methods of field theory, which I shall discuss towards the end of this chapter. The recognition of these similarities has proved enormously fruitful. For example, the methods of relativistic field theory have shed considerable light on certain problems in condensed matter physics, especially those involving phase transitions, as we shall see in the next chapter, while techniques developed originally for statistical mechanics provide alternative methods of calculation in relativistic field theories, when perturbation theory is not applicable.

In this book, I shall consider, for the most part, only *equilibrium statistical mechanics*. The assumption of thermal equilibrium, that is of a state in which all macroscopic properties of the system have settled down to constant values, leads to great simplifications, provided we accept that the measured values of these quantities are to be compared with suitably weighted averages over the microscopic states of our theoretical model system. For we then have only to establish what weight should be attached to a given state and are absolved from considering how the system passes from one state to another. The mathematical foundations of statistical mechanics have been developed rather more fully for classical systems than for quantum-mechanical ones. I shall begin by considering the kinds of justification which have been suggested for the use of particular statistical weight functions for classical systems and then examine the relationship between statistical mechanics and thermodynamics. Finally, I shall describe the adaptation of these ideas to quantum mechanics and quantum field theory.

10.1 Ergodic Theory and the Microcanonical Ensemble

It will probably strike readers as intuitively obvious that macroscopic measurements generally yield some kind of average value of the measured quantity. This is because of the limited resolution of our measuring apparatus, but there are at least two different aspects to this, both of which are called upon to justify different theoretical steps. Consider, for example, a largish amount of a fluid in a transparent container. Suppose, for the sake of argument, that we know, with negligible error, the total mass of fluid and the volume of the container. Then the ratio of the two gives us a value for the overall density. By passing a beam of light through the container, we can measure the refractive index, and hence the density, of that region of the fluid which the beam intersects. There are two reasons for expecting the density measured in this way to coincide with the overall density. One is that the measurement process takes much longer than the timescales which characterize the microscopic motions, for example the mean time between two collisions of a single particle, or the time taken for a particle to cross the beam. Therefore, although the number of particles in the volume defined by the beam fluctuates with time, we would expect the measured density to be a *long time average* of instantaneous densities and, further, that this average should coincide with the overall density. The second reason is that, even though the volume defined by the beam may be only a small part of the total volume, it will normally contain a large number of particles. Averaged over all possible configurations of the particles, the density should certainly be equal to the overall density, and probability theory tells us to expect relative fluctuations about this average which depend inversely on the square root of the mean number of particles. Because our measurement is *coarse grained*, in the sense that it probes distances much greater than the average separation of two particles, we would expect even an instantaneous measurement to give a value very close to the average.

The statistical description of systems in thermal equilibrium is based on the idea that the measured value of a quantity is a long time average. We further assume that, during the time taken to perform the measurement, the system passes through a sequence of instantaneous states which is representative of the whole set of states available to it. In classical mechanics, the instantaneous state of a system can be represented as a point in *phase space*. For a system of N particles, phase space Γ is the $6N$-dimensional manifold whose points correspond to the values of the $3N$ coordinates and $3N$ momenta. For the moment, it will be convenient to lump the coordinates and momenta together into a $6N$-dimensional coordinate X. A weighted average of a quantity $f(X)$ is of the form

$$\langle f \rangle_t = \int_\Gamma \mathrm{d}^{6N}X \rho(X, t) f(X) \tag{10.1}$$

where $\rho(X, t)$ is a probability density for finding the system in a state close to X at time t. The probability density can be visualized in terms of a *Gibbs ensemble* of very many identical systems, $\rho(X, t)\mathrm{d}^{6N}X$ being the fraction of these whose state at time t is in a phase-space volume element $\mathrm{d}^{6N}X$ containing X.

An equation governing the rate of change of the probability distribution with time can be deduced from Hamilton's equations (3.16). In fact, we have already derived this equation, namely (3.22), for the particular distribution (3.20). To show that the same equation is valid for any other distribution, we consider the points representing members of the ensemble as a 'probability fluid' in phase space. The current density of this fluid has components $j_i = \dot{X}_i \rho(X, t)$ and, since we are not going to change the probability artificially by adding or removing systems from the ensemble, the equation of continuity must hold:

$$\frac{\partial}{\partial t} \rho(X, t) = -\sum_{i=1}^{6N} \frac{\partial}{\partial X_i} [\dot{X}_i \rho(X, t)]. \tag{10.2}$$

From Hamilton's equations we find

$$\sum_{i=1}^{6N} \frac{\partial \dot{X}_i}{\partial X_i} = \sum_{i=1}^{3N} \left(\frac{\partial \dot{q}_i}{\partial q_i} + \frac{\partial \dot{p}_i}{\partial p_i} \right) = \sum_{i=1}^{3N} \left(\frac{\partial}{\partial q_i} \frac{\partial H}{\partial p_i} - \frac{\partial}{\partial p_i} \frac{\partial H}{\partial q_i} \right) = 0 \quad (10.3)$$

and therefore

$$\frac{\partial}{\partial t} \rho(X, t) = -\sum_{i=1}^{6N} \dot{X}_i \frac{\partial}{\partial X_i} \rho(X, t) = -\mathrm{i}\mathcal{H}\rho(X, t) \qquad (10.4)$$

where $\mathcal{H}$ is the Liouville operator defined in (3.19). This is the *Liouville equation*. It gives the rate of change of the probability density at a fixed point in phase space. We could also fix our attention on a particular member of the ensemble, whose state is $X(t)$, and ask how the probability density in its neighbourhood, $\rho(X(t), t)$ changes with time. The answer is

$$\frac{\mathrm{d}}{\mathrm{d}t} \rho(X(t), t) = \frac{\partial}{\partial t} \rho(X(t), t) + \sum_{i=1}^{6N} \dot{X}_i \frac{\partial}{\partial X_i} \rho(X(t), t) = 0. \quad (10.5)$$

This result is usually described by saying that the probability density behaves as an incompressible fluid. It does not imply, however, that ρ has a uniform value over that part of phase space where it is non-zero, as would be true for an ordinary incompressible fluid.

For a system in equilibrium, all averages of the form (10.1) should be constant in time, which means that $\partial\rho/\partial t = 0$. According to (10.4), this will be true if ρ depends on X only through quantities whose Poisson bracket with the Hamiltonian H is zero, which are conserved quantities. For simplicity, I shall assume that the only relevant conserved quantity is the energy. The probability density which describes a system in equilibrium depends, as we shall see, on how the system is permitted to interact with its environment. Once this interaction is specified, it is quite straightforward to construct the appropriate probability density. Ideally, however, we would like to have some reassurance on a number of points. First, we would like to know whether the ensemble average (10.1) is indeed equal to the long time average which, by hypothesis, corresponds to an experimental measurement. If so, we would like to be sure that the time-independent probability density we have constructed is unique, for if more than one could be found we would have no good reason for preferring any particular one. Finally, we would like to understand theoretically why a system which starts in a non-equilibrium state usually does settle down into a state of thermal equilibrium. The theory which tries to answer these questions in a mathematically rigorous manner is called *ergodic theory*. It is unfortunately true that, while many elegant mathematical results have been obtained, the effort required to derive them is out of all proportion to their practical utility in applications to actual physical systems. I shall therefore not attempt to do more than convey the flavour of what is involved.

We consider a system which is completely isolated from its environment. It is therefore *closed*, which means that no particles enter or leave it, and *isoenergetic*, which means that its energy is fixed at a definite value E. The probability density must be zero except on the $(6N - 1)$-dimensional surface where $H(X) = E$. A candidate for the equilibrium probability density, which depends on the phase-space point X only through $H(X)$, is

$$\rho_{\text{micro}}(X, E) = \frac{\delta[H(X) - E]}{\Sigma(E)} \tag{10.6}$$

where, to ensure the correct normalization,

$$\Sigma(E) = \int \mathrm{d}^{6N}X\delta[H(X) - E]. \tag{10.7}$$

The Gibbs ensemble corresponding to this probability density is called the *microcanonical ensemble*. It is uniformly distributed over the constant energy surface.

The microcanonical ensemble is likely to be relevant to experimental observations if the averages we calculate with it are equal to the corresponding long time averages. A system is said to be *ergodic* if, for any smooth function $f(X)$,

$$\int_\Gamma \mathrm{d}^{6N}X\rho_{\text{micro}}(X)f(X) = \lim_{T\to\infty}\frac{1}{T}\int_0^T \mathrm{d}tf(X(t)) \tag{10.8}$$

and if this is true for almost all starting points $X(t = 0)$ for the trajectory on the right-hand side. The phrase 'almost all' has the mathematical sense of 'except on a set of measure zero', which means that the set of exceptional starting points makes no contribution to the ensemble average on the left. The way this might come about is as follows. Imagine the constant energy surface to be divided into small cells. In the course of its motion over a very long time, the point $X(t)$ representing an ergodic system will pass through every cell, provided that we wait long enough, and the fraction of time it spends in each cell is equal to the weight of that cell in the ensemble average. This is true for any cells of finite size, which means that the trajectory will eventually pass arbitrarily close to any point of the energy surface. The stronger statement that it will eventually pass *through* every point is actually not true. The application of the microcanonical ensemble to averages in thermal equilibrium is justified by the *ergodic theorem* due to G D Birkhoff and A I Khinchin, which states that, for an ergodic system, the microcanonical ensemble is the only time-independent probability density on the energy surface. The converse, that a system for which the only time-independent distribution is the microcanonical one is ergodic, is also true. The drawback of this approach lies in the extreme difficulty of proving that any system of real physical interest actually is ergodic. One such proof, given by Y Sinai, applies to a gas of hard spheres, that is a gas of spherical molecules which do not deform or penetrate each other, but exert no other forces. Given that this admittedly idealized model system is ergodic, we might expect that other, more realistic models would also have this property.

Although ergodicity ensures that the microcanonical ensemble correctly describes thermal equilibrium, it does not ensure that an isolated system will eventually settle into equilibrium if it starts in some other state. In other words, a Gibbs ensemble which initially does not have the uniform microcanonical distribution over the energy surface will not necessarily approach such a distribution with the passage of time. On the face of it, indeed, it seems unlikely that this could ever happen. From (10.5), we know that the density in the neighbourhood of any particular member of the ensemble is constant, and therefore any initial inhomogeneities in $\rho(X)$ cannot be smoothed out with time, although they will move around the energy surface. The kind of thing which might happen is illustrated schematically in figure 10.1, where ρ is zero, except in the shaded region. The fraction of the energy surface where ρ is non-zero is constant in time, but the shape of this region may evolve in a complicated way, developing strands which spread out over the entire energy surface. If the surface is divided into small cells, and we define a *coarse-grained* probability density by averaging over each cell, then this coarse-grained density may well become uniform. Since our experimental measurements are in any case coarse grained, the actual probability density would, for practical purposes, become indistinguishable from the microcanonical one, because we would only want to average functions $f(X)$ which vary very little within a coarse-graining cell.

This kind of behaviour is somewhat analogous to the mixing of two immiscible liquids, such as oil and water, stirred together in a container, producing a mixture which is homogeneous in the coarse-grained sense. Systems whose trajectories in phase space lead to this kind of development of a probability density are called *mixing*. There is, of course, a precise mathematical definition, but we shall not be making any use of it. It can be shown that all mixing systems are also ergodic, but the converse is not true. The hard-sphere gas was in fact shown by Sinai to be mixing.

A simple example of the use of the microcanonical ensemble is provided by an ideal gas with Hamiltonian

$$H = \sum_{i=1}^{N} \frac{1}{2m} \boldsymbol{p}_i^2 \tag{10.9}$$

confined to a volume V. The area of the energy surface $\Sigma(E)$ can be expressed as

$$\Sigma(E) = \int \mathrm{d}^{3N}p\mathrm{d}^{3N}x\delta\left(E - \frac{1}{2m}\sum_{i=1}^{3N} p_i^2\right)$$
$$= \frac{\partial}{\partial E}\int \mathrm{d}^{3N}p\mathrm{d}^{3N}x\theta\left(E - \frac{1}{2m}\sum_{i=1}^{3N} p_i^2\right) \tag{10.10}$$

where $\theta(E - H)$ is the step function. The integral over coordinates gives V^N and the momentum integral is the volume of a $3N$-dimensional sphere of radius $(2mE)^{1/2}$, which can be evaluated as in appendix 1. The final answer is

$$\Sigma(E) = \frac{V^N(2\pi m)^{(3N/2)}E^{(3N/2)-1}}{(\frac{3}{2}N - 1)!} \tag{10.11}$$

and we shall see shortly that it is related to the entropy of the gas.

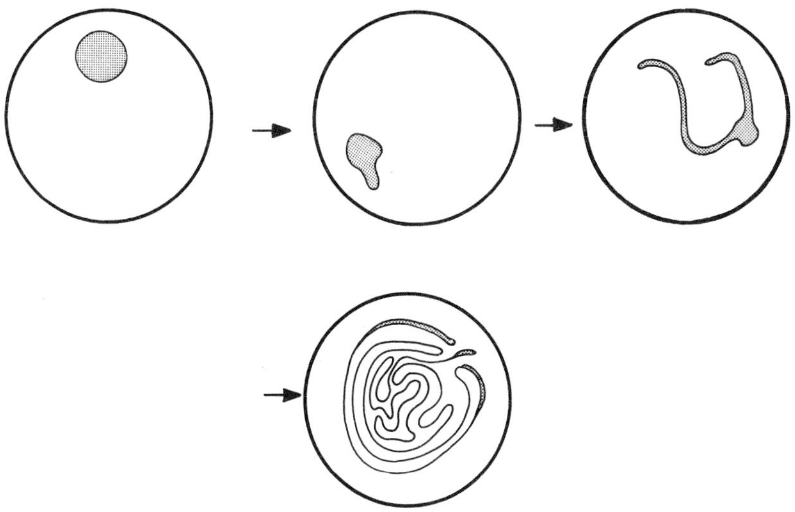

Figure 10.1 Schematic illustration of the evolution in time of the phase-space probability density of a mixing system. The probability density is non-zero only in the shaded region, whose area is constant.

10.2 The Canonical Ensemble

If our system is permitted to exchange heat energy with its surroundings, we need a somewhat different statistical description. So long as we restrict ourselves to equilibrium conditions, we need not be very precise about the mechanism which allows this exchange to take place. The simplest course is to suppose that the surroundings constitute a *heat bath*. Ideally, the heat bath is an infinite system which can exchange finite amounts of energy with the system of interest without any change occurring in its own properties. Experimentally, this situation can be accurately simulated by using thermostatic feedback techniques. Normally, we describe these as techniques for maintaining a constant temperature, but we have yet to establish a precise notion of temperature within statistical mechanics. We shall still take the total number of particles in the system to be fixed, in which case we are dealing with a *closed isothermal system*. The Gibbs ensemble for such a system is called the *canonical ensemble* and our first objective is to find the appropriate probability density $\rho_{\text{can}}(X)$. The question of what this probability density should be has not been investigated with the same degree of mathematical rigour as for the microcanonical ensemble, but the following simple argument produces what is universally accepted as the correct answer.

Consider two systems, A and B, which are both in equilibrium with the same heat bath but do not interact directly with each other. Individually, they have probability densities $\rho_{\text{can}}(X_{\text{A}})$ and $\rho_{\text{can}}(X_{\text{B}})$, which depend on the coordinates and momenta only through $H_{\text{A}}(X_{\text{A}})$ and $H_{\text{B}}(X_{\text{B}})$ respectively. Equally, we can regard A and B as a single system AB, with Hamiltonian $H_{\text{AB}}(X_{\text{AB}}) = H_{\text{A}}(X_{\text{A}}) + H_{\text{B}}(X_{\text{B}})$, whose probability density $\rho_{\text{can}}(X_{\text{AB}})$ depends only on H_{AB}. Since A and B do not interact, their probability densities should be independent, and the joint probability density is

$$\rho_{\text{can}}(H_{\text{AB}}) = \rho_{\text{can}}(H_{\text{A}} + H_{\text{B}}) = \rho_{\text{can}}(H_{\text{A}})\rho_{\text{can}}(H_{\text{B}}). \qquad (10.12)$$

This relation determines the form of ρ_{can} up to a single parameter. For a function of a single variable, $f(x)$, which has the property $f(x+y) = f(x)f(y)$, we can first deduce that $f(0) = 1$ by setting $x = y = 0$. Then, by choosing y to be a small increment δx and defining $\beta = -f'(0)$, we obtain the differential equation $\mathrm{d}f(x)/\mathrm{d}x = -\beta f(0)$. Since $f(0) = 1$, the unique solution is $f(x) = \exp(-\beta x)$. In (10.12), the analogue of x is the function $H(X)$, and this allows some extra freedom in the normalization. It is easy to see that the normalized probability density

$$\rho_{\text{can}}(X, \beta) = \mathrm{e}^{-\beta H(X)}\left(\int \mathrm{d}^{6N}X \mathrm{e}^{-\beta H(X)}\right)^{-1} \qquad (10.13)$$

satisfies (10.12). The undetermined constant β is the same for any system in contact with the same heat bath and must therefore be a property of the heat bath itself. Thermodynamically, the only relevant property is its temperature. Thus, β must be a function of temperature, and we can clearly relate it to the ideal gas scale of temperature by taking the system to be an ideal gas.

For a gas or liquid consisting of N identical molecules, we define the *canonical partition function* $Z_{\text{can}}(\beta, V, N)$ in terms of the normalizing factor in (10.13) by

$$Z_{\text{can}}(\beta, V, N) = \frac{1}{h^{3N}N!}\int \mathrm{d}^{6N}X\mathrm{e}^{-\beta H(X)}. \tag{10.14}$$

By including the $1/N!$, we get a sum over all distinct states of the system, counting any two states which differ only by the interchange of a pair of particles as indistinguishable. The factor h^{-3N} has no physical significance and is included as a matter of theoretical convenience to make Z_{can} dimensionless. The constant h is arbitrary but must have the dimensions of action. It is convenient to take it to be Planck's constant, since this allows a direct comparison between corresponding classical and quantum-mechanical systems. Many quantities of thermodynamic interest can be expressed as derivatives of the partition function. In particular, the average internal energy U is obviously given by

$$\begin{aligned} U(\beta, V, N) &= \int \mathrm{d}^{6N}XH(X)\mathrm{e}^{-\beta H(X)}\left(\int \mathrm{d}^{6N}X\mathrm{e}^{-\beta H(X)}\right)^{-1} \\ &= -\frac{\partial}{\partial\beta}\ln Z_{\text{can}}(\beta, V, N). \end{aligned} \tag{10.15}$$

For an ideal monatomic gas, we easily obtain

$$Z_{\text{can}}(\beta, V, N) = \frac{V^N}{N!}\left(\frac{2\pi m}{\beta h^2}\right)^{3N/2} \tag{10.16}$$

and the internal energy is found to be $U = 3N/2\beta$. Comparing this with the value $U = 3Nk_{\text{B}}T/2$ obtained from elementary kinetic theory, where $k_{\text{B}} = 1.380\,54 \times 10^{-23}\ \text{J K}^{-1}$ is Boltzmann's constant and T the absolute temperature, we identify

$$\beta = 1/k_{\text{B}}T. \tag{10.17}$$

10.3 The Grand Canonical Ensemble

A system which can exchange both heat energy and particles with its surroundings is called an *open isothermal system*. Exactly what this means depends to some extent on the particular physical situation we want to investigate. Most straightforwardly, we can think of a very large homogeneous system, within which we draw an imaginary boundary enclosing a small part of the whole, which still contains a very large number of particles. Our earlier example of a light beam intersecting a large container of gas would be a case in point. The small subsystem constitutes 'the system' while the remainder of the original large system acts as an (ideally infinite) heat bath and particle reservoir. The Gibbs ensemble which describes an open isothermal system is the *grand canonical ensemble*.

The grand canonical probability density allows for the possibility of the system's containing any number of particles. It must have the general form

$$\rho(X) = g_N \exp[-\beta H_N(X)]\left(\sum_{N=0}^{\infty} g_N \int \mathrm{d}^{6N}X \exp[-\beta H_N(X)]\right)^{-1} \quad (10.18)$$

where H_N is the Hamiltonian of the system when it contains exactly N particles and g_N is related to the probability that it does contain N particles. This probability is obtained by integrating over the coordinates and momenta which the N particles might have:

$$P_N = g_N \int \mathrm{d}^{6N}X \exp[-\beta H_N(X)]\left(\sum_{N=1}^{\infty} g_N \int \mathrm{d}^{6N}X \exp[-\beta H_N(X)]\right)^{-1}. \quad (10.19)$$

If a particular particle can find itself, with equal probability, anywhere in the system or reservoir, and the reservoir is very much larger than the system, then the probabilities P_N should form a Poisson distribution,

$$P_N = \frac{\bar{N}^N \mathrm{e}^{-\bar{N}}}{N!} \quad (10.20)$$

where $\bar{N}$ is the average number of particles in the system.

In the case of non-interacting particles, the N-particle Hamiltonian is just the sum of single-particle Hamiltonians, and

$$\int \mathrm{d}^{6N}X \exp[-\beta H_N(X)] = \left(\int \mathrm{d}^3x\, \mathrm{d}^3p \exp[-\beta H_1(x, p)]\right)^N = (h^3 Z_1)^N \quad (10.21)$$

where Z_1 is the canonical partition function for a single particle. The two expressions (10.19) and (10.20) are then consistent if we set

$$g_N = \frac{1}{N!}\left(\frac{\bar{N}}{h^3 Z_1}\right)^N. \quad (10.22)$$

In general, the grand canonical probability density is defined as

$$\rho_{\mathrm{gr}}(N, X, \beta, \mu) = \frac{1}{h^{3N}N!} z^N \exp[-\beta H_N(X)]\left(\sum_{N=0}^{\infty} \frac{1}{h^{3N}N!} z^N \int \mathrm{d}^{6N}X \exp[-\beta H_N(X)]\right)^{-1} \quad (10.23)$$

where the *fugacity* z is

$$z = \mathrm{e}^{\beta\mu} \quad (10.24)$$

and μ is called the *chemical potential*. The chemical potential is taken to be a property of the particle reservoir and so, while it controls the average number $\bar{N}$ of particles in the system, it is independent of the number N which characterizes a particular configuration of the system.

From the derivation of (10.22), it is clear that the general expression (10.24) is strictly valid only when the integral $Y_N = \int d^{6N}X \exp[-\beta H_N(X)]$ can be written as Y^N, where Y is a quantity independent of N. This is usually not true when particles interact, but it is an excellent approximation when we consider a large system and interactions which are appreciable only over a distance which is small compared with the dimensions of the system. In that case, we can divide the volume of the system into a large number of cells, each of a size greater than the range of interactions, and ignore interactions between particles in different cells. The integral Y then factorizes into a product of single-cell terms, and the number of these terms is proportional to the number of particles in the system. Finally, since the relative fluctuations in the number of particles in a large system are small, only those terms in (10.23) for which N is large will be important.

The *grand canonical partition function* is defined as the normalizing denominator in (10.23):

$$Z_{\mathrm{gr}}(\beta, V, \mu) = \sum_{N=0}^{\infty} \exp(\beta\mu N) Z_{\mathrm{can}}(\beta, V, N)$$
$$= \sum_{N=0}^{\infty} \frac{1}{h^{3N}N!} \exp(\beta\mu N) \int d^{6N}X \exp[-\beta H_N(X)]. \tag{10.25}$$

For an ideal gas, we easily find

$$Z_{\mathrm{gr}}(\beta, V, \mu) = \exp\left[e^{\beta\mu} V \left(\frac{2\pi m}{\beta h^2}\right)^{3/2}\right]. \tag{10.26}$$

The average internal energy and number of particles are

$$U = -\left.\frac{\partial \ln Z_{\mathrm{gr}}}{\partial \beta}\right|_{\beta\mu} = \frac{3}{2}\beta^{-1} e^{\beta\mu} V \left(\frac{2\pi m}{\beta h^2}\right)^{3/2} \tag{10.27}$$

$$\bar{N} = \left.\frac{\partial \ln Z_{\mathrm{gr}}}{\partial (\beta\mu)}\right|_{\beta} = e^{\beta\mu} V \left(\frac{2\pi m}{\beta h^2}\right)^{3/2} \tag{10.28}$$

from which we recover the relation $U = 3\bar{N}/2\beta$, now involving the average particle number.

10.4 Relation Between Statistical Mechanics and Thermodynamics

The highly successful science of thermodynamics deals with large systems in terms of macroscopic observable quantities alone. Equilibrium thermodynamics is derived, for the most part, from three basic principles, known as the zeroth, first and second laws, which summarize the phenomenological results of countless experiments. These principles are so well established by observation as to stand in no real need of further justification. However, our theoretical understanding would be seriously incomplete if we could not recover the results of thermodynamics from the microscopic laws of motion for the particles which constitute a macroscopic system. Moreover, once we can identify thermodynamic functions in statistical mechanical terms, we can set about obtaining predictions for their properties which cannot be obtained from thermodynamics alone. I am going to assume that readers are familiar with the principles of thermodynamics, but I shall first give a short summary of the points which will particularly concern us. For simplicity, I shall deal explicitly only with fluid systems, but other systems, such as magnets and superconductors, which we shall need to consider later, can be dealt with by using straightforward analogies.

If two systems which are internally in equilibrium, their macroscopic properties having reached steady values, are brought into thermal contact, allowing heat energy to pass between them, their individual equilibria may be disturbed. If we wait long enough, however, the combined system will settle into a new equilibrium state, and we say that the two systems are in equilibrium with each other. The zeroth law of thermodynamics asserts that if two systems are simultaneously in equilibrium with a third, then they will be found to be in equilibrium with each other also. This implies that the systems share a common property, which has the same value for any two systems which are in equilibrium with each other. The property in question is *temperature*, and our discussion of the canonical ensemble indicates that β is indeed a measure of thermodynamic temperature. The zeroth law does not, however, provide a means of assigning numerical values to temperature. Indeed, any appropriately varying property of a chosen standard system—a thermometer—could be used to define a numerical scale of temperature. Two scales do not necessarily agree with each other, although there must obviously be a unique correspondence between temperatures measured on different scales.

The first law is essentially a statement of the conservation of energy, which explicitly recognizes that a change in the internal energy of a system can result equally from a flow of heat or from the performance of an equivalent amount of work. In a rudimentary way, we can distinguish a higher temperature from a lower one by agreeing that the temperature of a system increases if heat flows into it and no work is done in the process.

The second law has been formulated in many different ways. The simplest, due in slightly different forms to Clausius and Kelvin, asserts that no process is possible whose only effect is the transfer of heat from a colder body to a hotter one. On the face of it, this is a purely qualitative statement, and it is quite remarkable that two, precise, quantitative results follow from it. These are derived in every self-respecting textbook on thermodynamics. The first is that we can define an *absolute thermodynamic scale of temperature*. This scale is independent, in principle, of the properties of any specific system, but it coincides with the ideal gas scale defined by (10.17). The units of temperature, like any other units, are not determined by any physical principle. For a monatomic ideal gas, temperature is just a measure of the average translational kinetic energy of its molecules, and the value of k_B simply converts units of energy to the conventional units of temperature. The second result is that every equilibrium state of a system can be assigned an *entropy*, S, in such a way that, if an amount of heat ΔQ flows into the system at a fixed temperature T, the change in entropy is $\Delta S = \Delta Q/T$. This actually defines the difference in entropy between any two equilibrium states, but not its absolute value.

Combining the first and second laws, we obtain the fundamental equation of the thermodynamics of fluids

$$dU = TdS - pdV \tag{10.29}$$

which expresses any change in internal energy as the sum of heat flow into the system and work done on it. In thermodynamic terms, this serves to define the pressure p. Because of this equation, the internal energy is naturally expressed in terms of the two quantities S and V, $U = U(S, V)$. This means that the partial derivatives $(\partial U/\partial S)_V$ and $(\partial U/\partial V)_S$ have recognizable physical interpretations as T and $-p$ respectively. While it is perfectly possible to write U as a function of, say, T and p, its partial derivatives with respect to these variables have no simple significance. If we wish to consider the possibility of particles entering or leaving the system, we extend (10.29) to read

$$dU = TdS - pdV + \mu dN \tag{10.30}$$

where μ is the increase in internal energy due to the addition of a particle, when no heat flow or performance of work accompanies the change. This provides the thermodynamic definition of the chemical potential.

The last two equations exemplify a general feature of thermodynamics, namely that a system can be characterized by a *thermodynamic potential*. This is a function of several independent variables which together specify the macroscopic state of the system, and whose partial derivatives produce other quantities of physical interest. Several different functions may be used as potentials, and the criterion for a specific choice is that its natural independent variables should be quantities over which we exert experimental control. In statistical mechanics, we consider various idealized experimental situations in which the system is constrained in different ways and, as we have seen, these lead to different statistical ensembles. For a closed isoenergetic system, described by the microcanonical ensemble, the energy E (which for the moment I shall consider as identical to U), volume V and particle number N are all fixed and we need a potential for which these are natural independent variables. By rearranging (10.30), we find

$$\mathrm{d}S = (1/T)\,\mathrm{d}E + (p/T)\,\mathrm{d}V - (\mu/T)\,\mathrm{d}N \qquad (10.31)$$

which shows that the entropy $S(E, V, N)$ is a suitable choice.

For a closed isothermal system, described by the canonical ensemble, the variables are T, V and N. The appropriate potential is the *Helmholtz free energy* $F = U - TS$. Using $\mathrm{d}(TS) = T\,\mathrm{d}S + S\,\mathrm{d}T$, we get

$$\mathrm{d}F = -S\,\mathrm{d}T - p\,\mathrm{d}V + \mu\,\mathrm{d}N \qquad (10.32)$$

so indeed F is naturally expressed as $F(T, V, N)$. It is important to notice that we have done more than subtract TS from U. In (10.30) it is implied that both U and its partial derivatives T, p and μ are regarded as functions of S, V and N. In (10.32), it is similarly implied that F, S, p and μ are functions of T, V and N. This demands that we re-express S as a function of these variables by solving the equation

$$T = \left(\frac{\partial}{\partial S}\, U(S, V, N)\right)_{V,N} \qquad (10.33)$$

for S. This whole process is called a *Legendre transformation*.

For an open isothermal system, described by the grand canonical ensemble, the independent variables are T, V and μ. By another Legendre transformation, we identify the appropriate potential as

$$\Omega(T, V, \mu) = F - \mu N = U - TS - \mu N \qquad (10.34)$$

which is called the *grand potential*.

These three potentials can be identified in terms of the statistical partition functions $\Sigma(E, V, N)$, $Z_{\mathrm{can}}(\beta, V, N)$ and $Z_{\mathrm{gr}}(\beta, V, \mu)$. To do this safely, however, it is necessary to consider the *thermodynamic limit*, in which N and V are taken to infinity, with the number of particles per unit volume N/V held fixed. The reason for this is that, in thermodynamics, it is assumed that all of the quantities U, T, μ and N have definite values. In statistical mechanics this is not true. In an isothermal system, for example, the temperature is fixed by an infinite heat bath, but the energy fluctuates and U can be identified only as an average energy. Because the interpretation of the variables varies from one ensemble to another, the entropy, Helmholtz free energy and grand potential obtained from the appropriate ensembles will not be related by the thermodynamic Legendre transformations unless the effect of fluctuations is negligible. I said earlier that relative fluctuations are expected to be proportional to $N^{-1/2}$, and readers are invited to investigate this in the exercises. If so, then we can expect to obtain a unique correspondence between statistical mechanics and thermodynamics in the thermodynamic limit. Experimentally, we deal with systems of finite size, but typical numbers of particles are of the order of Avogadro's number 6.02×10^{23} which is, to a fair approximation, infinite!

Let us start with the grand canonical ensemble and define

$$\Omega_{\mathrm{gr}}(\beta, V, \mu) = -k_{\mathrm{B}} T \ln Z_{\mathrm{gr}}(\beta, V, \mu). \tag{10.35}$$

We would like to identify this as the grand canonical version of the thermodynamic potential $\Omega(T, V, \mu)$. If we can identify its partial derivatives with respect to T, V and μ as $-S$, $-p$ and $-N$ respectively, then the two functions can differ only by an additive constant, which can be determined by direct calculation if necessary. Using (10.17), (10.27) and (10.28) we find

$$\begin{aligned}\frac{\partial \Omega_{\mathrm{gr}}}{\partial T} = -\frac{1}{k_{\mathrm{B}} T^2}\frac{\partial \Omega_{\mathrm{gr}}}{\partial \beta} &= \frac{1}{T}\left(\Omega_{\mathrm{gr}} + \left.\frac{\partial \ln Z_{\mathrm{gr}}}{\partial \beta}\right|_{\beta\mu} + \left.\frac{\mu \partial \ln Z_{\mathrm{gr}}}{\partial(\beta\mu)}\right|_{\beta}\right) \\ &= \frac{1}{T}(\Omega_{\mathrm{gr}} - U + \mu \bar{N}).\end{aligned} \tag{10.36}$$

We must now argue self-consistently that if, indeed, (10.35) is the correct grand canonical version of Ω, then by (10.34) this must be $-S$. It follows from (10.28) that the μ derivative is $-N$, but the identification of $\partial\Omega_{\mathrm{gr}}/\partial V$ as $-p$ is problematic, since we have no grand canonical definition of pressure. What we do in practice is to adopt the definition $p = -\partial\Omega_{\mathrm{gr}}/\partial V$. We can check that this is, at least, sensible in the case of an ideal gas, by using (10.26)–(10.28) to recover the equation of state $p = Nk_{\mathrm{B}}T/V$.

Readers may like to develop similar arguments to show that, for the canonical ensemble,

$$F_{\mathrm{can}}(T, V, N) = -k_{\mathrm{B}} T \ln Z_{\mathrm{can}}(\beta, V, N) \tag{10.37}$$

and for the microcanonical ensemble

$$S_{\mathrm{micro}}(E, V, N) = k_{\mathrm{B}} \ln\left(\frac{\Sigma(E, V, N)}{h^{3N} N!}\right). \tag{10.38}$$

I shall follow the alternative course of showing that, in the thermodynamic limit, these functions are obtained from (10.35) by the thermodynamical Legendre transforms. Consider equation (10.25). In the thermodynamic limit, we expect fluctuations in N to be small relative to $\bar{N}$, so only those terms in the sum for which $N \approx \bar{N}$ should make significant contributions. We can therefore make the estimate

$$Z_{\mathrm{gr}}(\beta, V, \mu) = K e^{\beta\mu\bar{N}} Z_{\mathrm{can}}(\beta, V, \bar{N}) \tag{10.39}$$

where K represents the number of important terms. We now use (10.35) and (10.37) to write

$$\frac{\Omega_{\text{gr}}}{\bar{N}} = \frac{F_{\text{can}}}{\bar{N}} - \mu - \frac{\ln K}{\bar{N}}. \tag{10.40}$$

In the thermodynamic limit, we expect the potentials to be *extensive*, or proportional to the size of the system. The quantity K is not precisely defined, but it should depend only weakly on N. In the thermodynamic limit, therefore, the last term in (10.40) vanishes and the remaining equation coincides with (10.34). Both potentials can now be obtained from either ensemble with the same result, so we have a unique correspondence with thermodynamics and the ensemble subscripts can be dropped.

A relation between the canonical and microcanonical ensembles can be derived in a similar manner. Using (10.7) and (10.14), we can write

$$\begin{aligned} Z_{\text{can}}(\beta, V, N) &= \frac{1}{h^{3N}N!}\int \mathrm{d}E \int \mathrm{d}^{6N}X\mathrm{e}^{-\beta E}\delta(E - H_N(X)) \\ &= \frac{1}{h^{3N}N!}\int \mathrm{d}E\mathrm{e}^{-\beta E}\Sigma(E, V, N). \end{aligned} \tag{10.41}$$

Then, treating fluctuations in energy in the same way as those of the number of particles, and using (10.37) and (10.38), we recover the thermodynamic relation $F = U - TS$. In this way, we see that, in the thermodynamic limit, all three statistical ensembles are equivalent and their partition functions can be identified in a unique manner in terms of thermodynamic potentials. Mathematically, it is interesting to note that the Legendre transformations which relate these potentials correspond to *Laplace transforms* which relate the partition functions. The arguments I used to derive these relations are, of course, by no means rigorous. In principle, assumptions such as the extensivity of the potentials should be checked for each system to which the theory is applied. Indeed, it is possible to invent theoretical models for which the arguments do not work. For example, in models where long-range forces act between particles, the thermodynamic limit may not exist. As far as I know, the arguments are sound for all systems of physical interest. Readers may like to check for themselves that everything goes through smoothly for the ideal gas. They should find that the entropy is given by the *Sackur–Tetrode equation*

$$\frac{S}{N} = k_{\mathrm{B}}\left\{\frac{5}{2} + \ln\left[\frac{V}{N}\left(\frac{2\pi m k_{\mathrm{B}} T}{h^2}\right)^{3/2}\right]\right\}. \tag{10.42}$$

Factors of $N!$ should be treated using Stirling's approximation

$$\ln(N!) = N\ln(N) - N + \tfrac{1}{2}\ln(2\pi N) + \ldots \tag{10.43}$$

valid for large N.

10.5 Quantum Statistical Mechanics

When dealing with a large quantum-mechanical system, we need to estimate the expectation values of operators in states which we are unable to specify exactly at a microscopic level. We therefore have to take two averages, one over the uncertainties inherent in a definite quantum state and one to take account of our ignorance of what the state actually is. For the time being, I shall work in the Schrödinger picture. Suppose we have a complete orthonormal set of states $|\psi_n(t)\rangle$ for which

$$\langle\psi_m(t)|\psi_n(t)\rangle = \delta_{mn} \quad \text{and} \quad \sum_n |\psi_n(t)\rangle\langle\psi_n(t)| = \hat{I}. \tag{10.44}$$

For simplicity, I am assuming that these states can be labelled by a discrete index n: there will be no difficulty converting the sums into integrals where necessary. Suppose further that we can specify for each state the probability P_n of finding the system in that state. As long as the system is left undisturbed, P_n does not change with time. Using (10.44), we can write the expectation value of an observable A at time t as

$$\bar{A}(t) = \sum_n \langle\psi_n(t)|\hat{A}|\psi_n(t)\rangle P_n = \sum_{m,n} \langle\psi_m(t)|\hat{A}|\psi_n(t)\rangle P_n \langle\psi_n(t)|\psi_m(t)\rangle. \tag{10.45}$$

The object

$$\hat{\rho}(t) = \sum_n |\psi_n(t)\rangle P_n \langle\psi_n(t)| \tag{10.46}$$

can be regarded as an operator, called the *density operator*, which acts on a bra or ket vector to produce another:

$$\langle\Psi|\hat{\rho} = \sum_n \langle\Psi|\psi_n(t)\rangle P_n \langle\psi_n(t)| \quad \text{or} \quad \hat{\rho}|\Psi\rangle = \sum_n |\psi_n(t)\rangle P_n \langle\psi_n(t)|\Psi\rangle. \tag{10.47}$$

The expectation value (10.45) is the sum of diagonal matrix elements of $\hat{A}\hat{\rho}$, which is the trace of $\hat{A}\hat{\rho}$:

$$\bar{A}(t) = \sum_n \langle\psi_n(t)|\hat{A}\hat{\rho}|\psi_n(t)\rangle = \mathrm{Tr}[\hat{A}\hat{\rho}]. \tag{10.48}$$

It is easily verified that

$$\mathrm{Tr}[\hat{\rho}\hat{A}] = \mathrm{Tr}[\hat{A}\hat{\rho}] \tag{10.49}$$

and, on account of the normalization of probabilities, that

$$\mathrm{Tr}[\hat{\rho}] = 1. \tag{10.50}$$

The density operator behaves rather differently from the operators which represent observable quantities. Because it is constructed from state vectors which represent possible histories of the system, it is time dependent in the Schrödinger picture and time independent in the Heisenberg picture. In the Schrödinger picture, we can use (5.31) with (5.32) to obtain the equation of motion

$$\frac{\mathrm{d}}{\mathrm{d}t}\hat{\rho}(t) = \frac{\mathrm{i}}{\hbar}[\hat{\rho}(t), \hat{H}] \tag{10.51}$$

which is the quantum-mechanical version of the Liouville equation (10.4). It differs by a minus sign from the equation of motion (5.35) for time-dependent operators which represent observables in the Heisenberg picture.

The arguments we used to derive the ensembles of classical statistical mechanics can be taken over to the quantum case. To describe thermal equilibrium, we want the density operator to be time independent in the Schrödinger picture. According to (10.51) it must therefore be constructed from operators which commute with the Hamiltonian, including the Hamiltonian itself. For a system of N particles, confined to a volume V, we obtain the canonical density operator as

$$\hat{\rho}_{\text{can}} = Z_{\text{can}}^{-1}\exp(-\beta\hat{H}_N) \tag{10.52}$$

where the partition function is given by

$$Z_{\text{can}}(\beta, V, N) = \operatorname{Tr}\exp(-\beta\hat{H}_N). \tag{10.53}$$

No factor of h^{-3N} is required because this expression is already dimensionless, and no factor of $1/N!$ because the indistinguishability of identical particles is taken into account in the definition of the quantum states. The grand partition function may be defined by analogy with (10.25) as

$$Z_{\text{gr}}(\beta, V, \mu) = \sum_N \exp(\beta\mu N) Z_{\text{can}}(\beta, V, N). \tag{10.54}$$

Alternatively, we can resort to second quantization and define the grand canonical density operator and partition function by

$$\hat{\rho}_{\text{gr}} = Z_{\text{gr}}^{-1}\exp[-\beta(\hat{H} - \mu\hat{N})] \tag{10.55}$$

$$Z_{\text{gr}}(\beta, V, \mu) = \operatorname{Tr}\exp[-\beta(\hat{H} - \mu\hat{N})]. \tag{10.56}$$

Here, of course, the trace includes states with any number of particles. When the number of particles is not conserved, it makes no sense to speak of a fixed number. Moreover, $\hat{N}$ does not then commute with the Hamiltonian and cannot appear in the equilibrium density operator. In that case, we must use (10.55) and (10.56) with $\mu = 0$. It is a matter of taste whether this is regarded as a grand canonical description of a system of particles or, on the other hand, as a canonical description of the underlying system of quantum fields.

Quantum-mechanical ideal gases are most conveniently treated in the grand canonical ensemble. Since the particles do not interact, eigenstates of the operator $\hat{H} - \mu\hat{N}$ can be built from single-particle energy eigenstates. If we consider a gas confined to a cubical box of side L, the single-particle momentum eigenstates have momenta

$$\boldsymbol{p} = \left(\frac{h}{L}\right)\boldsymbol{i} \tag{10.57}$$

where $\boldsymbol{i}$ is a triplet of integers, each of which can have any positive or negative value. For each momentum value, with single-particle energy $\varepsilon_i = p_i^2/2m$, there are $(2s + 1)$ independent spin polarization states for particles of spin s. We now take the states $|\psi_n\rangle$ to be the basis states of the occupation number representation, with $n_{i\sigma}$ particles in the state with momentum labelled by $\boldsymbol{i}$ and spin polarization σ. The grand partition function is

$$Z_{\text{gr}} = \sum_{\{n_{i\sigma}\}} \exp\left[-\beta \sum_{i\sigma} (\varepsilon_i - \mu)n_{i\sigma}\right] = \sum_{\{n_{i\sigma}\}} \prod_{i\sigma} \exp[-\beta(\varepsilon_i - \mu)n_{i\sigma}]. \tag{10.58}$$

For bosons, each $n_{i\sigma}$ ranges from 0 to ∞, while for fermions, it takes only the values 0 or 1. In either case, all the sums can be carried out, giving

$$Z_{\text{gr}} = \prod_i \{1 \pm \exp[-\beta(\varepsilon_i - \mu)]\}^{\pm(2s+1)} \tag{10.59}$$

where the upper signs refer to fermions and the lower ones to bosons. The average occupation numbers of single-particle momentum states are easily found:

$$\bar{n}_i = \sum_\sigma \bar{n}_{i\sigma} = -\frac{\partial}{\partial(\beta\varepsilon_i)} \ln Z_{\text{gr}} = (2s + 1)\{\exp[\beta(\varepsilon_i - \mu)] \pm 1\}^{-1}. \tag{10.60}$$

Under all circumstances of practical interest, sums over momentum states can be replaced by integrals, and (10.57) leads to the replacement $\Sigma_i \to (V/h^3)\int \mathrm{d}^3p$, where $V = L^3$ is the volume. The energy becomes $\varepsilon = p^2/2m$ and, since this depends only on the magnitude of $\boldsymbol{p}$, the angular integrals over the direction of $\boldsymbol{p}$ can be carried out. After defining $x = (\beta p^2/2m)^{1/2}$, we find for the logarithm of the partition function

$$\ln Z_{\text{gr}} = \pm 4\pi V(2s + 1)\left(\frac{2m}{\beta h^2}\right)^{3/2} \int_0^\infty \mathrm{d}x\, x^2 \ln(1 \pm z\mathrm{e}^{-x^2}) \tag{10.61}$$

and for the average number of particles per unit volume

$$\frac{\bar{N}}{V} = 4\pi(2s + 1)\left(\frac{2m}{\beta h^2}\right)^{3/2} z \int_0^\infty \mathrm{d}x\, x^2\mathrm{e}^{-x^2}(1 \pm z\mathrm{e}^{-x^2})^{-1} \tag{10.62}$$

where z is the fugacity (10.24). At low temperatures, quantum ideal gases behave very differently from classical ones. The case of bosons will be discussed in the next chapter. The case of fermions, which I shall not discuss, is particularly important when applied to the gas of electrons in a metal and is dealt with extensively in most textbooks of solid state physics. At high temperatures, on the other hand, quantum gases differ very little from classical ones. From (10.62), we see that if β becomes very small with $\bar{N}/V$ fixed, the fugacity must also become small. In that case, (10.61) can be approximated as

$$\ln Z_{\mathrm{gr}} \approx 4\pi V(2s+1)\left(\frac{2m}{\beta h^2}\right)^{3/2} z\int_0^\infty \mathrm{d}x\, x^2 \mathrm{e}^{-x^2} = (2s+1)zV\left(\frac{2\pi m}{\beta h^2}\right)^{3/2}. \tag{10.63}$$

This agrees exactly with (10.26), apart from the spin multiplicity factor $(2s+1)$. For spin-0 particles, which can be compared most directly with their classical counterparts, this factor is 1. For particles with higher spin, the familiar relations $U = 3Nk_\mathrm{B}T/2$ and $pV = Nk_\mathrm{B}T$ are unaffected.

10.6 Field Theories at Finite Temperature

Although we have found it possible to treat ideal quantum gases without any detailed use of second quantization, field-theoretic methods are more or less essential for the systematic study of large systems of interacting particles. We have seen, moreover, that relativistic particles can be correctly described only by a field theory. It is therefore necessary to find methods of evaluating quantities of the form (10.49) or (10.56) when $\hat{H}$ and $\hat{N}$ are second-quantized operators. A useful technique comes about from realizing that each of the matrix elements in the trace in (10.56) is analogous to that on the left-hand side of (9.23), if we replace $\hat{H}$ by $\hat{H} - \mu\hat{N}$ and $t_\mathrm{f} - t_\mathrm{i}$ by $-\mathrm{i}\beta$. This leads to the *imaginary time formalism*, in which the diagrammatic perturbation theory we discussed in chapter 9 can be taken over more or less intact, simply by replacing real time t with an imaginary time $\tau = \mathrm{i}t$. This imaginary time takes values between 0 and β. Here, I shall discuss only the case of a relativistic scalar field ϕ, but other relativistic and non-relativistic field theories can be treated by similar methods.

Since we are considering a many-particle system in thermal equilibrium, its rest frame is a preferred frame of reference. Therefore, even in a relativistic theory, there is a preferred measure of time, namely that measured in the rest frame, which provides a natural means of distinguishing Heisenberg and Schrödinger pictures. For simplicity, I shall take the chemical potential to be zero. If $\phi(\boldsymbol{x})$ is the Schrödinger-picture field operator, then we define the *imaginary time Heisenberg picture* by

$$\hat{\phi}(\boldsymbol{x}, \tau) = \mathrm{e}^{\hat{H}\tau}\hat{\phi}(\boldsymbol{x})\mathrm{e}^{-\hat{H}\tau} \quad \text{and} \quad \hat{\phi}^{\dagger}(\boldsymbol{x}, \tau) = \mathrm{e}^{\hat{H}\tau}\hat{\phi}^{\dagger}(\boldsymbol{x})\mathrm{e}^{-\hat{H}\tau}. \tag{10.64}$$

It should be noticed that $\hat{\phi}^{\dagger}(\boldsymbol{x}, \tau)$ is *not* the adjoint of $\hat{\phi}(\boldsymbol{x}, \tau)$ in the usual sense. By analogy with (9.48), we define an imaginary time propagator by

$$G(\boldsymbol{x} - \boldsymbol{x}', \tau - \tau') = \mathrm{Tr}\,\{\hat{\rho}T_{\tau}[\hat{\phi}(\boldsymbol{x}, \tau)\hat{\phi}^{\dagger}(\boldsymbol{x}', \tau')]\} \tag{10.65}$$

where T_{τ} is the latest-on-the-left ordering operator for imaginary times. This propagator will indeed depend only on $\boldsymbol{x} - \boldsymbol{x}'$ if the equilibrum state is homogeneous, as intuitively it must be. By using the identity $\mathrm{Tr}\,(\hat{A}\hat{B}) = \mathrm{Tr}\,(\hat{B}\hat{A})$, valid for any $\hat{A}$ and $\hat{B}$, it is easy to show that it depends only on $\tau - \tau'$.

The same identity may be used to derive a vital property of the propagator, namely that it is periodic in $\tau - \tau'$, with period β:

$$G(\boldsymbol{x} - \boldsymbol{x}', \tau - \tau' + \beta) = G(\boldsymbol{x} - \boldsymbol{x}', \tau - \tau'). \tag{10.66}$$

Since τ and τ' are both between 0 and β, their difference is between $-\beta$ and β. Therefore, (10.66) is meaningful only when $\tau < \tau'$. On the other hand, $\tau + \beta$ must be greater than τ', so we have

$$\begin{aligned} G(\boldsymbol{x} - \boldsymbol{x}', \tau - \tau' + \beta) &= Z_{\mathrm{gr}}^{-1}\,\mathrm{Tr}\,[\mathrm{e}^{-\beta\hat{H}}\mathrm{e}^{(\tau+\beta)\hat{H}}\hat{\phi}(\boldsymbol{x})\mathrm{e}^{-(\tau+\beta)\hat{H}}\mathrm{e}^{\tau'\hat{H}}\hat{\phi}^{\dagger}(\boldsymbol{x}')\mathrm{e}^{-\tau'\hat{H}}] \\ &= Z_{\mathrm{gr}}^{-1}\,\mathrm{Tr}\,[\mathrm{e}^{\tau\hat{H}}\hat{\phi}(\boldsymbol{x})\mathrm{e}^{-\tau\hat{H}}\mathrm{e}^{-\beta\hat{H}}\mathrm{e}^{\tau'\hat{H}}\hat{\phi}^{\dagger}(\boldsymbol{x}')\mathrm{e}^{-\tau'\hat{H}}] \\ &= Z_{\mathrm{gr}}^{-1}\,\mathrm{Tr}\,[\mathrm{e}^{-\beta\hat{H}}\mathrm{e}^{\tau'\hat{H}}\hat{\phi}^{\dagger}(\boldsymbol{x}')\mathrm{e}^{-\tau'\hat{H}}\mathrm{e}^{\tau\hat{H}}\hat{\phi}(\boldsymbol{x})\mathrm{e}^{-\tau\hat{H}}]. \end{aligned} \tag{10.67}$$

For $\tau < \tau'$, this is indeed equal to $G(\boldsymbol{x} - \boldsymbol{x}', \tau - \tau')$. For $\tau > \tau'$, the corresponding relation

$$G(\boldsymbol{x} - \boldsymbol{x}', \tau - \tau' - \beta) = G(\boldsymbol{x} - \boldsymbol{x}', \tau - \tau') \tag{10.68}$$

can be established in the same way. In the case of fermions, the propagator is antiperiodic, which means $G(\boldsymbol{x} - \boldsymbol{x}', \tau - \tau' \pm \beta) = -G(\boldsymbol{x} - \boldsymbol{x}', \tau - \tau')$.

The expectation value of any operator constructed from the fields can, in principle, be calculated from the propagator or from other imaginary time Green functions. For example, to obtain the expectation value of $\hat{\phi}^{\dagger}(\boldsymbol{x})\hat{\phi}(\boldsymbol{x})$, we would use

$$\begin{aligned} \langle\hat{\phi}^{\dagger}(\boldsymbol{x})\hat{\phi}(\boldsymbol{x})\rangle = \mathrm{Tr}\,[\hat{\rho}\hat{\phi}^{\dagger}(\boldsymbol{x})\hat{\phi}(\boldsymbol{x})] &= \lim_{\varepsilon\to 0}\mathrm{Tr}\,\{\hat{\rho}T_{\tau}[\hat{\phi}(\boldsymbol{x}, \tau)\hat{\phi}^{\dagger}(\boldsymbol{x}, \tau + \varepsilon)]\} \\ &= \lim_{\varepsilon\to 0} G(\mathbf{0}, -\varepsilon). \end{aligned} \tag{10.69}$$

The Green functions in turn can be represented by functional integrals of the form (9.30), except that these must be converted to imaginary time. The result, derived by a similar method to that of §9.3, is

$$\mathrm{Tr}\{\hat{\rho}T_\tau[\hat{\phi}(x_1)\ldots\hat{\phi}^\dagger(x_n)]\}$$
$$= Z_{\mathrm{gr}}^{-1}\int \mathcal{D}\phi(x)\phi(x_1)\ldots\phi^*(x_n)\exp[-S_\beta(\phi)] \tag{10.70}$$

where $\phi(x)$ means $\phi(\boldsymbol{x}, \tau)$ and the symbol $\mathcal{D}\phi$ includes a normalizing factor to make (10.50) true. The finite temperature action S_β is found by replacing t with $-\mathrm{i}\tau$. For the self-interacting scalar field we studied in chapter 9, it is given by

$$S_\beta(\phi) = \int_0^\beta \mathrm{d}\tau \int \mathrm{d}^3x\left(\frac{\partial\phi^*}{\partial\tau}\frac{\partial\phi}{\partial\tau} + \boldsymbol{\nabla}\phi^*\cdot\boldsymbol{\nabla}\phi + m^2\phi^*\phi + \frac{\lambda}{4}(\phi^*\phi)^2\right). \tag{10.71}$$

Proceeding as in chapter 9, we find that the unperturbed propagator $G_0(\boldsymbol{x} - \boldsymbol{x}', \tau - \tau')$ satisfies the equation

$$\left(\frac{\partial^2}{\partial\tau^2} + \nabla^2 - m^2\right)G_0(\boldsymbol{x} - \boldsymbol{x}', \tau - \tau') = -\delta(\tau - \tau')\delta(\boldsymbol{x} - \boldsymbol{x}'). \tag{10.72}$$

Because of the periodicity in imaginary time, we express it in terms of a Fourier transform as

$$G_0(\boldsymbol{x} - \boldsymbol{x}', \tau - \tau')$$
$$= \int \frac{\mathrm{d}^3k}{(2\pi)^3}\exp[\mathrm{i}\boldsymbol{k}\cdot(\boldsymbol{x} - \boldsymbol{x}')]\beta^{-1}\sum_{n=-\infty}^{\infty}\exp[\mathrm{i}(2\pi n/\beta)(\tau - \tau')]\widetilde{G}_0(\boldsymbol{k}, n). \tag{10.73}$$

The frequencies $\omega_n = 2\pi n/\beta$ are known as *Matsubara frequencies*. On substituting in (10.72), we find

$$\widetilde{G}_0(\boldsymbol{k}, n) = [k^2 + \omega_n^2 + m^2]^{-1}. \tag{10.74}$$

To see how the finite temperature field theory fits in with our earlier discussion of quantum gases, let us evaluate $\ln Z_{\mathrm{gr}}$ for the case of an ideal relativistic gas, with $\lambda = 0$. According to (10.50) and (10.70), the partition function is given by

$$Z_{\mathrm{gr}} = \int \mathcal{D}\phi(x)\exp[-S_\beta(\phi)]. \tag{10.75}$$

This, however, is slightly ambiguous because of an ill-defined constant which appears in the definition of the functional integral (see (9.27), for example). To avoid this difficulty, we can calculate the quantity $-\partial\ln Z_{\mathrm{gr}}/\partial m^2$ which, according to (10.71), is given by

$$-\frac{\partial}{\partial m^2}\ln Z_{\mathrm{gr}} = \int_0^\beta \mathrm{d}\tau \int \mathrm{d}^3x \langle \phi^*(\boldsymbol{x}, \tau)\phi(\boldsymbol{x}, \tau)\rangle = \int_0^\beta \mathrm{d}\tau \int \mathrm{d}^3x G_0(\boldsymbol{0}, 0). \tag{10.76}$$

Since $G_0(\boldsymbol{0}, 0)$ is independent of $\boldsymbol{x}$ and τ, the two integrals just give an overall factor of $V\beta$.

To evaluate $G_0(\boldsymbol{0}, 0)$ itself, we use the identity

$$\sum_{n=-\infty}^{\infty} \frac{1}{n^2 + a^2} = \frac{\pi}{a}\left(\frac{\mathrm{e}^{\pi a} + \mathrm{e}^{-\pi a}}{\mathrm{e}^{\pi a} - \mathrm{e}^{-\pi a}}\right) = \frac{\pi}{a}\coth(\pi a) \tag{10.77}$$

which readers are invited to prove in exercise 10.6. We obtain

$$\begin{aligned} -\frac{\partial}{\partial m^2}\ln Z_{\mathrm{gr}} &= V \int \frac{\mathrm{d}^3k}{(2\pi)^3}\frac{\beta}{2\omega(\boldsymbol{k})}\coth\left(\tfrac{1}{2}\beta\omega(\boldsymbol{k})\right) \\ &= -\frac{\partial}{\partial m^2}\left(2V\int\frac{\mathrm{d}^3k}{(2\pi)^3}\ln\{\exp[\beta\omega(\boldsymbol{k})/2] \right. \\ &\quad \left. - \exp[-\beta\omega(\boldsymbol{k})/2]\}\right). \end{aligned} \tag{10.78}$$

Up to a possible constant of integration, this gives

$$\begin{aligned} -\ln Z_{\mathrm{gr}} &= 2V\frac{1}{2\pi^2\beta^3}\int_0^\infty \mathrm{d}x\, x^2 \ln\{1 - \exp[-(x^2 + \beta^2 m^2)^{1/2}]\} \\ &\quad + V\beta\int\frac{\mathrm{d}^3k}{(2\pi)^3}\,\omega(\boldsymbol{k}). \end{aligned} \tag{10.79}$$

Remembering that the internal energy is $U = -\partial \ln Z_{\mathrm{gr}}/\partial\beta$, we recognize the last term as the infinite vacuum energy encountered in (7.19), as long as we identify $(2\pi)^3\delta(0) = \int \mathrm{d}^3x = V$. The first term, in which $x = \beta|\boldsymbol{k}|$, is obviously similar to (10.61) with $z = 1$. The field theory describes particles of spin $s = 0$, and the overall factor of 2 represents the two equal contributions from particles and antiparticles. Other differences arise from the relativistic energy relation $\omega(\boldsymbol{k}) = (k^2 + m^2)^{1/2}$ and the use of natural units, in which $h = 2\pi$. The non-relativistic limit of (10.79) is explored in exercise 10.7.

10.7 Black-body Radiation

Black-body radiation is most simply conceived of as an ideal gas of photons in thermal equilibrium with the walls of a cavity which contains it. According to quantum electrodynamics, photons can scatter from each other by way of intermediate states containing virtual charged particles. Under almost all circumstances, however, this interaction is entirely negligible. Because photons are massless, there is no lower limit to the energy change involved in the emission or absorption of a photon by the cavity walls. There is therefore no constraint on the total number of photons in the gas and its chemical potential is zero. It is possible to derive the partition function from QED, but problems are again encountered with redundant gauge degrees of freedom. In particular, treatment of the component A_0 of the vector potential in the imaginary time formalism requires careful consideration. For want of space, I shall not discuss these questions in detail. It should come as no surprise to readers that we obtain the correct result simply by setting $m = 0$ in (10.79). Since photons are their own antiparticles, the overall factor of 2 arises in this case from the two independent spin polarization states.

At very high temperatures, such as we shall later encounter in connection with the early universe, a modified version of black-body radiation arises, in which any particle species whose mass is much smaller than $k_B T$ can be considered effectively massless and treated on the same footing as photons. As long as an ideal gas description remains appropriate, we simply add the contributions to $\ln Z_{gr}$ from each species. If we drop the unobservable vacuum energy, then, for each bosonic species, the contribution is

$$-\ln Z_{gr} = g_b V \frac{1}{2\pi^2 \beta^3} I_b \tag{10.80}$$

where g_b is the number of independent spin polarization states of particles and antiparticles and

$$I_b = \int_0^\infty dx\, x^2 \ln(1 - e^{-x}) = -\frac{\pi^4}{45}. \tag{10.81}$$

For fermions, this integral is modified in the same way as that in (10.61). It is given by

$$I_f = -\int_0^\infty dx\, x^2 \ln(1 + e^{-x}) = \tfrac{7}{8} I_b. \tag{10.82}$$

In view of this relation (which is readily verified by showing that $I_b - I_f = I_b/8$), we can treat the gas as a whole by defining

$$g = \sum_{\text{boson species}} g_b + \tfrac{7}{8} \sum_{\text{fermion species}} g_f. \tag{10.83}$$

To return to laboratory units, we must divide $\ln Z_{gr}$ by $(\hbar c)^3$ to make it dimensionless. We then have

$$\Omega = -k_B T \ln Z_{gr} = -V \frac{2g\sigma}{3c} T^4 \tag{10.84}$$

where

$$\sigma = \frac{\pi^2 k_B^4}{60 \hbar^3 c^2} = 5.6698 \times 10^{-8}\ \mathrm{W\,m^{-2}\,K^{-4}} \tag{10.85}$$

is the Stefan–Boltzmann constant. It is a simple matter to derive the following expressions for the energy and entropy densities and the pressure:

$$\frac{U}{V} = -\frac{1}{V}\frac{\partial \ln Z_{\mathrm{gr}}}{\partial \beta} = \frac{2g\sigma}{c}T^4 \tag{10.86}$$

$$\frac{S}{V} = -\frac{1}{V}\frac{\partial \Omega}{\partial T} = \frac{8g\sigma}{3c}T^3 \tag{10.87}$$

$$p = -\frac{\partial \Omega}{\partial V} = \frac{2g\sigma}{3c}T^4 = \frac{1}{3}\frac{U}{V}. \tag{10.88}$$

10.8 The Classical Lattice Gas

Our explicit examples have so far been restricted to ideal gases, because the approximation methods needed to treat non-ideal gases and liquids require quite lengthy development, for which there is no space in this book. I shall, however, describe a straightforward, if somewhat crude, approximation to a non-ideal classical gas, which is of some importance in the theory of phase transitions. This is the *lattice gas*. We consider a gas whose molecules interact through a pair potential $W(r)$, so the Hamiltonian for N molecules is

$$H_N = \sum_{i=1}^{N}\frac{1}{2m}\boldsymbol{p}_i^2 + \tfrac{1}{2}\sum_{i,j=1}^{N} W(|\boldsymbol{x}_i - \boldsymbol{x}_j|). \tag{10.89}$$

Inserting this into (10.14), we find that the momentum integrals can be carried out, so the canonical partition function is

$$Z_{\mathrm{can}}(\beta, V, N) = \left(\frac{2\pi m}{\beta h^2}\right)^{3N/2}\frac{1}{N!}\int \mathrm{d}^{3N}x \exp\left(-\tfrac{1}{2}\beta\sum_{i,j=1}^{N} W(|\boldsymbol{x}_i - \boldsymbol{x}_j|)\right). \tag{10.90}$$

The remaining integral is a sum over all instantaneous configurations of the positions of the molecules, and there is some advantage to re-expressing this sum in the following approximate manner. Real molecules exhibit a strong repulsion at short distances, so it makes sense to divide the total volume occupied by the gas into a large number of cells, each with a volume v comparable with the volume of a single molecule, and to suppose that there can be at most one molecule in any one cell. The mid-points of the cells will usually be taken to form a regular lattice in space. To the ith cell, we assign an occupation number n_i, which is either 0 or 1, and the sum of these over all cells is N. We now take the potential energy of a pair of molecules to depend only upon the cells occupied by the molecules, but not upon their precise locations within the cells, which will be a reasonable approximation for potentials which vary little over the size of a cell. For a given set of N occupied cells, the integral in (10.90) now gives v^N for each of the $N!$ distributions of N molecules in the N cells. By summing over all such sets, we obtain

$$Z_{\mathrm{can}}(\beta, V, N) = \left(\frac{2\pi m}{\beta h^2}\right)^{3N/2} v^N \sum_{\{n_i\}}{}^{(N)} \exp\left(-\tfrac{1}{2}\beta \sum_{i,j} W_{ij} n_i n_j\right) \quad (10.91)$$

where i and j now label cells and W_{ij} is the potential between particles in cells i and j when these cells are occupied. The configuration sum is over all sets of values $n_i = 0, 1$ consistent with their sum being N.

For reasons I shall explain below, it is convenient to write $n_i = (1 + s_i)/2$, where the new variables s_i take the values ± 1. Also, if interactions are appreciable only over distances much shorter than the size of the whole system, then we can write

$$\sum_j W_{ij} = \sum_j W_{ji} = W_0 \quad (10.92)$$

where W_0 is independent of the location of cell i. This will be true except for cells close to the edge of the system, which can be neglected in the thermodynamic limit. We now use (10.25) to obtain the grand canonical partition function. It is

$$Z_{\mathrm{gr}}(\beta, V, \mu) = \exp\left[-\left(\frac{\beta V}{2v}\right)(\bar{\mu} + \tfrac{1}{4} W_0)\right] \sum_{\{s_i\}} \exp \beta \left(\tfrac{1}{2}\bar{\mu} \sum_i s_i - \tfrac{1}{8} \sum_{i,j} W_{ij} s_i s_j\right) \quad (10.93)$$

where the modified chemical potential is given by

$$\beta\bar{\mu} = \beta\mu + \ln\left[\left(\frac{2\pi m}{\beta h^2}\right)^{3/2} v\right] - \tfrac{1}{2}\beta W_0. \quad (10.94)$$

The special value of this result is that, apart from the leading factor, it has the same form as the partition function of a well known model for ferromagnetism, the *Ising model*, which we shall encounter in the next chapter. In that model, the s variables represent atomic spins (or magnetic dipole moments) situated at the sites of a crystal lattice. That this analogy between a ferromagnet and an imperfect gas can be made is, as we shall see, both theoretically important and experimentally well verified.

10.9 Analogies Between Field Theory and Statistical Mechanics

Since both quantum mechanics and statistical mechanics require us to calculate suitably weighted averages of physical quantities, it is not too surprising that formal analogies can be made between them. Under appropriate circumstances, however, these analogies can be closer than we might have expected, and it is interesting to see how they work out.

Consider first of all the imaginary time action (10.71) for a scalar field theory at finite temperature. In the integrand, the imaginary time variable appears on an equal footing with spatial coordinates so, in effect, $\phi(\boldsymbol{x}, \tau)$ lives in a $(d + 1)$-dimensional Euclidean space, d being the original number of spatial dimensions, which is of finite extent β in the extra dimension. The extra dimension is sometimes regarded as being of quantum-mechanical origin, in the following sense. The Hamiltonian of the scalar field theory may be written as

$$\hat{H} = \int \mathrm{d}^3x[\hat{\Pi}^\dagger\hat{\Pi} + (\boldsymbol{\nabla}\hat{\phi}^\dagger)\cdot(\boldsymbol{\nabla}\hat{\phi}) + m^2\hat{\phi}^\dagger\hat{\phi} + \tfrac{1}{4}\lambda(\hat{\phi}^\dagger\hat{\phi})^2] \qquad (10.95)$$

and may loosely be compared with (10.89) for a classical gas. In the classical case, the momenta can be trivially integrated out leaving, as in (10.90), a configurational integral involving the potential energy part of the Hamiltonian with its original number of dimensions. By contrast, we could regard (10.71) as being obtained from (10.95) by again dropping the momentum term, but now also adding an extra spatial dimension. If (10.95) were to be interpreted not as the Hamiltonian of a quantum field theory but as a classical Hamiltonian (with ϕ being, say, the displacement of a continuous vibrating medium), then the configurational integral would be weighted with the exponential of $-\beta$ times its potential energy part. It would, in other words, be similar to (10.75), but with a factor β in the exponent instead of the integral over an extra dimension. While we must obviously be cautious when arguing in this way, it is frequently true that the properties of a quantum-mechanical system in d spatial dimensions can be related to those of a $(d + 1)$-dimensional classical system.

There is clearly also an analogy between the configurational integrals of classical statistical mechanics and the functional integrals of chapter 9, which represent purely quantum-mechanical expectation values. If, in a functional integral such as (9.30), we make the replacement $t = -\mathrm{i}x_4$, the weight function becomes $\exp(-S_\mathrm{E})$, where the Euclidean action is

$$S_\mathrm{E}(\phi) = \int \mathrm{d}^4x[\boldsymbol{\nabla}\phi^*\cdot\boldsymbol{\nabla}\phi + m^2\phi^*\phi + \tfrac{1}{4}\lambda(\phi^*\phi)^2] \qquad (10.96)$$

the gradient operator now being the four-dimensional Euclidean one. The introduction of an imaginary time here has nothing to do with temperature—the fourth Euclidean dimension being of infinite extent—and is, in fact, equivalent to the Wick rotation we used to evaluate Feynman integrals such as (9.76). The original Lorentz invariance of the action has been replaced by invariance under rotations in four-dimensional Euclidean space. There is clearly a rough correspondence between the Euclidean functional integral and the configurational integrals or sums of classical statistical mechanics, if we make S_E correspond to βW, W being the potential energy.

For a sum like (10.93), this correspondence can be made more precise by a change of variables known as the *Hubbard–Stratonovitch transformation*. The energies W_{ij} arise from attractive forces between gas molecules and are negative if we take them to decay to zero at large distances. If we denote by Γ_{ij} the inverse of the matrix $(-W_{ij}/4)$, we can prove the identity

$$\exp\left[\tfrac{1}{2}\beta\sum_{i,j}(-\tfrac{1}{4}W_{ij})s_i s_j\right] = Q\int_{-\infty}^{\infty}\prod_i \mathrm{d}\phi_i \exp\left(-\frac{1}{2\beta}\sum_{i,j}\Gamma_{ij}\phi_i\phi_j + \sum_i \phi_i s_i\right) \tag{10.97}$$

by completing the square on the right-hand side, that is by making the shift $\phi_i \to \phi_i - (\beta/4)\Sigma_j W_{ij}s_j$. Obviously, Q is the appropriate normalizing factor. Using this identity in (10.93), it becomes easy to carry out the sum over the s:

$$\sum_{s=\pm 1}\exp(\tfrac{1}{2}\beta\bar{\mu} + \phi)s = 2\cosh(\tfrac{1}{2}\beta\bar{\mu} + \phi). \tag{10.98}$$

Thus, the partition function of the lattice gas becomes

$$Z_{\mathrm{gr}} = \bar{Z}\int_{-\infty}^{\infty}\prod_i \mathrm{d}\phi_i \exp\left(-\frac{1}{2\beta}\sum_{i,j}\Gamma_{ij}\phi_i\phi_j + \sum_i \ln\cosh(\tfrac{1}{2}\beta\bar{\mu} + \phi_i)\right) \tag{10.99}$$

where $\bar{Z}$ denotes the collection of normalizing factors we have accumulated. Now, the matrix Γ_{ij} depends on the distance $|\boldsymbol{r}_i - \boldsymbol{r}_j|$ between cells i and j of the lattice gas. Under circumstances I shall discuss in the next chapter, it is permissible to expand its Fourier transform as

$$\Gamma_{ij} = v\int\frac{\mathrm{d}^3k}{(2\pi)^3}\exp[\mathrm{i}\boldsymbol{k}\cdot(\boldsymbol{r}_i - \boldsymbol{r}_j)][\Gamma_0 + \Gamma_1 k^2 + \ldots] \tag{10.100}$$

and keep only the first two terms. If we finally approximate the sum over cells of volume v by an integral, we obtain

$$Z_{\mathrm{gr}} = \bar{Z}\int\mathcal{D}\phi\exp\left\{-\int\mathrm{d}^3x\left[\frac{1}{2}\left(\frac{\Gamma_1}{\beta v}\right)\boldsymbol{\nabla}\phi\cdot\boldsymbol{\nabla}\phi + \frac{1}{2}\left(\frac{\Gamma_0}{\beta v}\right)\phi^2 - \frac{1}{v}\ln\cosh(\tfrac{1}{2}\beta\bar{\mu} + \phi)\right]\right\}. \tag{10.101}$$

The effective Hamiltonian in this expression is very similar to (10.96), except that here ϕ is real and the form of the self-interaction is different. Obviously, we can set the coefficient of $(\boldsymbol{\nabla}\phi)^2$ equal to $\frac{1}{2}$ (as is appropriate for a real field) simply by rescaling ϕ. We can also recover the ϕ^4 form of self-interaction by expanding $\ln\cosh(\beta\bar{\mu}/2 + \phi)$ and discarding higher-order terms. It will transpire in the next chapter that this is indeed a useful and legitimate procedure in the neighbourhood of a *critical point*, where the analogy is most useful. Of course, (10.96) and

(10.101) involve different numbers of spatial dimensions and we shall see that this has important consequences.

Exercises

10.1. Consider a classical one-dimensional harmonic oscillator, with Hamiltonian $H = p^2/2m + m\omega^2x^2/2$. What are the curves of constant energy in its two-dimensional phase space? Show that $\Sigma(E) = 2\pi/\omega$. Show that both the long time average and the microcanonical average of a function $f(x, p)$ are given by

$$\frac{1}{2\pi}\int_0^{2\pi} \mathrm{d}\theta f((2E/m\omega^2)^{1/2}\sin\theta,\ (2mE)^{1/2}\cos\theta).$$

This system is therefore ergodic. By considering the flow of an ensemble of points on the energy surface, show that it is not mixing.

10.2. For an open system, define the fluctuation ΔN in the number of particles by $(\Delta N)^2 = \langle (N - \bar{N})^2 \rangle$. Show that $(\Delta N)^2 = \partial^2 \ln Z_{\mathrm{gr}}/\partial(\beta\mu)^2$. For a classical ideal gas, show that $\Delta N/\bar{N} = \bar{N}^{-1/2}$. In the same way, show that relative fluctuations in the internal energy U are proportional to $\bar{N}^{-1/2}$.

10.3. The partition function for the *pressure ensemble* is

$$Z_{\mathrm{pr}}(\beta, p, N) = \int_0^\infty \mathrm{d}V \mathrm{e}^{-\beta pV} Z_{\mathrm{can}}(\beta, V, N).$$

Calculate this partition function for a classical ideal gas. Suggest an expression, in terms of Z_{pr} and its derivatives, for the mean volume of a system maintained at constant pressure p, and check it by recovering the ideal gas equation of state in the thermodynamic limit. Show that, in the thermodynamic limit, the quantity $G = -k_{\mathrm{B}}T \ln Z_{\mathrm{pr}}$ is the Gibbs free energy, $G = F + pV$.

10.4. The four quantities S, U, V, and N are all extensive. Show that this implies the relation $S(\lambda U, \lambda V, \lambda N) = \lambda S(U, V, N)$, where λ is an arbitrary number. By differentiating with respect to λ and setting $\lambda = 1$, derive the relation $U = TS - pV + \mu N$. Hence show that $\Omega = -pV$ and $G = \mu N$.

10.5. Show that the density operator (10.46) is Hermitian and that the trace in (10.48) does not depend on which complete orthonormal set of states is used to compute it.

10.6. In the complex z plane, let C be the closed contour which runs from $-\infty$ to $+\infty$ just below the real axis and returns to $-\infty$ just above the real axis. Show that, for any sufficiently well behaved function $f(z)$,

$$\lim_{\eta\to 0}\oint_C \mathrm{d}z\,\frac{\mathrm{e}^{\mathrm{i}\eta z}f(z)}{\mathrm{e}^{2\pi\mathrm{i}z}-1} = \sum_{n=-\infty}^{\infty} f(n).$$

Verify (10.77) by choosing $f(z) = (z^2 + a^2)^{-1}$ and deforming the contour in an appropriate manner.

10.7. Consider the field-theoretic partition function (10.79) in the limit that βm is very large. By making the change of variable $x \to (2\beta m)^{1/2}x$, show that (10.79) reduces to the non-relativistic partition function (10.61) for spin-0 particles, with $\hbar = 1$ and a chemical potential $\mu = m$.

10.8. Consider a gas of N hydrogen atoms in a container of volume V, at a temperature sufficiently high for all H_2 molecules to be dissociated and some atoms to be ionized. Using classical non-relativistic statistical mechanics, work out the canonical partition function for $N - \nu$ indistinguishable atoms, ν indistinguishable protons and ν indistinguishable electrons. For each ionized atom, include a potential energy I, equal to the ionization potential. Assume that the masses of a hydrogen atom and a proton are equal. By finding the most probable value of ν, show that the fraction $x = \nu/N$ of ionized atoms is given by the *Saha equation*

$$\frac{x^2}{1-x} = \frac{1}{n}\left(\frac{2\pi m}{\beta h^2}\right)^{3/2} \mathrm{e}^{-\beta I}$$

where m is the electron mass and $n = N/V$. Note that this result depends on h, which is an arbitrary parameter in the classical theory. Why is this? Why would you expect to obtain the correct answer by taking h to be Planck's constant?

10.9. From equations (10.83) and (10.86)–(10.88), it might appear that a fermion simply counts as $\frac{7}{8}$ of a boson as far as black-body radiation is concerned, but this is not so. By direct calculation or informal arguments, satisfy yourself that the number density of particles of species i is given by

$$\frac{N}{V} = \frac{g_i}{2\pi^2\beta^3}\int_0^\infty \mathrm{d}x\, x^2(\mathrm{e}^x \pm 1)^{-1}.$$

Show that the fermionic integral is $\frac{3}{4}$ of the bosonic one. The value of the bosonic integral is $2\zeta(3)$, where ζ is the Riemann zeta function.

11

Phase Transitions

Among the many applications of statistical mechanics, some of the most intriguing and challenging theoretical problems arise in connection with *phase transitions*. These are abrupt changes of state such as occur, for example, when a liquid is transformed into vapour, a ferromagnet loses its magnetization upon heating to its Curie temperature, or at the onset at sufficiently low temperatures of superfluidity or superconductivity. It is within the theory of phase transitions, too, that the mathematical relationships between statistical mechanics and relativistic field theories are most powerful. Indeed, the idea of *spontaneous symmetry breaking*, which lies at the heart of the theory of phase transitions, is the crucial ingredient which turns the gauge theories of chapter 8 into a real working model of the fundamental forces of nature, to be discussed in the next chapter.

It is not possible in the space of a single chapter to cover adequately the wide and diverse range of phenomena which theoretical and experimental ingenuity have uncovered. I shall therefore discuss only a few standard examples and the key theoretical arguments which have been devised to deal with them. In almost all cases, phase transitions can occur only by virtue of interactions between particles. This, indeed, is what gives rise to the greatest technical challenges. The one exception to this rule is the case of *Bose–Einstein condensation* in an ideal Bose gas, which I shall discuss first. The greater part of the chapter will deal with the gas–liquid and ferromagnetic transitions, which illustrate most of the essential theory, and I shall end by describing the Ginzburg–Landau theory of superconductivity, which provides the closest analogy with the gauge theories of particle physics.

11.1. Bose–Einstein Condensation

Consider an ideal gas of spin-0 particles. According to (10.60), the number of particles in the ith momentum state, with momentum given by (10.57) is

$$n_i = z[\exp(\beta\varepsilon_i) - z]^{-1}. \tag{11.1}$$

For a given number of particles per unit volume, the fugacity z is determined implicitly in terms of $\bar{N}/V$ and temperature by an equation of the form (10.62). By its definition (10.24), z is positive. On the other hand, since the occupation numbers (11.1) cannot be negative, z cannot be greater than $\exp(\beta\varepsilon_0)$, where ε_0 is the smallest single-particle energy. For a large volume, we can take this energy to be zero, so $0 < z < 1$, which means that the chemical potential μ must be negative. The interesting question is, what happens as z approaches 1? We see from (11.1) that the occupation number of the zero-energy state can become indefinitely large. In fact, the growth of this number is limited by the total number of particles available, but it can be a significant fraction of the total number. This phenomenon, known as *Bose–Einstein condensation*, is the basic cause of superfluidity and superconductivity.

When the zero-energy state is macroscopically occupied, we have to reconsider equations such as (10.61) and (10.62), where we replaced a sum over momentum states by an integral. This is normally valid because the momentum eigenvalues are very closely spaced, but it assumes that the number of particles per unit volume with momentum in the infinitesimal range d^3p is infinitesimal. This will not be true for the element d^3p which includes the zero-energy state when there is condensation. In fact, the integrals in (10.61) and (10.62) do assign only an infinitesimal number of particles to this element, so we can correct them simply by adding on the contributions of the condensed particles. For the grand potential and particle number per unit volume, we obtain

$$\frac{\Omega}{V} = \frac{1}{\beta V}\ln(1 - z) + 4\pi\beta^{-5/2}\left(\frac{2m}{h^2}\right)^{3/2}\int_0^\infty dx\, x^2 \ln(1 - z e^{-x^2}) \tag{11.2}$$

$$\frac{\bar{N}}{V} = \frac{\bar{n}_0}{V} + 4\pi\left(\frac{2m}{\beta h^2}\right)^{3/2} z\int_0^\infty dx\, x^2\, e^{-x^2}(1 - z e^{-x^2})^{-1} \tag{11.3}$$

where $\bar{n}_0$ is the average number of condensed particles. These equations are to be understood as applying to the thermodynamic limit. When $V \to \infty$, the condensation terms go to zero, unless z is infinitesimally close to 1 and the number of condensed particles per unit volume is finite.

The conditions under which condensation occurs can be investigated as follows. Suppose first that condensation does occur. Then z is infinitesimally close to 1. The second term in (11.3) is proportional to the integral

$$4\pi^{-1/2}\int_0^\infty dx\, x^2(e^{x^2} - 1) = \zeta(\tfrac{3}{2}) = 2.612\ldots \tag{11.4}$$

where ζ is the Riemann zeta function, and we have

$$\frac{\bar{N}}{V} = \frac{\bar{n}_0}{V} + 2.612\left(\frac{2\pi m k_B}{h^2}\right)^{3/2} T^{3/2}. \tag{11.5}$$

For a given number density, there is a *critical temperature* T_c at which the number of condensed particles just vanishes:

$$T_c = \frac{h^2}{2\pi m k_B}\left(\frac{\bar{N}}{2.612V}\right)^{2/3}. \tag{11.6}$$

At lower temperatures than this, $\bar{n}_0$ is a non-zero fraction of $\bar{N}$. At higher temperatures, on the other hand, (11.5) cannot be true, since n_0 cannot be negative. We must then have $n_0 = 0$, so there is no condensation and z must be less than 1.

In the condensed phase (i.e. the low-temperature state in which condensation occurs), it is easy to see that the fraction of particles in the condensate is given by

$$\frac{\bar{n}_0}{\bar{N}} = 1 - \left(\frac{T}{T_c}\right)^{3/2}. \tag{11.7}$$

Under the influnce of an applied force, such as gravity or the attraction of container walls, this condensate moves as a coherent whole and is responsible for the frictionless flow characteristic of superfluid helium. (Helium is the only substance known to exhibit superfluidity. It is not an ideal gas, and intermolecular forces are essential for understanding its properties in detail.) The condensate can be described by a macroscopic wavefunction, ϕ, whose magnitude is proportional to $(\bar{n}_0)^{1/2}$. The temperature dependence of quantities like $|\phi|$ in the immediate neighbourhood of a critical temperature will be a recurring theme. If we expand (11.7) in powers of $T_c - T$, we find

$$|\phi| \sim (T_c - T)^{\beta} \tag{11.8}$$

where β, an example of what are known as *critical exponents*, has the value $\frac{1}{2}$. The symbol $\sim$ indicates both that a constant of proportionality is missing and that this is only the leading behaviour when $T_c - T$ is small.

Another important feature which is common to all phase transitions is that the transition is sharply defined only in the thermodynamic limit. When $V \to \infty$ in (11.3), we can draw a sharp distinction between the condensed phase, in which $\bar{n}_0/V$ has a non-zero limit and the normal phase in which it goes to zero. When the volume is large but finite, there is a narrow range of temperature in which $\bar{n}_0/V$ decreases from being a significant fraction of $\bar{N}/V$ to being extremely small, but no precise dividing line between the two phases. Although experiments deal with finite systems, these systems do occupy a volume which is extremely large compared with average intermolecular distances. Under these circumstances the theoretical ambiguity as to the precise location of a critical temperature may well be much smaller than the resolution in temperature which an experimenter can achieve. Thus, to all intents and purposes, well defined transitions can indeed be seen in practice.

11.2 Critical Points in Fluids and Magnets

Much of the theoretical interest in phase transitions has to do with *critical points*. The exact nature of a critical point will emerge as we study examples, but one essential feature is already apparent from the case of Bose–Einstein condensation. The condensed phase, which exists below T_c, is distinguished from the normal high-temperature phase by a non-zero value of $\bar{n}_0/V$. On approaching the critical temperature, this quantity goes continuously to zero, and so, exactly *at* the critical temperature, the condensed and normal phases are identical. This behaviour is distinctive of critical points, which may also be described as *continuous* or *second-order* phase transitions, the terminology depending somewhat upon its context. Had n_0/V dropped discontinuously to zero at T_c, it would have been possible for distinct condensed and normal phases to coexist with each other at T_c, which is characteristic of a *first-order* phase transition. A classification of phase transitions due to P Ehrenfest defines a phase transition to be of nth order if an nth derivative of the appropriate thermodynamic potential is discontinuous, while all its $(n - 1)$th derivatives are continuous. If we introduce a separate chemical potential μ_0 for particles in the zero-energy state, then $\bar{n}_0$ is the first derivative of Ω with respect to μ_0. It is continuous at T_c, but $\partial n_0/\partial T$ is not, so the condensation is indeed second order according to this classification. However, the singularities found at phase transitions are often more complicated than simple discontinuities, so the general classification scheme has fallen out of common use.

Two, standard, easily studied examples of critical points are those which occur in simple fluids and in ferromagnets, and I shall deal first with ferromagnetism. As readers are no doubt aware, a permanently magnetized sample of, say, iron typically contains a number of domains, the directions of magnetization being different in neighbouring domains. The physical factors which control the size of a domain have no direct bearing on the phase transitions we are discussing, and I shall simplify matters by assuming that the magnetization of the sample is completely uniform. In practice, our considerations will apply to the interior of a single domain. The magnetization M_S which exists in the absence of any applied magnetic field is called the *spontaneous magnetization* and its magnitude depends on temperature in the manner sketched in figure 11.1. Upon heating to the critical or Curie temperature T_c, the spontaneous magnetization vanishes continuously. In the immediate neighbourhood of T_c, called the *critical region*, we find

$$M_S \sim (T_c - T)^{\beta}. \tag{11.9}$$

The exponent β varies rather little from one ferromagnetic material to another and is typically about $\frac{1}{3}$.

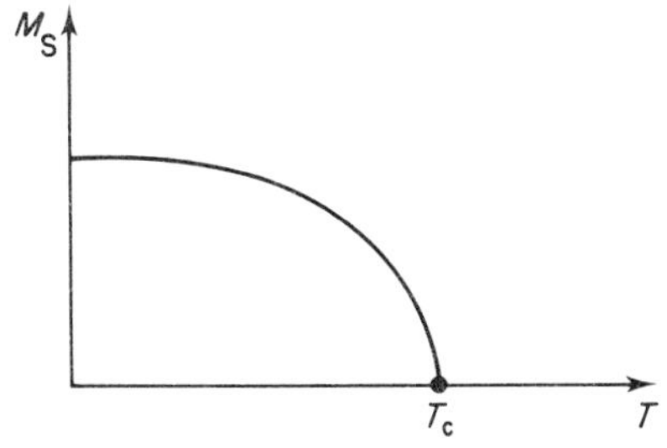

Figure 11.1 The spontaneous magnetization of a ferromagnet as a function of temperature.

The direction in which the spontaneous magnetization points usually lies along one of several *easy axes*, defined by the crystal structure of the material. For simplicitly, I shall consider only *uniaxial* materials in which there is only one easy axis. Then the magnetization can point in one of two opposite directions along this axis. Consider what happens when a magnetic field H is applied in a direction parallel to the easy axis. To be specific, let us suppose that the magnet is heated to a temperature above T_c, at which point a large magnetic field is applied. There will be a magnetization parallel to H, which would decrease to zero if the field were removed. If, in the presence of H, the magnet is cooled to below T_c, the magnetization continues to be parallel to H, but reduces to M_S when the field is removed. If the process is repeated, but with the direction of H reversed, we end up with a magnetization of magnitude M_S pointing in the opposite direction.

Ideally, given a temperature T and magnetic field H, there is a magnetization $M(T, H)$ which has a unique value, except along the line $H = 0$ for $T < T_c$, where the limit as $H \to 0$ of $M(T, H)$ is either $M_S(T)$ if H is positive, or $-M_S(T)$ if H is negative. We may therefore draw an idealized *phase diagram* as in figure 11.2. As far as the line $H = 0$ is concerned, we can identify three different phases, namely two ferromagnetic phases, distinguished by oppositely directed magnetizations, which exist below T_c, and the *paramagnetic* phase, with $M = 0$, above T_c. At the critical point $(T, H) = (T_c, 0)$, the two ferromagnetic phases become identical and also indistinguishable from the paramagnetic phase. The line $H = 0$, $T < T_c$ is a line of two-phase coexistence, where oppositely magnetized domains can coexist in the same sample. Ideally, it is a line of first-order phase transitions, since the magnetization changes discontinuously from M_S to $-M_S$ as the magnetic field decreases through zero. We see, however, that any two states, say A and B in figure 11.2, can be connected by a path along which no phase transition occurs. The essential definition of a critical point is that it marks the end of a line where two or more phases coexist, and that these phases become identical in a continuous manner. In practice, the way in which the magnetization of a sample varies with temperature and magnetic field is more complicated and it is necessary to consider, for example, the motion of domain walls, which gives rise to hysteresis. The actual magnetization of a sample is not given by a single-valued function, but depends on its history. Nevertheless, the function $M(T, H)$ can be found by careful experimental procedures and it is this function which we hope to be able to calculate, at least approximately, from equilibrium statistical mechanics.

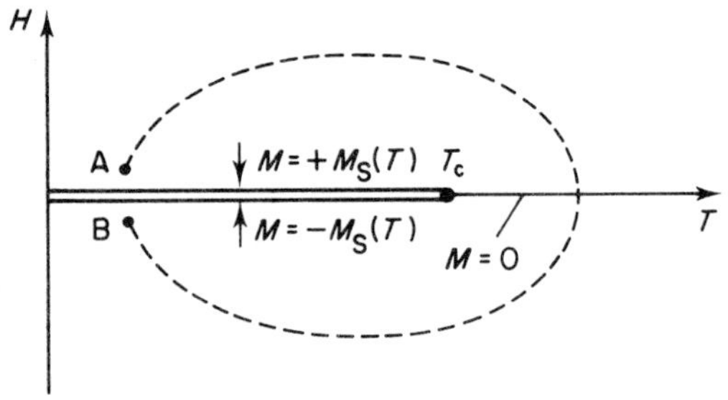

Figure 11.2 Phase diagram of a ferromagnet in the temperature–magnetic field plane. Below T_c, the magnetization is discontinuous at $H = 0$. However, by varying H and T along the broken curve, we can pass from state A to state B without encountering a phase transition.

The magnetic susceptibility is defined by

$$\chi = \partial M/\partial H. \tag{11.10}$$

The zero-field susceptibility, sometimes called the *initial susceptibility*, is found to diverge at the critical point. That is, it becomes infinite, and it does so as a power of $|T - T_c|$:

$$\chi(T, 0) \approx \chi_0 \, |T - T_c|^{-\gamma}. \tag{11.11}$$

The critical exponent γ, which has similar values of about 1.3 for all ferromagnets, is found to be the same whether the critical temperature is approached from above or below, but the amplitude χ_0 may be different in the two cases.

The behaviour of simple fluids is quite analogous to that of ferromagnets. Figure 11.3 represents the vapour pressure curve $p = p_v(T)$, which ends at a critical point (T_c, p_c). By speaking of a 'simple' fluid, I mean that additional complications, such as the possibility of solidification, will be ignored. Although most real substances have more complicated phase diagrams than that shown in figure 11.3, these complications do not affect the critical properties we are discussing. Along the vapour pressure curve, the liquid and vapour phases of the same substance can coexist in the same container. By varying the pressure at a fixed temperature below T_c, we can transform liquid into vapour or vice versa. This is a first-order transition, because the density ρ changes discontinuously. If we plot the densities of the liquid and vapour, both measured at the vapour pressure, as functions of temperature, the result is that sketched in figure 11.4. It is obviously analogous to the spontaneous magnetization curve of figure 11.1, if we include the oppositely directed magnetization, except that it is not symmetrical. Near the critical point, the difference in density between the liquid and vapour is found to vary as

$$\rho_l - \rho_v \sim (T_c - T)^\beta. \tag{11.12}$$

Measured values of the exponent β are very similar for all fluids. Remarkably, they are also very similar to the values obtained for ferromagnets, being in the neigbourhood of $\frac{1}{3}$. Indeed, it is found that all *critical phenomena*, that is the properties of systems in the neighbourhoods of their critical points, are substantially independent of the detailed microscopic constitution of the system considered. This *universality* of critical phenomena is, of course, one of the principal features we should like to understand theoretically.

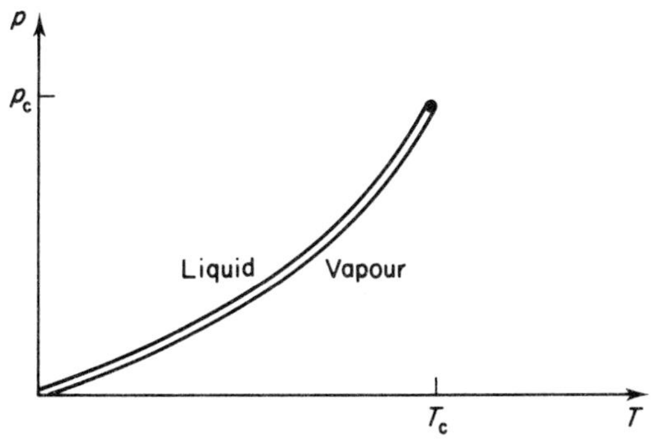

Figure 11.3 Phase diagram of a simple fluid in the temperature–pressure plane.

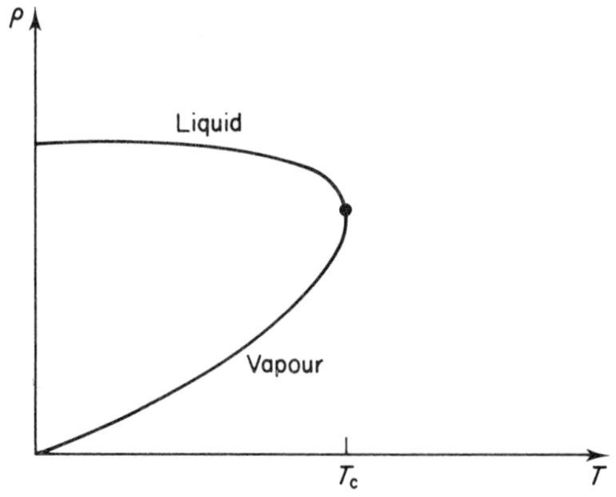

Figure 11.4 Variation with temperature of the densities of liquid and vapour phases of a simple fluid at the vapour pressure.

It is convenient to focus theoretical discussions on magnetic systems because, as is evident from figures 11.2 and 11.3, they possess a greater degree of symmetry. The magnetization of a macroscopic sample is a magnetic dipole moment per unit volume, which may have contributions from the intrinsic dipole moments of fixed atoms or ions and mobile electrons and also from the orbital motion of electrons. When the major contribution is from mobile electrons, the magnetism is said to be *itinerant*. When the major contribution is from atoms or ions fixed at the sites of a crystal lattice or from electrons which, though mobile, tend to congregate near these lattice sites, the magnetism is said to be *localized*. The exact degree of itineracy or localization is not easy to establish, but it appears that the three common metallic ferromagnets, namely iron, cobalt and nickel, are predominantly itinerant. Theoretically, it is somewhat easier to deal with localized magnets and, because of universality, this does not, in the end, make much difference as far as their critical properties are concerned. I shall therefore regard ferromagnetism as arising from localized magnetic moments situated at the sites of a lattice. Each of these magnetic moments is proportional to the intrinsic spin of an atom or ion, and the basic constituents of the magnet are conventionally referred to as *spins*.

To understand the origin of universality, it is necessary to consider correlations between the directions of spins at different sites. Our sample will exhibit a net magnetization if, on average, all the spins tend to point in the same direction. In a large sample, the average of a spin variable $\boldsymbol{s}_i$ at the ith lattice site will be independent of the particular site. The magnetization per spin is

$$\boldsymbol{M} = m\langle \boldsymbol{s}_i \rangle \tag{11.13}$$

where $m\boldsymbol{s}_i$ is the magnetic moment associated with the spin. The fluctuations of a given spin away from its average value are measured by

$s_i - \langle s_i \rangle$. What particularly concerns us is the correlation between such fluctuations at two different sites. We define the *correlation function* $G(\boldsymbol{r}_i - \boldsymbol{r}_j)$ as

$$G(\boldsymbol{r}_i - \boldsymbol{r}_j) = \langle (s_i - \langle s_i \rangle) \cdot (s_j - \langle s_j \rangle) \rangle = \langle s_i \cdot s_j \rangle - \langle s_i \rangle \cdot \langle s_j \rangle \quad (11.14)$$

where $\boldsymbol{r}_i$ is the position of the ith lattice site. Analogous correlation functions can be defined in terms of magnetization density for itinerant magnets or density fluctuations in a fluid. Assuming that only short-ranged forces act between spins, we would expect this correlation function to decay to zero at large distances, and so it does. Under most circumstances, we find

$$G(\boldsymbol{r}_i - \boldsymbol{r}_j) \sim \exp(-|\boldsymbol{r}_i - \boldsymbol{r}_j|/\xi) \quad (11.15)$$

where ξ is a characteristic distance called the *correlation length*. The correlation length depends on temperature and on the applied magnetic field. In the absence of an applied field, it diverges at the critical point, and this divergence is governed by a new critical exponent ν:

$$\xi \approx \xi_0 |T - T_c|^{-\nu}. \quad (11.16)$$

As with the susceptibility, the same exponent governs the divergence as the critical temperature is approached from above or below, but the amplitudes ξ_0 may be different. Typically, we find $\nu \approx 0.6$–0.7.

This divergence of the correlation length is at the root of the universality of critical phenomena. Because fluctuations are strongly correlated over large distances, they, and the critical properties which depend on them, are insensitive to details of the forces which act over microscopic distances. Experimentally, the correlation function can be investigated by scattering. In the case of magnets, the scattering of neutrons is affected by magnetic forces, while the scattering of light by fluids depends on density correlations. When the correlation length is large, the scattered waves from points widely separated in the sample are coherent, and so strong scattering results. This is visible to the naked eye in a fluid near its critical point. The strong scattering by a substance which is normally transparent gives it a foggy or milky appearance known as *critical opalescence*. From a theoretical point of view, we might expect that quantities such as critical exponents could be calculated on the basis of quite highly idealized models, which take little account of the detailed microscopic constitution of real materials, and this appears to be borne out in practice. Later on, we shall see in rather more detail why this is so.

11.3 The Ising Model and its Approximation by a Field Theory

The forces which tend to align spins in a ferromagnet have electrostatic and quantum-mechanical origins. For example, if the spins of two electrons in neighbouring atoms in a triplet state, which roughly means that they are parallel, then, to maintain the overall antisymmetry of the two-electron state, their orbital motion must be described by an antisymmetric combination of atomic orbitals. Conversely, if their spins are in a singlet antiparallel state, then their orbital state must be symmetric. The expectation value of the electrons' electrostatic energy is different in the symmetric and antisymmetric orbital states, and therefore also in the singlet and triplet spin states. This leads to an effective interaction between spins, called the *exchange* interaction. To study magnetic effects, we would like to use an effective Hamiltonian which depends only on spin degrees of freedom. It was shown by Heisenberg that such a Hamiltonian must have the form

$$\hat{H} = -\sum_{ij} J_{ij} \hat{\boldsymbol{S}}_i \cdot \hat{\boldsymbol{S}}_j \tag{11.17}$$

where $\hat{\boldsymbol{S}}_i$ is the spin operator for the ith lattice site and J_{ij} is a symmetric matrix of constants representing the exchange energies. Usually, these energies will be appreciable only when sites i and j are close together. The exchange energies can have either sign. If the J_{ij} are predominantly positive, then parallel spins have the lower energy and ferromagnetism will result; if they are negative, then we shall have antiferromagnetism.

In a uniaxial ferromagnet, the spins point preferentially along one crystallographic axis, say the z axis, so we can delete the x and y spin components in (11.17). In that case, all the remaining operators commute with each other, and we can choose a set of basis states in which they are all diagonal. If we take the $\hat{S}$ to be spin-$\frac{1}{2}$ operators, their eigenvalues are $\pm \hbar/2$. For theoretical purposes, it is useful to imagine that an independent magnetic field H_i can be applied at each lattice site. It is also convenient to absorb factors of $\hbar/2$ into the definitions of J_{ij} and H_i and the magnetic moment m in (11.13) into the definition of H_i. Then the partition function may be written as

$$Z(\beta, \{H_i\}) = \operatorname{Tr} e^{-\beta \hat{H}} = \sum_{\{s_i = \pm 1\}} e^{-\beta H_{\mathrm{I}}} \tag{11.18}$$

where

$$H_{\mathrm{I}} = -\sum_{ij} J_{ij} s_i s_j - \sum_i H_i s_i . \tag{11.19}$$

This is the *Ising model*. As a model of a ferromagnet, it is clearly rather idealized, taking into account only the configurational average of spin degrees of freedom. Thus, the free energy $F = -k_{\mathrm{B}} T \ln Z$ obtained from (11.18) represents not the whole free energy of a ferromagnetic material, but only that contribution to it which is directly involved in the ferromagnetic transition.

The partition function (11.18) is obviously of the same form as the configuration sum in (10.93) for the lattice gas, so long as we identify the adjusted chemical potential (10.94) with a uniform magnetic field. We see from (10.93) that the grand potential of the lattice gas receives a contribution from the factor multiplying the configuration sum, but that this contribution is a smooth function of temperature and chemical potential and cannot be directly involved in the gas–liquid transition.

From (11.18), the correlation function (11.14) can be expressed as

$$G(\boldsymbol{r}_i - \boldsymbol{r}_j) = \beta^{-2}\frac{\partial^2}{\partial H_i \partial H_j}\ln Z = \beta^{-1}\frac{\partial}{\partial H_i}\langle s_j\rangle. \tag{11.20}$$

For a uniform field, the magnetic susceptibility can be written in terms of the correlation function as

$$\chi = \frac{\partial}{\partial H}\langle s_i\rangle = \beta\sum_j G(\boldsymbol{r}_i - \boldsymbol{r}_j). \tag{11.21}$$

We saw in the last chapter that, by means of the Hubbard–Stratonovich transformation, the spin variables s_i can be replaced by a new set of variables ϕ_i, in terms of which the Ising model takes on an appearance similar to a relativistic scalar field theory. If, in (10.99), we replace $\bar{\mu}/2$ with the site-dependent magnetic field H_i, we find that, when H is zero, the averages of s_i and ϕ_i are related by

$$\langle s_i\rangle = \langle\tanh(\phi_i)\rangle. \tag{11.22}$$

In the neighbourhood of the critical point, the magnetization is small and since, for small ϕ, $\tanh(\phi) \approx \phi$, the magnetization can be approximately represented by $\langle\phi\rangle$. Moreover, since critical phenomena are associated with strong correlations over large distances, the expansion (10.100) in powers of the wavevector k should be justified near the critical point. When ϕ, H and $|T - T_c|$ are all small, the partition function of the Ising model can be approximated, via (10.101), by an expression of the form

$$Z(T, h) = \int \mathcal{D}\phi \exp[-H_{\text{eff}}(\phi)] \tag{11.23}$$

where the effective Hamiltonian is

$$H_{\text{eff}}(\phi) = \int \mathrm{d}^d x(\tfrac{1}{2}\boldsymbol{\nabla}\phi\cdot\boldsymbol{\nabla}\phi + \tfrac{1}{2}r_0\phi^2 + \tfrac{1}{4!}u_0\phi^4 - h\phi). \tag{11.24}$$

To arrive at this form, we have rescaled ϕ so as to make the coefficient of $\boldsymbol{\nabla}\phi\cdot\boldsymbol{\nabla}\phi$ equal to $\frac{1}{2}$. The parameter r_0 is approximately linear in temperature,

$$r_0 \propto (T - T_0) \tag{11.25}$$

where T_0 is an approximation to the critical temperature, u_0 is a constant and h is proportional to the original magnetic field. It is useful to allow for d spatial dimensions rather than just three.

A large number of approximations stand between the field theory (11.23) and any realistic model of a ferromagnet. Nevertheless, it is believed to embody exact information about universal quantities such as the exponents β, γ and ν. However, the information we need to calculate non-universal quantities such as the critical temperature itself or the amplitude χ_0 in (11.11) has largely been lost.

11.4 Order, Disorder and Spontaneous Symmetry Breaking

The phase transition which takes place at the Curie temperature in zero magnetic field can be described as an *order–disorder transition*. The high-temperature paramagnetic phase is one in which fluctuations in the orientation of each spin variable are entirely random, so that the configurational average is zero: the state is a disordered one. In the ferromagnetic phase, on the other hand, the spins point preferentially in one particular direction and, in this sense, the state is ordered. On the face of it, it is hard to understand how such an ordered state can come about. The Ising Hamiltonian (11.19) with $H_i = 0$ has a symmetry: if we reverse the sign of every spin, it remains unchanged. Therefore, for each configuration in which a given spin s_i has the value $+1$, there is another one, obtained by reversing all the spins, in which it has the value -1, and these two configurations have the same statistical weight. Thus, the magnetization per spin defined by (11.13) ought to be zero, and this argument is apparently valid for any temperature. Indeed, for any *finite* system, the conclusion is inescapable. Within the theoretical framework of the Ising model, the only way to obtain a non-zero spontaneous magnetization is to consider an infinite system, that is to take the thermodynamic limit.

The way in which this comes about is illustrated in figure 11.5. If we apply a uniform magnetic field, then the symmetry of the Hamiltonian is broken, and there is a magnetization in the direction of the field. Figure 11.5(*a*) shows the variation of M with H at a fixed low temperature for a large but finite system. It is indeed zero at $H = 0$, but increases rapidly when a small field is applied. As the size of the system is taken to infinity, the slope at $H = 0$ increases and eventually becomes a discontinuity as shown in figure 11.5(*b*). In the limit, the value of M at $H = 0$ is not well defined, but the limit of M as $H \rightarrow 0$ from above or below is $\pm M_S(T)$. I cannot reproduce here the detailed calculations which support this picture. Interested readers may like to consult, for example, Reichl (1980) for some further explanation and references to the original literature. In fact, even an infinite Ising model does not always show a spontaneous magnetization. Whether it does or not depends on the number of spatial dimensions, d. It is possible to obtain an exact solution only in one and two dimensions. (By 'solution' is meant a method of calculating actual values for thermodynamic quantities like the magnetization, susceptibility and specific heat.) In one dimension, there is no ferromagnetic state. For two dimensions, the solution was given for the case of zero magnetic field in a celebrated paper by L Onsager (1944) and for a non-zero field by C N Yang (1952) and there is a spontaneous magnetization at low temperatures. For three dimensions, no exact solution has been found, but all approximate calculations indicate that there is a ferromagnetic phase. There is, in fact, very little doubt that the ferromagnetic state exists for all $d \geqslant 2$.

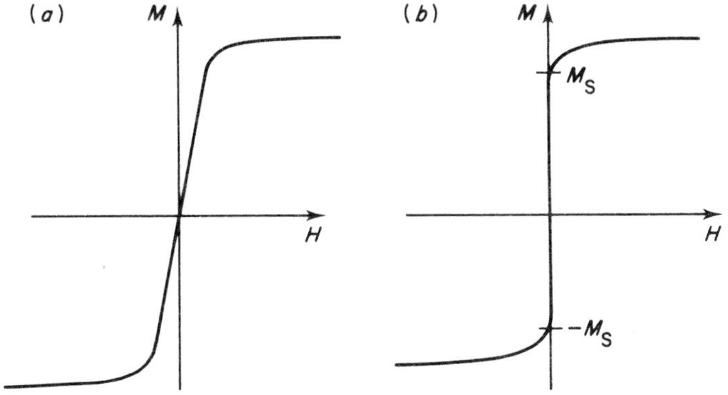

Figure 11.5 Variation of magnetization with magnetic field in a ferromagnet at a fixed temperature below T_c: (*a*) finite system; (*b*) infinite system.

Real ferromagnets do, of course, exhibit a spontaneous magnetization and, though they may be very large, they certainly are not infinite. To understand this, we must think back to our discussion of ergodic theory. There we saw that the ensemble averages of equilibrium statistical mechanics correspond to long time averages of the instantaneous states of an experimentally observed system. In the case of a uniaxial ferromagnet, the low-energy states with positive magnetization and those with negative magnetization constitute two separate regions of phase space. Either there are very few trajectories which can connect these two regions without passing through states of much higher energy, or there are no such trajectories at all. In a large system below its Curie temperature, fluctuations in energy large enough to surmount the energy barrier between the two regions will be sufficiently rare that only one region is explored during the finite time over which the system is observed. (Just how long we would have to wait for a suitable fluctuation to occur is hard to estimate in a definitive manner, but estimates greater than the present age of the universe are sometimes quoted for systems of everyday size!) Thus, the occurrence of a spontaneous magnetization indicates a partial breakdown of ergodicity. In effect, what should be compared with observations is not the complete equilibrium ensemble average but rather an average over half of the configurations, namely those which have a net magnetization in the same direction. This is achieved by the thermodynamic limit in the following way. When a magnetic field is applied, the statistical weight of the 'wrong' configurations is reduced relative to those of the 'right' ones. If we now take the infinite volume limit, the 'wrong' configurations are suppressed entirely, so, if we subsequently remove the field, a spontaneous magnetization remains.

Spontaneous symmetry breaking may be defined as a situation in which the Hamiltonian of a system possesses a symmetry but the equilibrium state does not have the same symmetry. Ferromagnetism is obviously a case in point. The effective Hamiltonian (11.24) in the field-theoretic approximation to the Ising model clearly inherits the same symmetry, $\phi \rightarrow -\phi$, when h is zero, and we shall shortly see how this symmetry may be spontaneously broken. Evidently, the same phenomenon must be possible in genuine relativistic field theories. For zero-temperature field theories of the kind discussed in chapter 9, the analogue of two (or more) possible states of magnetization is the existence of several possible vacuum states, one of which has been spontaneously chosen by our universe. As we shall see in chapter 12, this symmetry breaking may be invoked to explain the different strengths of the fundamental interactions. Alternatively, it could be that the universe, like a ferromagnet, possesses many domains in which the symmetry is broken in different ways.

11.5 The Ginzburg–Landau Theory

The field-theoretic approximation (11.23) and (11.24) to the Ising model is similar to the self-interacting scalar field theory we studied in chapter 9. As we saw, it is not possible to compute exact values for the expectation values of any functions of the field, and some approximations must be made. One useful approximation is obtained when we evaluate the integral (11.23) by the method of steepest descent. In its simplest form, this means finding the value of ϕ at which the integrand $\exp[-H_{\text{eff}}(\phi)]$ has its maximum value and replacing the integral by a constant times this maximum value of the integrand. The maximum value of the integrand corresponds to a minimum value of H_{eff}, so we have

$$Z(T, h) \approx \text{constant} \times e^{-F(T,h)} \tag{11.26}$$

where $F(T, h)$ is the minimum value of H_{eff}. Apart from a constant and a factor of $1/k_BT$ which, since we are considering only the critical region, can be replaced by $1/k_BT_c$, $F(T, h)$ is our approximation to the free energy. The value $M(T, h)$ of ϕ which minimizes H_{eff} is our approximation to the magnetization. This approximation constitutes the *Ginzburg–Landau* theory of phase transitions, although Ginzburg and Landau did not arrive at it in quite this way.

In general, we can allow M to depend also on position $\boldsymbol{x}$. To be a minimum of H_{eff}, it must satisfy what amounts to an Euler–Lagrange equation

$$-\nabla^2 M(\boldsymbol{x}) + r_0 M(\boldsymbol{x}) + \tfrac{1}{6}u_0 M^3(\boldsymbol{x}) = h(\boldsymbol{x}). \tag{11.27}$$

When h is independent of position, it is not hard to see that the minimum of H_{eff} occurs for a position-independent value of M. This value minimizes the potential

$$V(\phi) = \tfrac{1}{2}r_0\phi^2 + \tfrac{1}{4!}u_0\phi^4 - h\phi. \tag{11.28}$$

According to (11.25), r_0 is positive if $T > T_0$ and negative if $T < T_0$, and in figure 11.6 $V(\phi)$ is sketched for these two cases. In the high-temperature case, there is a single minimum which is $M = 0$ when $h = 0$. In the low-temperature case, there are two minima (if h is not too large). When $h = 0$, these two minima are at the same depth; otherwise, one or other of them is lower, according to the sign of h. This evidently corresponds, at least qualitatively, to the behaviour of a ferromagnet, if we identify T_0 as the critical temperature in this approximation. It is a simple matter to obtain the value of the critical exponent β in (11.9). When $h = 0$ and r_0 is negative, the solution of (11.27) for the spontaneous magnetization is

$$M_S = \left(-\frac{6r_0}{u_0}\right)^{1/2} \propto (T_0 - T)^{1/2} \tag{11.29}$$

so $\beta = \frac{1}{2}$.

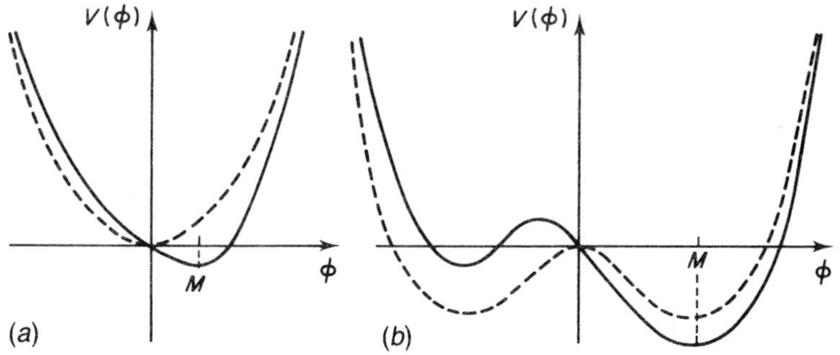

Figure 11.6 The Ginzburg–Landau potential (a) for $r_0 > 0$ and (b) for $r_0 < 0$. The symmetrical curves (broken) are for $h = 0$ and the asymmetrical ones (full) for $h > 0$.

The correlation function may be defined by analogy with (11.20), using the functional derivative discussed in appendix 1, as

$$G(\boldsymbol{x}-\boldsymbol{y}) = \delta M(\boldsymbol{x})/\delta h(\boldsymbol{y}). \tag{11.30}$$

By differentiating (11.27), we find that it satisfies the equation

$$(-\nabla^2 + r_0 + \tfrac{1}{2}u_0 M^2)\, G(\boldsymbol{x}-\boldsymbol{y}) = \delta(\boldsymbol{x}-\boldsymbol{y}) \tag{11.31}$$

which is, not too surprisingly, the Euclidean version of (9.34) for the propagator of a scalar field. The solution, analogous to (9.38), is

$$G(\boldsymbol{x}-\boldsymbol{y}) = \int \frac{\mathrm{d}^d k}{(2\pi)^d} \frac{\exp[\mathrm{i}\boldsymbol{k}\cdot(\boldsymbol{x}-\boldsymbol{y})]}{(k^2+m^2)} \tag{11.32}$$

where $m^2 = r_0 + u_0 M^2/6$. When $\boldsymbol{x}$ and $\boldsymbol{y}$ are far apart, this gives

$$G(\boldsymbol{x}-\boldsymbol{y}) \sim \exp(-m|\boldsymbol{x}-\boldsymbol{y}|) \tag{11.33}$$

so we identify the correlation length as

$$\xi = 1/m. \tag{11.34}$$

When $h = 0$, we have $m^2 = r_0$ above the critical temperature or, using (11.29), $m^2 = -2r_0$ below the critical temperature, and so the critical exponent for the correlation length is $\nu = \frac{1}{2}$. The susceptibility is given by

$$\chi = \partial M/\partial h = \int \mathrm{d}^d x G(\boldsymbol{x}-\boldsymbol{y}) = 1/m^2 \tag{11.35}$$

and so its critical exponent is $\gamma = 2\nu = 1$.

We see that the Ginzburg–Landau theory does indeed predict critical exponents which are universal: they do not depend, for example, on u_0 or on the constant which relates r_0 to $T - T_c$. It might be thought that this is an artificial result, arising from the quite drastic approximations we used to get from a real magnet or fluid to the effective Hamiltonian (11.24). This is not so, however. We could systematically improve upon these approximations by adding higher powers of ϕ and higher derivatives, and by allowing coefficients such as r_0 and u_0 to contain additional higher powers of $T - T_c$ and h. Readers may easily convince themselves that, when M, $T - T_c$ and h are sufficiently small, all these additional terms become negligible in comparison with those we retained. The only proviso is that u_0 should remain positive. If u_0 becomes zero or negative then, in order for the potential to have a minimum, a higher power of ϕ with a positive coefficient must be added, and new types of critical behaviour result (see, for example, Lawrie and Sarbach (1984)).

The critical exponents of the Ginzburg–Landau theory are the same as those obtained from a variety of simple approximations known collectively as *classical* or *mean field* theories. Other examples of such approximations are the van der Waals theory of imperfect gases and the Weiss molecular field theory of ferromagnetism. The reason for this is that, in all such approximations, the appropriate free energy can be written in the Ginzburg–Landau form when we are close enough to the critical point. Although the classical exponents are universal, they are only in very rough numerical agreement with the typical experimental values I quoted earlier on. The fault lies not with the idealized model defined by (11.23) and (11.24), but with the approximation we used to estimate the functional integral. Numerous methods are available for improving upon this approximation. We can, for example, return to the original Ising model (11.18) and attempt to evaluate its thermodynamic properties directly. One method of approximation is the *high-temperature series expansion* in powers of β. Since this is most accurate at very high temperatures, careful methods of extrapolation are needed to obtain results valid at the critical temperature, but good agreement with experimental values can be obtained. Another approach is the *Monte Carlo* method, which carries out configuration sums directly by generating a set of configurations with the correct statistical weight, which should be representative of the whole ensemble. In the next section, I shall discuss an alternative approach, called the *renormalization group*, which yields rather more insight and further illustrates the analogy with relativistic quantum field theory.

11.6 The Renormalization Group

We have seen that the distinctive behaviour of a system near a critical point derives from the fact that the correlation length ξ becomes very large or, in the ideal case of an infinite system whose temperature can be adjusted to be exactly T_c, infinite. A somewhat highbrow way of expressing this is to say that the system becomes *scale invariant* at the critical point. To see what this means, consider first the case of a finite correlation length and, to be specific, a magnet. If we examine a part of the system whose diameter is much smaller than ξ, we find that fluctuations in all the spins are strongly correlated. If, on the other hand, we examine a region whose diameter is much larger that ξ, we find that there are strong correlations only within what now count as small subregions of diameter ξ, but not over the whole region. Thus, the appearance of the system depends upon the *length scale*, or characteristic size of the region we choose to examine. By contrast, when ξ is infinite, the appearance of the system is much the same, at whatever length scale we choose to examine it. It turns out that much valuable information about critical phenomena, including improved approximations to the values of critical exponents, can be obtained by investigating how the appearance of the system changes with the scale of length on which we examine it. That this might be possible was first suggested by L P Kadanoff, and detailed techniques for putting the idea into practice have been developed by many others, notably by K G Wilson and M E Fisher.

These techniques, known as *renormalization group* techniques exist nowadays in many varied forms. Some of them are described in the books by Amit (1978), Domb and Green (1976), Ma (1976) and Reichl (1980). Here, I shall discuss one particular method which is well suited to field-theoretic models like (11.23). In chapter 9, we found that interacting relativistic field theories require renormalization, because parameters such as masses and coupling constants appearing in the action which defines the field theory do not correspond directly to measurable quantities. Here, the situation is quite similar. The parameters r_0 and u_0 in (11.24) are related to forces which act at a microscopic level, and are not best suited for describing the large-scale phenomena associated with critical points. In quantum electrodynamics, we saw that the net effect of an electric charge on, for example, the collisions of charged particles varies with the energy of the collision. This could be expressed in terms of a modified Coulomb potential or, as in (9.94), of an energy-dependent charge. Since the energy of a virtual photon exchanged in the collision can be expressed in terms of its wavelength, we can regard the energy dependence of the effective charge as a dependence on a characteristic length scale of the collision process. Furthermore, according to (9.93), the energy dependence can be related to the dependence on an arbitrary 'mass' parameter μ which may be introduced in the renormalization process. Indeed, the earliest version of the renormalization group was invented by M Gell-Mann and F E Low in just this context.

The Ginzburg–Landau theory is more or less equivalent to the lowest order of perturbation theory (for which, see chapter 9), which involves no closed-loop diagrams with momentum integrals. To obtain improved approximations for the critical exponents, it is necessary to consider higher-order contributions. Readers will recall that the momentum integrals contained in these higher-order contributions are frequently infinite, but that these infinities disappear (at least in the case of a renormalizable theory) when the results are expressed in terms of renormalized measurable quantities. In chapter 9, it appeared that these infinities were an embarrassment. Here, as we shall see, they actually work to our advantage. I shall describe only one particular calculation, that of the susceptibility exponent γ. As in (11.35), the susceptibility is the integral over all space of the correlation function, which is the Fourier transform of this function evaluated at $k = 0$. It is actually convenient to deal with the inverse of the susceptibility, which I shall denote by $\Gamma = \chi^{-1}$. At the first order of perturbation theory, it is given by an obvious modification of equations (9.68) and (9.75), with $p = 0$:

$$\Gamma = r_0 + \tfrac{1}{2}u_0 \int \frac{\mathrm{d}^d k}{(2\pi)^d} \frac{1}{(k^2 + r_0)}. \tag{11.36}$$

The dummy integration variable k is, of course, not the one we just set to zero.

As it stands, the integral in (11.36) is infinite if $d \geqslant 2$, and this infinity arises from the upper limit $k \to \infty$. However, if our model field theory is regarded as an approximation to a condensed matter system such as a magnet or a fluid, then infinite values of k are not really allowed. For a magnet or lattice gas, the field ϕ existed originally only at the sites of a regular lattice, and $\boldsymbol{k}$ should take values only within the first Brillouin zone of the lattice. It is an adequate approximation to take this Brillouin zone to be a sphere of radius Λ, approximately equal to π/a, where a is the spacing between lattice sites. The integrand in (11.36) depends only on the magnitude of $\boldsymbol{k}$, so as in (9.76), angular integrations can be carried out, leaving

$$\Gamma = r_0 + \tfrac{1}{2}u_0 S_d \int_0^\Lambda \mathrm{d}k \frac{k^{d-1}}{(k^2 + r_0)}. \tag{11.37}$$

The factor S_d is $(2\pi)^{-d}$ times the surface area of a unit sphere in d dimensions and can be shown to be given by $S_d = 2\pi^{d/2}/(2\pi)^d\Gamma(d/2)$, where $\Gamma(x) = (x-1)!$ is Euler's gamma function. At the critical temperature, the inverse susceptibility should be zero, and we see that this now occurs when r_0 takes a value $r_{0\mathrm{c}}$ which is of order u_0. Up to corrections of order u_0^2, we find that

$$r_{0\mathrm{c}} = -\tfrac{1}{2}u_0 S_d \int_0^\Lambda \mathrm{d}k\, k^{d-3}. \tag{11.38}$$

If we now define a new variable

$$t_0 = r_0 - r_{0\mathrm{c}} \tag{11.39}$$

which is proportional to $T - T_\mathrm{c}$, then (11.37) can be rewritten, again up to corrections of order u_0^2, as

$$\Gamma = t_0\left(1 - \tfrac{1}{2}u_0 S_d \int_0^\Lambda \mathrm{d}k \frac{k^{d-1}}{k^2(k^2 + t_0)}\right). \tag{11.40}$$

This certainly vanishes when $t_0 = 0$, but we want to know how it behaves for small t_0. The answer depends crucially on the number of spatial dimensions d. Mathematically, it is perfectly possible to take $d > 4$. In that case, the integral in (11.40) approaches a finite constant value at $t_0 = 0$. For small t_0, we then find that Γ is approximately a constant times t_0. Since Γ is χ^{-1}, this means that $\gamma = 1$, which is the classical value given by the Ginzburg–Landau theory. Indeed, further arguments along these lines show that in more than four dimensions, all the critical exponents of the Ginzburg–Landau theory should be exactly correct. For practical purposes, we are, of course, more interested in dimensions less than four. Below four dimensions, the integral in (11.40) is infinite when $t_0 = 0$, but now the infinity comes from the limit $k \to 0$. This is called an *infrared* divergence and, unlike the *ultraviolet* divergences at infinite values of k, it has a genuine physical significance, being associated with the singular behaviour of thermodynamic quantities at the critical point. To deal with the infrared divergence, we may rescale k by a factor of $t_0^{1/2}$, which gives

$$\Gamma = t_0 \left(1 - \tfrac{1}{2} u_0 t_0^{(d-4)/2} S_d \int_0^{\Lambda/t_0^{1/2}} \mathrm{d}k \, \frac{k^{d-1}}{k^2(k^2+1)}\right). \qquad (11.41)$$

In the limit $t_0 \to 0$, the upper limit $\Lambda/t_0^{1/2}$ becomes infinite, but the integral is finite. However, the factor $u_0 t_0^{(d-4)/2}$ now becomes infinite. In terms of the dimensional analysis introduced in §9.6, this quantity is dimensionless, u_0 itself having dimension $4 - d$. Thus, if the expansion in (11.41) were continued to higher orders in u_0, successive terms would be proportional to successively higher powers of $u_0 t_0^{(d-4)/2}$, each term becoming infinite more rapidly than the previous one as $t_0 \to 0$.

From this it is clear that perturbation theory does not give us a sensible answer for the dependence of the susceptibility on temperature near the critical point. The role of renormalization will be to reformulate perturbation theory in such a way that a sensible answer does emerge. This can be done in several ways. The principle of the method I am going to explain was put forward by Wilson and Fisher (1972). The crucial observation is that expressions like (11.41) can be evaluated, in principle, when d has any value, not necessarily an integer. As a purely mathematical device, therefore, we can consider d to be a continuous real variable. The value $d = 4$ clearly marks a borderline between different kinds of critical behaviour, and it will be convenient to define a variable ε by

$$\varepsilon = 4 - d. \qquad (11.42)$$

If we assume that the variation of the susceptibility with temperature can indeed by described by an exponent γ, then this exponent is likely to depend upon ε. Since it is equal to 1 for any negative value of ε, we may anticipate that for positive values of ε, it can be expressed as a power series

$$\gamma = 1 + \gamma_1 \varepsilon + \gamma_2 \varepsilon^2 + \dots. \qquad (11.43)$$

If we can calculate a few terms of this expansion then, by setting ε equal to 1, we obtain an estimate of the value of γ in three dimensions. The reason this works is that, the smaller ε is, the less rapidly $u_0 t_0^{(d-4)/2}$

diverges and the easier it becomes to extract sensible answers from perturbation theory. Clearly, any answer we obtain must be valid right up to $d = 4$, or $\varepsilon = 0$. In this limit, however, the ultraviolet divergence of integrals like that in (11.41) reappears when $t_0 \to 0$. The key to calculating γ is that these divergences can be removed, as we saw in chapter 9, by the process of renormalization. It should not now be too surprising that this process actually yields all the information we need to calculate γ.

As I described it in chapter 9, the object of renormalization was to express quantities like scattering amplitudes in terms of physically measurable masses and coupling constants. For our present purposes, the main object is to remove the ultraviolet divergences, and there are many different ways in which this can be achieved. Details may be found, for example, in the book by Amit (1978), and I shall just quote the results of one method. Since we have to deal with the limit $\Lambda/t_0^{1/2} \to \infty$, we might as well take Λ to infinity at the outset. The infinities at $\varepsilon = 0$ can all be removed if we express thermodynamic quantities in terms of renormalized variables u and t which, at the first order of perturbation theory, are related to u_0 and t_0 by

$$u_0 = \mu^\varepsilon u\left[1 + \frac{3}{2\varepsilon}S_4 u + \ldots\right] \tag{11.44}$$

$$t_0 = t\left[1 + \frac{1}{2\varepsilon}S_4 u + \ldots\right]. \tag{11.45}$$

The factor μ^ε in (11.44) makes the renormalized u dimensionless. As we discussed earlier, μ is an arbitrary parameter, and u and t are variables appropriate for describing phenomena on a length scale μ^{-1}. The inverse susceptibility can now be written as

$$\Gamma = t[1 + \tfrac{1}{4}S_4 u \ln(t/\mu^2) + \ldots] \tag{11.46}$$

where, to simplify matters, I have expanded in powers of ε as well as u and kept only the leading term. At higher orders, a wavefunction renormalization as in (9.71) also becomes necessary. Since critical phenomena are associated with very large length scales, we shall want μ to have a very small value. The way in which u and t vary with our choice of μ is expressed by differentiating (11.44) and (11.45), keeping u_0 and t_0 fixed. This leads to two functions, $\beta(u)$ and $\tau(u)$, defined by

$$\beta(u) = \mu \left.\frac{\partial u}{\partial \mu}\right|_{u_0, t_0} = -\varepsilon u + \tfrac{3}{2}S_4 u^2 + \ldots \tag{11.47}$$

$$\tau(u) = \frac{\mu}{t}\left.\frac{\partial t}{\partial \mu}\right|_{u_0, t_0} = \tfrac{1}{2}S_4 u + \ldots . \tag{11.48}$$

The function $\beta(u)$ is sketched in figure 11.7. It vanishes at two values of u, called *fixed points*, namely $u = 0$ and $u = u^*$, where

$$S_4 u^* = \tfrac{2}{3}\varepsilon + O(\varepsilon^2). \tag{11.49}$$

Since $\beta(u)$ is positive for $u > u^*$ and negative for $u < u^*$, a little thought shows that, whatever value u has for some fixed value of μ, it approaches the value u^* as $\mu \to 0$. In the renormalization group approach, this is the explanation of universality. Whatever the value of u_0, determined in principle by the nature of microscopic forces, the renormalized coupling appropriate to very-large-scale phenomena is u^*.

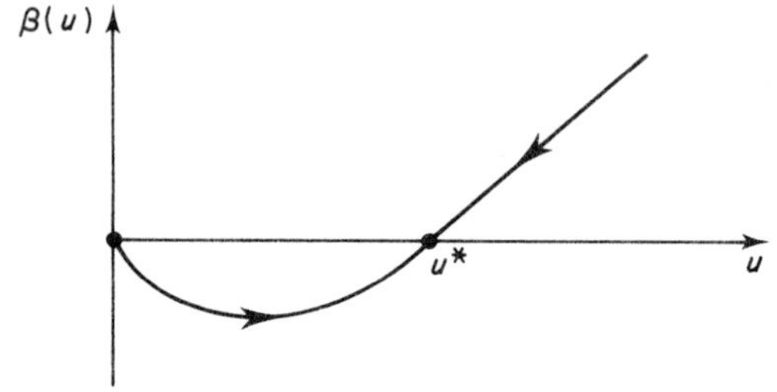

Figure 11.7 The renormalization group function $\beta(u)$. Arrows indicate the evolution of the running coupling constant as $\mu \to 0$.

Since u^* is of order ε, it is possible to calculate the coefficients in (11.43) from perturbation theory. Suppose, indeed, that we choose $\mu^2 = t$, which goes to zero at the critical point. Then (11.46) becomes just $\Gamma = t$. This might appear to imply that γ is 1, but in fact it does not. For a fixed value of μ, (11.44) shows that u is independent of temperature, since u_0 is, and then (11.45) shows that t is directly proportional to t_0 and therefore to $T - T_c$. If $\mu^2 = t$, however, the same is no longer true. To get round this, let us choose a fixed value $\hat{\mu}$, sufficiently small that u can be set equal to u^* with negligible error, and let $\hat{t}$ be the renormalized temperature variable corresponding to this choice. Then $\hat{t}$ *is* proportional to $T - T_c$. For a different choice of μ, still small enough to give $u = u^*$, we can relate the new variable t to $\hat{t}$ by solving (11.48) with the boundary condition that $t = \hat{t}$ when $\mu = \hat{\mu}$. We get

$$t = \hat{t}(\mu/\hat{\mu})^{\tau^*} \tag{11.50}$$

where

$$\tau^* = \tau(u^*) = \tfrac{1}{3}\varepsilon + O(\varepsilon^2). \tag{11.51}$$

From this, when we set $\mu = t^{1/2}$, we find

$$t = \hat{t}^{2/(2-\tau^*)}\hat{\mu}^{(-2\tau^*/(2-\tau^*))}. \tag{11.52}$$

Finally, therefore, we can identify the susceptibility exponent γ as

$$\gamma = 2/(2 - \tau^*) = 1 + \tfrac{1}{6}\varepsilon + O(\varepsilon^2). \qquad (11.53)$$

When $\varepsilon = 1$, this approximation gives $\gamma = 1.17$, which is certainly an improvement on the classical value of 1. The best available estimates from more extended calculations give a value of 1.24, in good agreement with other theoretical methods and with observations.

More important, perhaps, than the actual values of critical exponents is the insight the renormalization group provides as to how these universal values come about. We have seen that, although they do not depend on the detailed constitution of the system, as reflected, for example, in the value of u_0, they do depend on the number of spatial dimensions d. As it turns out, they also depend on some other general properties. We might, for example, generalize our field-theoretic model by taking ϕ to an n-component vector. For $n = 3$, this would correspond to an isotropic, rather than a uniaxial, ferromagnet. It is found that the critical exponents then vary slightly with n, for example $\gamma = 1 + (n + 2)\varepsilon/2(n + 8) + O(\varepsilon^2)$, which gives a value of 1.23 when $n = d = 3$.

11.7 The Ginzburg–Landau Theory of Superconductors

The phenomena of superconductivity are both theoretically interesting and of great technological importance. For want of space, I cannot describe them in anything like the detail they deserve, and I propose mainly to highlight some theoretical considerations which turn out to have implications beyond the science of superconductivity itself. From a microscopic point of view, superconductivity is a kind of Bose–Einstein condensation. The electrons which conduct electric currents in a metal are, of course, fermions, and a non-interacting gas of fermions cannot undergo condensation. The essence of the microscopic theory is that interactions between electrons and the positive ions which form a crystal lattice can result in a net weak attraction between electrons. By analogy with quantum electrodynamics, this force can be thought of as mediated by the exchange of *phonons*, which are quantized vibrations of the lattice, much as photons are quantized 'vibrations' of the electromagnetic field. Under the influence of this attraction, some electrons may form loosely bound pairs, known as *Cooper pairs*, whose net spin is zero and which behave as bosons. These boson pairs can then undergo condensation, and the condensed electrons can flow without friction, which means that their electrical resistance is zero. A simple experimental observation which supports this picture is that the metals which superconduct most readily tend to be rather poor conductors in the normal high-temperature state. This is because the interactions with the lattice which favour the formation of Cooper pairs cause relatively strong scattering in the normal state, which leads to a relatively high resistance. One reason for treating this qualitative picture with caution is that the mean separation of two electrons in a Cooper pair can be estimated, and it turns out to be comparable with, or greater than, the mean separation of the pairs themselves. A straightforward account of the microscopic theory, due largely to J Bardeen, L N Cooper and J R Schrieffer, is given by Reichl (1980).

11.7.1 Spontaneous breaking of continuous symmetries

In the Ginzburg–Landau theory, the phase transition which marks the onset of superfluidity or of superconductivity can be investigated in terms of an effective Hamiltonian similar to (11.24), in which ϕ is taken to be the macroscopic wavefunction of the condensate. This is a complex quantity which can be expressed as

$$\phi(\boldsymbol{x}) = \frac{1}{\sqrt{2}}[\phi_1(\boldsymbol{x}) + \mathrm{i}\phi_2(\boldsymbol{x})] \qquad \text{or} \qquad \phi(\boldsymbol{x}) = \psi(\boldsymbol{x})\mathrm{e}^{\mathrm{i}\alpha(\boldsymbol{x})}. \tag{11.54}$$

The effective Hamiltonian must be real, and therefore of the form

$$H_{\mathrm{eff}}(\phi) = \int \mathrm{d}^d x[\nabla\phi\cdot\nabla\phi + r_0\phi^*\phi + \tfrac{1}{4}u(\phi^*\phi)^2]. \tag{11.55}$$

I have not included a symmetry-breaking field h, because no such field exists physically, and I have chosen the normalization of the coefficients to coincide with those of the complex scalar field in chapter 9. Whereas (11.24) has a *discrete* symmetry, $\phi \to -\phi$, when $h = 0$, (11.55) has a *continuous* symmetry, in the sense that it is unchanged if we change the phase of ϕ by any constant angle θ. This is, in fact, a gauge symmetry of the kind we studied in chapter 8. Below the critical temperature, therefore, there are not just two possible minima but an entire circle of them, as sketched in figure 11.8. Any function of the form

$$M = v\mathrm{e}^{\mathrm{i}\alpha} \tag{11.56}$$

is a minimum if $v = (-2r_0/u_0)^{1/2}$ and α is any constant angle. Of course, M is not to be interpreted as a magnetization, but it plays a similar role. Such a quantity which, being non-zero in the ordered phase and zero in the disordered phase, serves to distinguish the two phases is called an *order parameter*.

It is interesting to examine fluctuations of ϕ about its mean value (11.56). Taking $\alpha = 0$ in (11.56), we write

$$\phi(\boldsymbol{x}) = \left(v + \frac{1}{\sqrt{2}}\chi(\boldsymbol{x})\right)\exp\left(\frac{\mathrm{i}\theta(\boldsymbol{x})}{\sqrt{2}v}\right) \tag{11.57}$$

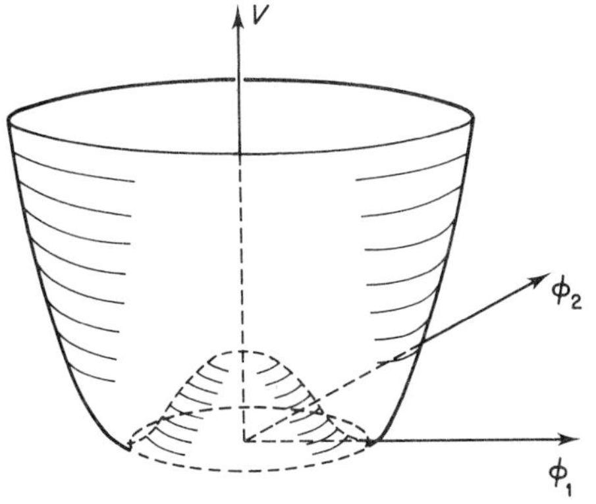

Figure 11.8 Potential for a complex scalar field with spontaneously broken symmetry. It is the surface of revolution of the symmetrical curve in figure 11.6(*b*), and its minima lie on the broken circle.

so that χ and θ measure fluctuations in the amplitude and phase, respectively, away from the mean values. Upon substituting this into the effective Hamiltonian (11.55), we obtain

$$H_{\text{eff}} = \int \mathrm{d}^d x[\tfrac{1}{2}\nabla\chi\cdot\nabla\chi + \tfrac{1}{2}(-2r_0)\chi^2 + \tfrac{1}{2}\nabla\theta\cdot\nabla\theta] + H_{\text{int}} \quad (11.58)$$

where H_{int} contains higher powers of χ and θ, and I have dropped a constant term. If this were to be interpreted as a quantum field theory, it would represent two species of particles, the χ particles with mass $(-2r_0)^{1/2}$, and the θ particles, with zero mass, interacting through the terms in H_{int}. In the same sense that states containing such particles would be *excitations* of the vacuum state, we can speak of statistical fluctuations about the mean value of ϕ as excitations. These excitations are wave-like disturbances which, in a quantum-mechanical system, will propagate in much the same way as particles. Phonons in a solid provide an example of this. The fact that θ excitations have zero 'mass' is easily understood from figure 11.8. A non-zero value of θ just moves ϕ around the circle of minima of the potential, which costs no potential energy. A χ fluctuation, on the other hand, moves ϕ in the radial direction, which requires an increase in potential energy. This is an example of *Goldstone's theorem*, which asserts that for any spontaneously broken continuous symmetry there is a massless particle (or 'massless' excitation), called a *Goldstone boson*.

These excitations are perhaps most easily visualized if we regard (11.55) as a model of a ferromagnet in which the spins can point with equal ease in any direction in a plane, their components being ϕ_1 and ϕ_2. The spontaneous magnetization points in one particular direction. The χ fluctuations are then fluctuations in the magnitude of this magnetization, while θ fluctuations are fluctuations in its direction. The latter are called *spin waves*, and the quantized spin waves are *magnons*. In a real ferromagnet, there are always preferred directions of magnetization, defined by the crystal lattice, and fluctuations away from these directions incur an increase in potential energy, so 'massless' magnons are not observed in practice. In superfluid helium-4, two kinds of excitations, called *phonons* and *rotons*, are found, but their detailed properties cannot be found from the condensate wavefunction alone and they do not correspond exactly to the χ and θ excitations. In superconductors, as we are about to see, the θ fluctuations have a very special effect.

11.7.2 Magnetic effects in superconductors

An important property of superconductors is the fact that they expel magnetic flux, that is the magnetic induction $\boldsymbol{B}$ is always zero inside a superconductor. This is called the *Meissner effect*. When magnetic fields are present, the effective Hamiltonian must be modified as follows:

$$H_{\text{eff}} = \int \mathrm{d}^3x[\tfrac{1}{2}B^2 + |(\boldsymbol{\nabla} - 2\mathrm{i}e\boldsymbol{A})\phi|^2 + r_0|\phi|^2 + \tfrac{1}{4}u_0|\phi|^4 - \boldsymbol{B}\cdot\boldsymbol{H}]. \quad (11.59)$$

The term $B^2/2$ represents, in a suitable system of units, the magnetic field energy. In the next term, the gradient has been replaced by the spatial components of the covariant derivative (8.8) and the vector potential rescaled, as in (8.16), but using the charge $-2e$ of a Cooper pair. In the last term, $\boldsymbol{H}$ is an externally applied magnetic field strength. In the macroscopic theory of magnetic materials, $\boldsymbol{H}$ is related to $\boldsymbol{B}$ by the equation $\boldsymbol{B} = \boldsymbol{H} + \boldsymbol{M}$, where $\boldsymbol{M}$ is the magnetization. Inside a superconductor, this implies that $\boldsymbol{M} = -\boldsymbol{H}$. It is found, though I shall not describe the details here, that the magnetic moment of a superconducting sample is generated by superconducting current flowing on its surface. The superconductor is said to exhibit *perfect diamagnetism*. The exact relation between the $\boldsymbol{H}$ which exists inside the superconductor and that which would be there if the sample were removed depends on the shape of the sample. For our purposes, it is sufficient to take $\boldsymbol{H}$ to be a uniform, constant field in the z direction. The magnetic induction will adjust itself to a value appropriate to the equilibrium state of the material, so the effective Hamiltonian is to be minimized with respect to both ϕ and $\boldsymbol{B}$ (which is equal to $\boldsymbol{\nabla} \times \boldsymbol{A}$). The appearance of $-\boldsymbol{B}\cdot\boldsymbol{H}$ in the thermodynamic potential is discussed in most textbooks on thermodynamics. The simplest way to understand it is to consider the normal (non-superconducting) state in which $\phi = 0$. Then, by minimizing (11.59), we get the correct result $\boldsymbol{B} = \boldsymbol{H}$.

To understand the Meissner effect, we must first find a vector potential whose curl gives a uniform magnetic induction of magnitude B

in the z direction. It is easy to verify that a suitable potential is

$$\boldsymbol{A}(\boldsymbol{x}) = \tfrac{1}{2}B(-y, x, 0). \tag{11.60}$$

Assuming that the mean value of ϕ is a constant, as in (11.56), the effective Hamiltonian becomes

$$H_{\text{eff}} = \int d^3x[\tfrac{1}{2}B^2 + e^2B^2(x^2 + y^2)|\phi|^2 + r_0|\phi|^2 + \tfrac{1}{4}u_0|\phi|^4 - BH] \tag{11.61}$$

It is the second term which leads to the Meissner effect. The integral of $(x^2 + y^2)$ over the volume V of the sample is proportional to $V^{5/3}$, the exact value depending on the shape of the sample. This gives a contribution to the free energy *per unit volume* proportional to $B^2|\phi|^2V^{2/3}$, which is infinite in the thermodynamic limit, or at least very large for a macroscopic sample, if neither B nor ϕ is zero. We therefore conclude that B cannot be non-zero in a region of macroscopic size within a superconductor. There are thus two possible minima of (11.61), namely a normal state with $B = H$ and $\phi = 0$, and a superconducting state with $B = 0$ and $|\phi|^2 = -2r_0/u_0$. The free energies per unit volume of these two states are

$$F_n/V = -\tfrac{1}{2}H^2 \quad \text{and} \quad F_s/V = -r_0^2/u_0. \tag{11.62}$$

The stable equilibrium state is the one with the lower free energy. At a fixed temperature below T_c, therefore, the superconducting state is stable, provided that the applied field is smaller than a critical value given by

$$H_c = (-2r_0^2/u_0)^{1/2}. \tag{11.63}$$

When a field larger than this is applied, the superconductivity is destroyed. Near T_c, the critical field varies as $(T_c - T)^{1/2}$. At lower temperatures, however, we should need more detailed information about the dependence of r_0 on T in order to find the temperature dependence of the critical field. This can be done empirically or by deriving the effective Hamiltonian as an approximation to a detailed microscopic theory.

11.7.3 The Higgs mechanism

The nature of fluctuations in a superconductor is different from that described earlier because the effective Hamiltonian (11.59) is invariant under *local* gauge transformations (see chapter 8) whereas (11.55) has only a *global* gauge symmetry. Indeed, the term $B^2/2$ in (11.59) is the three-dimensional analogue of the F^2 term in, for example, (8.17). The magnetic induction is unchanged if we add to $\boldsymbol{A}$ the gradient of any scalar function. We can again study fluctuations by substituting (11.57) into (11.59). The only place where the phase fluctuation θ appears is in the covariant derivative term, which becomes

$$\frac{1}{2}\left|\nabla\chi + \mathrm{i}(\sqrt{2}v + \chi)\left(\frac{1}{\sqrt{2}v}\nabla\theta - 2eA\right)\right|^2 . \qquad (11.64)$$

Therefore, if we add to A the quantity $\nabla(\theta/2\sqrt{2}ev)$, then θ disappears entirely, and the effective Hamiltonian can be written in the form

$$H_{\mathrm{eff}} = \frac{1}{2}\int \mathrm{d}^3x[|\nabla \times A|^2 + (2\sqrt{2}ev)^2A^2 + |\nabla\chi|^2 + (-2r_0)\chi^2] + H_{\mathrm{int}} \qquad (11.65)$$

where H_{int} contains higher-order terms describing self-interactions of χ and interactions between χ and A. We see that the excitations are χ fluctuations of 'mass' $(-2r_0)^{1/2}$ and 'photons' of mass $2\sqrt{2}ev$. In a superconductor, the 'mass' of the χ excitations is to be interpreted in terms of correlation length $\xi = (-2r_0)^{-1/2}$, which in this context is called the *coherence length*. By analogy wih (9.85), we can identify a second characteristic distance, $\lambda = 1/2\sqrt{2}ev$, called the *penetration depth*, which governs the rate of decay of magnetic forces inside the superconductor. Roughly speaking, when a magnetic field weaker than H_{c} is applied to a superconducting specimen, the magnetic flux penetrates the superconductor to a depth λ.

It is clear that exactly the same analysis will carry over to a genuine relativistic gauge field theory. At the simplest level, we might consider the action

$$S = \int \mathrm{d}^4x[-\tfrac{1}{4}F_{\mu\nu}F^{\mu\nu} + (\mathrm{D}_\mu\phi)^*\mathrm{D}^\mu\phi - m_0^2\phi^*\phi - \tfrac{1}{4}\lambda_0(\phi^*\phi)^2] \qquad (11.66)$$

where $\mathrm{D}_\mu = \partial_\mu + \mathrm{i}eA_\mu$ is the gauge-covariant derivative. When m_0^2 is positive, this describes scalar particles of charge e and their antiparticles of charge $-e$ interacting with massless photons. Bearing in mind that the photon has only two independent spin polarization states, this gives a total of four physical degrees of freedom. When m_0^2 is negative, the gauge symmetry is spontaneously broken. The theory then describes a single scalar χ particle interacting with a massive spin-1 particle, which is no longer recognizable as a photon. The massive spin-1 particle has three independent spin states, so there are again a total of four physical degrees of freedom. We may say that one of the scalar degrees of freedom, namely the phase angle which disappears, has combined with the redundant gauge degrees of freedom to produce the third physical polarization state of the spin-1 particle. In the context of particle physics, this is known as the *Higgs mechanism* (after P Higgs, who first described it). The Higgs mechanism solves the problem we encountered in chapter 8 of constructing a gauge theory in which the gauge quanta are massive and can be identified with observed particles such as the W and Z. This was the last barrier in the way of constructing a unified theory of strong, weak and electromagnetic interactions, and I shall describe this construction in the next chapter. The price to be paid is that we have to introduce scalar fields. Some of these fields, analogous to χ, should correspond to observable spin-0 particles, called *Higgs bosons*. Such particles have never been observed. The reason for this could be that their masses, whose values cannot be reliably predicted, are too large for them to be produced at current accelerator energies, but their non-appearance leaves some cause for dissatisfaction with an otherwise excellent theory.

Exercises

11.1 For a ferromagnet at its critical temperature, the magnetization is found to vary with magnetic field as $M \sim h^{1/\delta}$, where δ is a critical exponent. Show that the Ginzburg–Landau theory gives $\delta = 3$. It can often be shown that the free energy can be expressed near a critical point in the *scaling* form

$$F(t, h) = |t|^{2-\alpha} f(h/|t|^{\Delta})$$

where α and Δ are two further critical exponents. Thus, up to an overall factor, it depends only on the single variable $h/|t|^{\Delta}$ rather than on h and t independently. Show that if the scaling form is correct, then the specific heat at $h = 0$ diverges as $C \sim |t|^{-\alpha}$. Show that the free energy of the Ginzburg–Landau theory does have the scaling form, with $\alpha = 0$. For any free energy which can be expressed in scaling form, show that

(a) $\beta = 2 - \alpha - \Delta$ and $\gamma = \Delta - \beta$

(b) when $y = h/|t|^{\Delta} \to \infty$, $\mathrm{d}f(y)/\mathrm{d}y \sim y^{1/\delta}$

(c) $\Delta = \beta\delta$

(d) $\gamma = \beta(\delta - 1)$

and check these results for the Ginzburg–Landau theory. The scaling property and the relations between critical exponents which follow from it are an automatic consequence of the renormalization group analysis (see, for example, Amit (1978)).

11.2. When a ferromagnet contains two or more domains, or a liquid coexists with its vapour, there is a narrow region—a domain wall or interface—between the two phases in which the magnetization or density varies quite rapidly. Consider equation (11.27) with $h = 0$ and suppose that M depends on only one spatial coordinate, say z. Show that this equation has a *soliton* solution of the form

$$M(z) = M_S \tanh(\lambda z)$$

and identify the constant λ. Hence show that the thickness of the domain wall is approximately equal to the correlation length. Note that this applies to an *Ising* ferromagnet, in which the magnetization can point only in one of two opposite directions. In a *Bloch wall* the magnetization *rotates* as we pass through the wall, and the thickness depends on the anisotropy energy, which is the increase in a spin's potential energy as it rotates away from the easy axis. Can you develop a variant of the Ginzburg–Landau theory to investigate this possibility? (See Lawrie and Lowe (1981).)

12

Unified Theories of the Fundamental Interactions

We saw in chapter 8 that a special class of interacting field theories, the *gauge theories*, arise almost inevitably when we investigate the relationship between the 'internal spaces' in which fields or wavefunctions exist at different points of spacetime. We found that the simplest of these theories can be interpreted in terms of observed electromagnetic forces and, indeed, that quantum electrodynamics agrees with experimental measurements with extremely high precision. In this chapter, I shall describe how the weak and strong nuclear interactions can also be interpreted in terms of gauge theories. It would be most satisfying if the three interactions could be explained in terms not of three different gauge theories but of a single unified theory. Such theories have, as we shall see, been proposed. Just what is entailed in this unification will become clear as we proceed, but it is not entirely clear at the time of writing whether a completely unified theory can be achieved, or whether such a theory could be subjected to any very stringent experimental test.

It will, of course, be necessary to have some idea of the observed phenomena which need to be explained. High-energy particle physics is a large and rather technical subject, and it will be possible for me to give only a cursory description of some of the key facts which have emerged from many years of research. The weak interaction, because of its weakness, is amenable to theoretical treatment on the basis of perturbation theory and is now quite well understood. Strong interaction phenomena, on the other hand, can often not be adequately treated by perturbation theory and, because of the difficulty of devising alternative methods of approximation, are not really understood with the same degree of confidence.

It is worth considering briefly just what 'understanding' means in this context. It is generally agreed that fundamental processes can be described by some form of relativistic quantum field theory. Therefore, a large part of the problem is to be able to write down the action (or Lagrangian density) from which observed phenomena can, in principle, be seen to stem. At the present time, it seems that an action incorporating the weak, strong and electromagnetic interactions can be written down with some confidence. It is called the *standard model*. A second part of the problem is to be able actually to derive all the observable consequences of this model, and it is here that the strong interaction still present difficulties. To some extent, therefore, our confidence in the standard model is a matter of theoretical prejudice. A third aspect of understanding is to decide whether the model which accounts for our current observations is truly fundamental. We have seen that large-distance or low-energy phenomena can be well described on the basis of effective Hamiltonians which bear rather little resemblance to the models we believe to represent the microscopic physical constitution of the system we study. It is entirely possible that the standard model of particle physics is itself only an effective action, valid only for the range of energies which can be produced in present-day accelerators. We shall see that there are some theoretical reasons for believing that this is indeed the case.

12.1 The Weak Interaction

The simplest reason for distinguishing weak, electromagnetic and strong interactions is that a hierarchy is observed in the magnitudes of quantities such as scattering cross-sections and decay rates which, on the basis of formulae such as those given in appendix 4, we are inclined to attribute to a corresponding hierarchy of coupling constants. (We shall see, however, that the situation is more subtle than this.) For example, the neutral pion π^0 decays to two photons, an effect we naturally associate with electromagnetism, with a mean lifetime of about 10^{-16} s. A muon, on the other hand, lives for some 10^{-6} s before decaying, through what we identify as a weak interaction process, into an electron, a neutrino and an antineutrino. The beta decay of a free neutron into a proton, electron and antineutrino takes, on average, about 15 min, but this is exceptional even for weak interactions and is explained by the very low energy involved. The lifetimes of particles which decay by the strong interactions are typically of the order of 10^{-23} s.

In the early days of particle physics, the particles themselves were classified according to their masses into *leptons* (light particles), *baryons* (heavy particles) and *mesons* (particles of intermediate mass). In the light of improved understanding, a more detailed classification seems appropriate, which is the following. Particles which undergo strong interactions are called *hadrons*, and these can be subdivided into fermionic hadrons, the *baryons*, of which the most familiar examples are protons and neutrons, and bosonic hadrons, such as pions and kaons, which are *mesons*. Fermionic particles which have strong interactions are called *leptons*. They include the electron, the muon and the more recently discovered tau lepton, which are all negatively charged (their antiparticles being positive) and three species of neutrino, which appear to be associated with the three charged lepton species. All experimental evidence is consistent with the neutrinos being exactly massless and, although very small neutrino masses cannot be ruled out, this masslessness is assumed in most theoretical models. While the observed hadrons have complicated internal structure, consistent with their being composed or more fundamental particles, the *quarks*, there is no evidence that the leptons have any internal structure, and they are normally taken to be truly fundamental. The photon and the more recently discovered W^+, W^- and Z^0 particles occupy a distinguished position in this classification scheme, being (in theory) quanta of the gauge fields which mediate the electromagnetic and weak interactions. In the standard model, there are further gauge bosons, the *gluons*, associated with the strong interaction but these, like the quarks, have not been detected in isolation.

In the early 1970s, all known weak interaction phenomena could be reasonably well described by applying first-order perturbation theory to a field theory in which interactions were represented by a term in the Lagrangian density of the form

$$\mathscr{L}_{\mathrm{I}} = -\frac{1}{\sqrt{2}} G_{\mathrm{F}} \mathscr{J}_{\mu}^{\dagger} \mathscr{J}^{\mu}. \tag{12.1}$$

An interaction of this kind, known as the *current–current* interaction was first suggested by E Fermi. The current in question is given by

$$\mathscr{J}^{\mu} = \bar{\nu}_{\mathrm{e}} \gamma^{\mu}(1 - \gamma^5) e + \bar{\nu}_{\mu} \gamma^{\mu}(1 - \gamma^5) \mu + \text{hadronic terms} \tag{12.2}$$

where e and μ stand, respectively, for the electron and muon field operators, while ν_{e} and ν_{μ} are the field operators for the electron and muon neutrinos. The coupling constant G_{F} is called the *Fermi constant*, and its value is given by $G_{\mathrm{F}}/(\hbar c)^3 = 1.16\,\mathrm{GeV}^{-2}$. The interaction (12.1) contains contains a number of terms, each giving rise to a different kind of process. For example, muon decay ($\mu^- \to e^- + \bar{\nu}_{\mathrm{e}} + \nu_{\mu}$) is described by the vertex

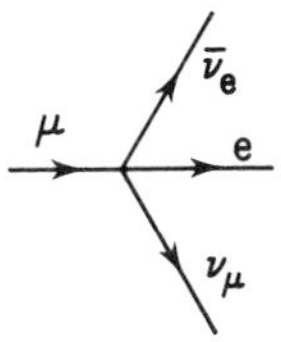

$$-\frac{1}{\sqrt{2}} G_{\mathrm{F}} \bar{e} \gamma_{\mu}(1 - \gamma^5) \nu_{\mathrm{e}} \cdot \bar{\nu}_{\mu} \gamma^{\mu}(1 - \gamma^5) \mu \tag{12.3}$$

where the field operator μ annihilates the decaying muon and the other three operators create the outgoing particles. The nature of the hadronic terms in (12.1) depends upon the kind of calculation we wish to undertake. For example, neutrino–neutron scattering ($\nu_e + n \rightarrow e^- + p$) could be decribed by a vertex of the form

$$-\frac{1}{\sqrt{2}} G_F \bar{e} \gamma_\mu (1 - \gamma^5) \nu_e \cdot \bar{p} \Gamma^\mu n. \qquad (12.4)$$

In this expression, n and p are to be treated as field operators for the neutron and proton and Γ^μ is a matrix, constructed from Dirac γ matrices, which represents strong interaction effects, involving the internal structure of the proton and neutron. In a theory of weak interactions only, Γ^μ is simply fitted to experimental data. In a theory which also purports to describe the strong interaction, we would instead construct contributions to the current (12.2) in terms of quark operators, of the same kind as those for the leptons. However, when we then calculate S-matrix elements as in (9.16), the 'in' and 'out' states still contain a proton or a neutron rather than free quarks, and we should have to find a means of calculating the effect of acting with quark operators on these states. This difficult task is equivalent to *calculating* Γ^μ from first principles.

The current (12.2) is called a *charged current*, because it has the net effect of raising by one unit the electric charge of a state on which it acts. For example, the electronic term annihilates a negative electron and creates a neutral neutrino. The form of this current and the interaction (12.1) is conjectured partly as a matter of theoretical prejudice and partly on the basis of experimental data. Since we believe the leptons to be truly fundamental particles, we expect that their interactions should be describable by a simple expression, involving a minimal number of adjustable parameters. The idea of using currents is motivated by the success of quantum electrodynamics, where the interactions of charged particles can indeed be expressed in terms of the electromagnetic current (9.79). The weak interaction currents are necessarily different, because they have to interconvert particles of different species. In principle, they might involve any or all of the bilinear covariants S, P, V^μ, A^μ and $T^{\mu\nu}$, discussed in chapter 7, with the obvious proviso that the two field operators do not necessarily refer to the same species. The particular form which is chosen summarizes a large amount of experimental data, of which I have space only to indicate a few important features.

The most significant feature is *parity violation*. Readers will recall from chapter 7 that the parity transformation is a change of coordinates which reverses the sign of all spatial coordinates. This is more or less equivalent to forming the mirror image of a physical state. (Strictly speaking, account must also be taken of the *intrinsic parity* of each particle species, as is explained in any particle physics textbook, but I shall not need to make use of this.) For a long time, it was believed that parity should be a symmetry of the fundamental interactions, in the sense that any state should evolve with time in the same way as its mirror image. This means that the Lagrangian density should be unchanged by a parity transformation. It was first suggested by T D Lee and C N Yang that this symmetry is in fact violated by the weak interaction. This was confirmed by C S Wu, who studied the beta decay of cobalt-60 and found an asymmetry in the numbers of electrons emitted parallel and antiparallel to the nuclear spin. In the mirror image system, this asymmetry would be reversed, and so parity is violated. Now, each of the leptonic terms in (12.2) has the form $V^\mu - A^\mu$, where V^μ is a vector current and A^μ an axial vector. If we consider the more general form

$$\mathcal{J}^\mu \propto (1 - \alpha^2)^{1/2} V^\mu + \alpha A^\mu \tag{12.5}$$

then for the interaction we have

$$\mathcal{J}^\dagger_\mu \mathcal{J}^\mu \propto [\alpha^2 V^\dagger_\mu V^\mu + (1 - \alpha^2) A^\dagger_\mu A^\mu] + \alpha(1 - \alpha^2)^{1/2}[V^\dagger_\mu A^\mu + A^\dagger_\mu V^\mu]. \tag{12.6}$$

According to the transformation rules given in chapter 7, the first term is unchanged by the parity transformation while the second changes sign. Thus, parity violation comes about through the interference between vector and axial vector currents and is a maximum when $\alpha = \pm 1/\sqrt{2}$. Thus, the $V^\mu - A^\mu$ form of the currents corresponds to *maximal parity violation*.

The reason for choosing $V^\mu - A^\mu$ rather than $V^\mu + A^\mu$ comes from observations of neutrinos. We saw in chapter 7 that, for massless particles, the *chiral projections* (7.74) correspond to helicity eigenstates. Experimentally, neutrinos are always observed in the left-handed polarization state while antineutrinos are always right handed. Readers should be able to convince themselves that only these states can be created by the $V^\mu - A^\mu$ current interaction.

It is, of course, possible to write down more general interactions involving the S, P and $T^{\mu\nu}$ covariants. When applied to muon decay, nuclear beta decay and neutrino–nucleus scattering, the various terms lead to different dependences on the angles between momenta and spins of the various particles involved. These place quite stringent limits on any possible contributions from scalar or tensor interactions. A sensitive test for the presence of pseudoscalar interactions is provided by the decay of charged pions. These decays almost always produce a muon and a neutrino, but a fraction of about 1.27×10^{-4} of pion decays produce instead an electron and a neutrino. Calculations show that if the interaction were entirely pseudoscalar, then the electronic decays would, on the contrary, be about five times more frequent than the muonic ones. Calculations based on the $V^\mu - A^\mu$ interaction, however, agree well with the observed ratio, so any pseudoscalar interaction must be extremely small. This close agreement also provides good evidence for *electron–muon universality*, which refers to the fact that the electron and muon currents appear in (12.2) with the same weight and therefore have weak interactions of the same strength. The value of G_F can be found by comparing calculated lifetimes both of muons and of nuclei which undergo beta decay with experimentally measured values, and consistent results are obtained by these two methods.

Because of the difficulty of carrying out reliable strong interaction calculations, there is less detailed information about the form of hadronic currents. If, in the vertex (12.4), it is assumed that

$$\Gamma^\mu = \gamma^\mu(C_V + C_A\gamma^5) \tag{12.7}$$

then it is found that $C_A/C_V \approx -1.26$. This can be taken as evidence for an underlying $V^\mu - A^\mu$ structure in the hadronic currents also.

Although the current–current interaction is able to account for quite a large body of observed phenomena, it has some important shortcomings. One is that there are some phenomena for which it cannot account, which I shall mention later. Theoretically, it has two highly undesirable features. One is that it does not satisfy the requirement of *unitarity*. Reduced to its simplest terms, this requirement means that, given an initial state, the total probability of observing *some* final state must be equal to 1. More technically, it means that the operator S which transforms 'in' states into 'out' states as in (9.5) must be unitary. From this it can be shown to follow that the total cross-section for, say, electron–neutrino scattering must decrease at high energies at least as fast as constant$/q^2$, where q is the total 4-momentum. When such cross-sections are calculated from the Fermi theory, they are found to *increase* as $G_F^2q^2$, as might be expected from dimensional analysis, so unitarity is violated. A related problem is that the theory is not renormalizable. Since the coupling constant G_F has the dimensions $(\text{energy})^{-2}$, the criterion for renormalizability discussed in chapter 9 is not satisfied. Therefore, at all orders of perturbation theory beyond the first, there are infinities which cannot be renormalized away and the theory does not make sense.

The accepted cure for these problems is to introduce an *intermediate vector boson*. If the field operator for this spin-1 particle is W^μ, then the current–current interaction is replaced by something like $g(\mathcal{J}^\dagger_\mu W^\mu + \mathcal{J}_\mu W^{\mu\dagger})$, since the action must be Hermitian. This is obviously similar to the electromagnetic interaction in (9.78) and, in particular, the new coupling constant g is dimensionless. The effect of this replacement upon processes of the kind we have been considering is to split four-fermion vertices like those in (12.3) and (12.4) into a pair of vertices of the kind which occur in QED, connected by a W propagator:

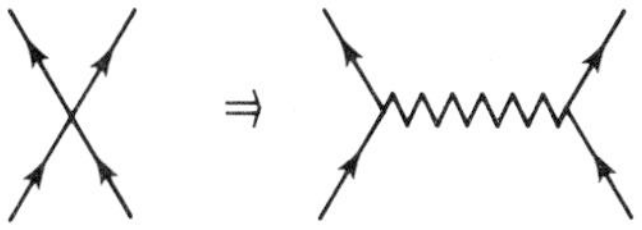

Ignoring technical details for the moment, this implies a corresponding replacement for the Fermi constant:

$$G_F \Rightarrow \frac{-g^2}{k^2 - M_W^2} \tag{12.8}$$

where k is the 4-momentum transferred between the two halves of the vertex. When the magnitude of this 4-momentum is much smaller than the mass M_W of the intermediate vector boson, this is just a constant, and we recover the Fermi theory with $G_F = g^2/M_W^2$. At high energies, however, the composite vertex behaves as $-g^2/k^2$ and this, other things being equal, solves the problem of unitarity and renormalizability.

Models of the weak interaction based on the idea of an intermediate vector boson were suggested by S Glashow (1961) and by A Salam and J C Ward (1964), but they lacked the crucial property of gauge invariance which, as we saw in chapter 9, is essential for a theory containing spin-1 particles to be renormalizable. The missing ingredient was the Higgs mechanism, discussed in the previous chapter, which allows masses for the spin-1 particles to be generated within a gauge-invariant theory by spontaneous symmetry breaking. A highly successful model which incorporates the Higgs mechanism was devised by S Weinberg (1967) and by Salam (1968). At the time, it was not entirely clear whether even this model would really be renormalizable, but its renormalizability was finally proved by G t'Hooft (1971).

12.2 The Glashow–Weinberg–Salam Model for Leptons

The Glashow–Weinberg–Salam model (which I shall abbreviate henceforth to GWS) is a non-Abelian gauge theory. As I explained it in chapter 8, these theories involve grouping observed particles into multiplets and regarding the members of a multiplet as different states of the same basic particle. Our problem is, of course, to decide which groups of particles nature actually does regard in this way. The groupings which have been found to work involve a further subtlety, which may appear strange at first sight. It will be convenient at the beginning to imagine that both the electron and its neutrino are massless and to endow the electron with a mass at a later stage. For massless particles, the right- and left-handed helicity states, whose field operators are the chiral projections (7.74), can be treated quite independently (see exercise 7.9). Since right-handed neutrinos are not observed, we can assume that they do not exist.

Consider now the electronic part of the current (12.2). It will be convenient to redefine it by inserting a factor of $\frac{1}{2}$. Because of the anticommutation relation (7.52), we see that it involves only the left-handed components of both the neutrino and the electron:

$$\mathcal{J}^{\mu}_{\mathrm{e}} = \bar{\nu}_{\mathrm{e}}\gamma^{\mu}\tfrac{1}{2}(1-\gamma^5)e = \bar{\nu}_{\mathrm{e}}\tfrac{1}{2}(1+\gamma^5)\gamma^{\mu}\tfrac{1}{2}(1-\gamma^5)e = \bar{\nu}_{\mathrm{eL}}\gamma^{\mu}e_{\mathrm{L}}. \quad (12.9)$$

These two left-handed components are assigned to a doublet, analogous to (8.18):

$$l_{\mathrm{e}} = \begin{pmatrix} \nu_{\mathrm{eL}} \\ e_{\mathrm{L}} \end{pmatrix}. \quad (12.10)$$

The notation indicates a doublet of left-handed electron-type particles. This commits us to an SU(2) × U(1) gauge theory like that discussed in §8.3, and the SU(2) property is called *weak isospin* to distinguish it from the nuclear isotopic spin. The doublet has, of course, a weak isospin of $t = \frac{1}{2}$, with $t^3 = +\frac{1}{2}$ for the neutrino and $-\frac{1}{2}$ for the electron. To get the correct electric charges from the Gell–Mann–Nishijima formula (8.51), we assign to the doublet a *weak hypercharge* of $y = -1$. As in §8.2, we now use the Pauli matrices to construct the current (12.9) as

$$\mathcal{J}^{\mu}_{\mathrm{e}} = \bar{l}_{\mathrm{e}}\gamma^{\mu}\tau^{+}l_{\mathrm{e}} \quad (12.11)$$

where

$$\tau^{+} = \tfrac{1}{2}(\tau^1 + \mathrm{i}\tau^2) = \begin{pmatrix} 0 & 1 \\ 0 & 0 \end{pmatrix}. \quad (12.12)$$

The coupling of this current to a gauge field must contribute to the Lagrangian density an Hermitian operator, consisting of the two terms

$$\mathcal{L}_{\mathrm{I}} = -\frac{g}{\sqrt{2}}\,\bar{l}_{\mathrm{e}}(\tau^{+}W^{+} + \tau^{-}W^{-})l_{\mathrm{e}} = -\frac{g}{2}\,\bar{l}_{\mathrm{e}}(\tau^1 W^1 + \tau^2 W^2)l_{\mathrm{e}} \quad (12.13)$$

where $\tau^{-} = (\tau^1 - \mathrm{i}\tau^2)/2$, $W^1_{\mu} = (W^{+}_{\mu} + W^{-}_{\mu})/\sqrt{2}$ and $W^2_{\mu} = \mathrm{i}(W^{+}_{\mu} - W^{-}_{\mu})/\sqrt{2}$. Since the current (12.11) annihilates a negative electron or creates a positron, W^{+}_{μ} must, to conserve charge, annihilate a positive gauge boson W^{+} or create its negative antiparticle, while its adjoint operator W^{-}_{μ} has the converse effect. This form of interaction will reproduce the Fermi theory of charged weak currents in the manner I described qualitatively in the previous section. It will not yet, however, lead to a gauge-invariant theory. By comparison with the SU(2) theory developed in chapter 8, we see that a third gauge field, W_μ^3, coupled to a new current, is required to make the interaction invariant under SU(2) rotations. Thus, we must enlarge (12.13) to read

$$\mathcal{L}_{\rm I} = -\frac{g}{2}\,\bar{l}_{\rm e}(\tau^1 \not{W}^1 + \tau^2 \not{W}^2 + \tau^3 \not{W}^3)l_{\rm e} = -g\bar{l}_{\rm e}\boldsymbol{t}\cdot\not{\boldsymbol{W}}l_{\rm e} \qquad (12.14)$$

where the three matrices $\boldsymbol{t} = \boldsymbol{\tau}/2$ are the generators of the isospin-$\frac{1}{2}$ representation. The extra current, given by

$$\bar{l}_{\rm e}\tau^3\gamma^\mu l_{\rm e} = \bar{\nu}_{\rm e}\gamma^\mu\nu_{\rm e} - \bar{e}\gamma^\mu e \qquad (12.15)$$

is a *neutral current*, which has no net effect on the charge of a state on which it acts. The second term is clearly proportional to the electromagnetic current. As we shall see, however, the gauge-invariant theory also involves a *weak* neutral current and thus predicts new interaction effects, which have indeed been observed.

In order to incorporate electromagnetism correctly, it is necessary to include a fourth gauge field B_μ associated with phase transformations. As we saw in chapter 8, the U(1) group of electromagnetism is not the same as the U(1) group of phase transformations. In accordance with the Gell-Mann–Nishijima formula, we shall find that the electromagnetic field A_μ is a linear combination of B_μ and W_μ^3. It is, of course, most gratifying that this leads to a description of both weak and electromagnetic forces within a unified framework, and it should be noted that we cannot treat the weak interaction in isolation by ignoring the phase transformations. (Readers should be able to satisfy themselves that this separation of weak and electromagnetic forces, with the electromagnetic field being just B_μ, would be possible if and only if the electron and neutrino had had the same charge.) At this point, the total Lagrangian density reads

$$\mathcal{L} = -\tfrac{1}{4}F_{\mu\nu}^{(W)}F^{(W)\mu\nu} - \tfrac{1}{4}F_{\mu\nu}^{(B)}F^{(B)\mu\nu} + \bar{l}_{\rm e}{\rm i}\gamma^\mu(\partial_\mu + {\rm i}g\boldsymbol{W}_\mu\cdot\boldsymbol{t} + {\rm i}g'\tfrac{1}{2}yB_\mu)l_{\rm e} \qquad (12.16)$$

where the field strength tensor $F^{(W)}$ is constructed from $\boldsymbol{W}_\mu$ in the same way as (8.37), with the SU(2) structure constants $C^{abc} = \varepsilon^{abc}$, and $F^{(B)}$ from B_μ as in (8.14). The two coupling constants g and g' associated with the two groups SU(2) and U(1) are independent. This Lagrangian density is invariant under the SU(2) × U(1) gauge transformations

$$\begin{aligned} l_{\rm e} &\to \exp[{\rm i}\tfrac{1}{2}y\theta(x) + {\rm i}\boldsymbol{\alpha}(x)\cdot\boldsymbol{t}]l_{\rm e} \equiv \exp[{\rm i}\tfrac{1}{2}y\theta(x)]U(\boldsymbol{\alpha})l_{\rm e} \\ W_\mu &\to U(\boldsymbol{\alpha})W_\mu U^{-1}(\boldsymbol{\alpha}) + ({\rm i}/g)(\partial_\mu U(\boldsymbol{\alpha}))U^{-1}(\boldsymbol{\alpha}) \qquad (12.17) \\ B_\mu &\to B_\mu - (1/g')\partial_\mu\theta. \end{aligned}$$

As in chapter 8, the matrix W_μ is defined as $\boldsymbol{W}_\mu\cdot\boldsymbol{t}$.

So, far, neither the gauge bosons nor the electron have masses. To put this right, without losing the gauge invariance, we must introduce a Higgs scalar field, as described in the previous chapter. In the simplest version of the GWS theory, it is an SU(2) doublet

$$\phi = \begin{pmatrix} \phi^+ \\ \phi^0 \end{pmatrix}. \tag{12.18}$$

The component ϕ^0 will be given a vacuum expectation value v

$$\langle 0|\phi|0\rangle = \begin{pmatrix} 0 \\ v \end{pmatrix} \tag{12.19}$$

so, since the vacuum contains no electric charge, the Higgs doublet must have hypercharge $y = 1$, making the ϕ^0 particles neutral. We add to the Lagrangian density (12.16) the quantity

$$\mathscr{L}_{\text{Higgs}} = (\mathrm{D}_\mu\phi)^\dagger(\mathrm{D}_\mu\phi) - \tfrac{1}{4}\lambda[(\phi^\dagger\phi)^2 - v^2]^2 \tag{12.20}$$

where the covariant derivative is

$$\mathrm{D}_\mu\phi = [\partial_\mu + \mathrm{i}g\boldsymbol{W}_\mu\cdot\boldsymbol{t} + \mathrm{i}g'\tfrac{1}{2}B_\mu]\phi. \tag{12.21}$$

Mathematically, of course, any constant value of ϕ such that $\phi^\dagger\phi = v^2$ is a minimum of the potential in (12.20). By making a gauge transformation, it is always possible to bring this expectation value into the form (12.19). This transformation will also redefine the electron–neutrino doublet. Physically, the particles we recognize as electrons and neutrinos are those annihilated by the field operators which appear in this doublet after the transformation has been made.

To find the masses of the gauge bosons, we set ϕ equal to its vacuum expectation value, which gives

$$\mathscr{L}_{\text{Higgs}} = \tfrac{1}{2}(gv)^2 W_\mu^+ W^{-\mu} + \tfrac{1}{4}v^2(gW_\mu^3 - g'B_\mu)^2. \tag{12.22}$$

From the first term, we identify the mass of the W^+ particle and its antiparticle, the W^-, as

$$M_{\mathrm{W}}^2 = \tfrac{1}{2}(vg)^2. \tag{12.23}$$

The second term contains a linear combination of W_μ^3 and B_μ, which is to be identified as the field operator for a third weak gauge boson, the Z^0, and the orthogonal combination to this, which does not appear in (12.23), will be the field operator A_μ for massless photons. To make Z_μ and A_μ orthogonal and properly normalized, they must be of the form

$$A_\mu = \cos\theta_{\mathrm{W}} B_\mu + \sin\theta_{\mathrm{W}} W_\mu^3 \qquad Z_\mu = -\sin\theta_{\mathrm{W}} B_\mu + \cos\theta_{\mathrm{W}} W_\mu^3. \tag{12.24}$$

The angle θ_{W} is called the *weak mixing angle* or the *Weinberg angle*. Its value is not known *a priori,* but it can be measured, by methods I shall mention shortly, and is found to be given by

$$\sin^2\theta_W \approx 0.22 \quad \text{or} \quad \theta_W \approx 28°. \tag{12.25}$$

By comparing (12.24) with (12.22), we see that the Weinberg angle is related to the two coupling constants by

$$\tan\theta_W = g'/g \tag{12.26}$$

and that the mass of the neutral weak gauge boson is given by

$$M_Z^2 = \frac{1}{2}\left(\frac{vg}{\cos\theta_W}\right)^2 = \frac{M_W^2}{\cos^2\theta_W}. \tag{12.27}$$

We should now check that the remaining gauge field A_μ really does correspond to the electromagnetic field. To do this, we consider a special gauge transformation of the form (8.47). Since the factor of $y/2$ is already included in the first of equations (12.17), we simply take $\alpha_1 = \alpha_2 = 0$ and $\alpha_3 = \theta$. Using (12.24), we find

$$A_\mu \to A_\mu - \left(\frac{\cos\theta_W}{g'} + \frac{\sin\theta_W}{g}\right)\partial_\mu\theta \qquad Z_\mu \to Z_\mu. \tag{12.28}$$

W_μ^+ and W_μ^- are also unchanged. Thus, the electromagnetic gauge transformation affects only A_μ, as it should, and the transformation law is correct provided that the coefficient of $\partial_\mu\theta$ is $-1/e$. Usng (12.26), therefore, we find that the fundamental electric charge is given in terms of the SU(2) and U(1) coupling constants by

$$e = gg'/(g^2 + g'^2)^{1/2}. \tag{12.29}$$

Finally, we must arrange for the electron to have a mass. This requires a term in the Lagrangian equal to $-m\bar{e}e = -m(\bar{e}_R e_L + \bar{e}_L e_R)$. In the standard version of the GWS model, the right-handed component e_R is treated on a separate footing from e_L. Since e_R is not involved in the weak currents, it is assumed to be unaffected by the SU(2) transformations and thus to have weak isospin $t = 0$ and, to get its charge right, a hypercharge $y = -2$. For this reason, the mass term quoted above is not gauge invariant. The electron mass can be generated in a gauge-invariant manner from spontaneous symmetry breaking. We add to $\mathcal{L}$ the gauge-invariant expression

$$\Delta\mathcal{L}_e = \bar{e}_R i\gamma^\mu(\partial_\mu - ig'B_\mu)e_R - f_e(\bar{l}_e\phi e_R + \bar{e}_R\phi^\dagger l_e) \tag{12.30}$$

where f_e is a constant. The contribution to this from the vacuum expectation value of ϕ gives the required mass term with

$$m = f_e v. \tag{12.31}$$

The muon, tau lepton and their associated neutrinos can now be incorporated by adding to $\mathcal{L}$ further terms of exactly the same form as those involving the electron and its neutrino.

12.3. Physical Implications of the Model for Leptons

As far as the electroweak interactions of leptons are concerned, the model is now complete. The easiest way to see its physical implications is to derive an effective Lagrangian density with an interaction term similar to that of the Fermi theory (12.1). To do this, we first recall that particles associated with the Higgs field have not been observed. They must, therefore, have large masses, which means that their propagators are very small and make a negligible contribution to observed processes. The first step, therefore, is to eliminate them by setting ϕ equal to its vacuum expectation value. We expect the Fermi theory to be a good approximation for processes in which the energies involved are much smaller than the W and Z masses. Under these circumstances, the only important terms in $\mathscr{L}$ which involve the weak gauge bosons and their interactions with the leptons can be written as

$$\hat{\mathscr{L}} = M_{\mathrm{W}}^2 W_\mu^+ W^{-\mu} + \tfrac{1}{2} M_{\mathrm{Z}}^2 Z_\mu Z^\mu - \frac{g}{\sqrt{2}} (W_\mu^+ \mathscr{J}^\mu + W_\mu^- \mathscr{J}^{\dagger\mu}) - \frac{g}{\cos\theta_{\mathrm{W}}} Z_\mu \mathscr{J}_0^\mu. \tag{12.32}$$

The first two terms come from the Higgs field Lagrangian (12.22) and the others from the electronic term in (12.16) and similar terms for the other lepton species. The charged current is (12.11), again with additional muon and tau terms. The neutral current which couples to Z_μ is

$$\mathscr{J}_0^\mu = \tfrac{1}{2}\bar{\nu}_{\mathrm{eL}}\gamma^\mu\nu_{\mathrm{eL}} + (\sin^2\theta_{\mathrm{W}} - \tfrac{1}{2})\bar{e}_{\mathrm{L}}\gamma^\mu e_{\mathrm{L}} + \sin^2\theta_{\mathrm{W}}\bar{e}_{\mathrm{R}}\gamma^\mu e_{\mathrm{R}} \tag{12.33}$$

with additional muon and tau terms.

As far as W_μ and Z_μ are concerned, (12.32) is a quadratic form. Remembering that the Lagrangian density we have constructed is to be used in a functional integral, the integral over W_μ and Z_μ can be carried out in much the same way that we used, for example, to obtain (9.39). Defining the effective Fermi interaction by

$$\int \mathscr{D}W\mathscr{D}Z \exp\left[\mathrm{i}\int \mathrm{d}^4x\, \hat{\mathscr{L}}\right] = \text{constant} \times \exp\left[\mathrm{i}\int \mathrm{d}^4x\, \mathscr{L}_{\mathrm{I,eff}}\right] \tag{12.34}$$

we find

$$\mathscr{L}_{\mathrm{I,eff}} = -\frac{g^2}{2M_{\mathrm{W}}^2}[\mathscr{J}_\mu^\dagger \mathscr{J}^\mu + \mathscr{J}_{0\mu}\mathscr{J}_0^\mu]. \tag{12.35}$$

The first, charged current, term has the same form as (12.1), except that the currents in the GWS theory differ from those in the Fermi theory by a factor of $\tfrac{1}{2}$. We can therefore identify the Fermi constant as

$$G_{\mathrm{F}} = g^2/4\sqrt{2}M_{\mathrm{W}}^2. \tag{12.36}$$

From (12.26) and (12.29), we find that $g = e/\sin\theta_W$, so this can be rearranged to express the W mass as

$$M_W^2 = e^2/4\sqrt{2}G_F\sin^2\theta_W. \tag{12.37}$$

The values of e and G_F are well known from experiment, so we can now *predict* the mass of the W and, from (12.27), the mass of the Z, provided that the Weinberg angle can be ascertained. This angle appears in the neutral current (12.33), which is an addition to the Fermi theory. The neutral current leads to new processes such as the elastic scattering of neutrinos by electrons. The neutrino beams needed to observe these processes first became available in the early 1970s, when the predicted neutral current effects were indeed found, giving the first experimental evidence in favour of the GWS theory. The value of $\sin^2\theta_W$ emerging from these experiments is 0.217 ± 0.014. From this value, we get the following predictions for the W and Z masses:

$$M_W = 80.2 \pm 2.6 \text{ GeV} \qquad M_Z = 90.6 \pm 2.1 \text{ GeV}. \tag{12.38}$$

When these particles were actually observed at CERN in 1982–3, their masses were found to be $M_W = 80.8 \pm 2.7$ GeV and $M_Z = 92.9 \pm 1.6$ GeV, giving strong support to the GWS theory.

12.4 Hadronic Particles in the Electroweak Theory

The idea that the hadrons are composed of *quarks* was first put forward by M Gell-Mann and G Zweig in the early 1960s. The species or *flavours* of quarks which are currently thought to exist, together with their electric charges Q in units of e, are

up (u)	charmed (c)	top (t)	$Q = \frac{2}{3}$
down (d)	strange (s)	bottom (b)	$Q = -\frac{1}{3}$.

At the time of writing, however, particles containing the top quark have been observed. There are several kinds of evidence for the existence of quarks. The masses and magnetic moments of all the observed hadrons can be reasonably well accounted for by modelling them as bound states of quarks, each baryon being composed of three quarks and each meson of a quark and an antiquark. A few examples are the proton (uud), neutron (udd), Ω^- (sss), π^+ (u$\bar{\text{d}}$) and K^0 (d$\bar{\text{s}}$). All the particles expected on this basis are observed and all observed particles fit into the scheme. The transformations of observed particle species which occur in scattering and decay processes are all consistent with rearrangements of their quark contents. Moreover, the energy and angular dependences of scattering cross-sections at high energies are characteristic of those expected for scattering by point-like constituent particles. This is quite analogous to the strong backscattering of α particles which led Rutherford to postulate the existence of atomic nuclei. For details of these matters, I must ask readers to consult more specialized textbooks, some of which are listed in the bibliography.

As is apparent from the above list, the quarks appear in pairs, (u, d), (c, s) and (t, b), whose charges differ by one unit, and these pairs, like the charged lepton–neutrino pairs, form weak isospin doublets. There is, however, a complication. The three gauge fields $\boldsymbol{W}_\mu$ form a weak isospin triplet (see the discussion following (8.31)) but, as we have seen, W_μ^3 cannot be directly identified as the field operator for a particle because the term in (12.22) which generates the gauge boson masses involves a linear combination of W_μ^3 and Z_μ. Now, the quark masses will be generated by a term in the Lagrangian density similar to (12.30), and the fields which appear in this term are linear combinations of those needed to form the weak isospin doublets. What these linear combinations are is a matter to be determined experimentally, and I shall shortly give a brief discussion of what is involved. The fact that no difficulty was encountered for the leptons can be traced to the fact that all the neutrinos were assumed to have the same mass, namely zero. As with the leptons, then, the left-handed components of the various quarks are grouped into weak isospin doublets, with $t = \frac{1}{2}$ and $y = \frac{1}{3}$, to give the correct charges:

$$\begin{pmatrix} u \\ d' \end{pmatrix}_{\mathrm{L}} \quad \begin{pmatrix} c \\ s' \end{pmatrix}_{\mathrm{L}} \quad \begin{pmatrix} t \\ b' \end{pmatrix}_{\mathrm{L}} \tag{12.39}$$

where d', s' and b' are linear combinations of d, s and b. All the right-handed components are SU(2) singlets with hypercharge $y = \frac{4}{3}$ for u, c and t and $y = -\frac{2}{3}$ for d, s and b.

To see some of the implications of all this, let us construct the hadronic contribution to the charged current. It is helpful to express d', s' and b' in terms of d, s and b as

$$\begin{pmatrix} d' \\ s' \\ b' \end{pmatrix} = U \begin{pmatrix} d \\ s \\ b \end{pmatrix} \tag{12.40}$$

where U is a unitary matrix called, after its inventors, the *Kobayashi–Maskawa matrix*. This matrix can be written in terms of four *weak mixing angles*, analogous to the Weinberg angle. Following the pattern of (12.9), the hadronic current is

$$\mathcal{J}_{\mathrm{h}}^{\mu} = \bar{u}_{\mathrm{L}}\gamma^{\mu}d'_{\mathrm{L}} + \bar{c}_{\mathrm{L}}\gamma^{\mu}s'_{\mathrm{L}} + \bar{t}_{\mathrm{L}}\gamma^{\mu}b'_{\mathrm{L}} = (\bar{u}_{\mathrm{L}}, \bar{c}_{\mathrm{L}}, \bar{t}_{\mathrm{L}})\gamma^{\mu}U\begin{pmatrix} d_{\mathrm{L}} \\ s_{\mathrm{L}} \\ b_{\mathrm{L}} \end{pmatrix} \tag{12.41}$$

and the second form indicates that, had we also considered linear combinations of u, c and t, this would simply have meant redefining U. (More detailed arguments are necessary to show that this particular version of U is unitary, however.)

The situation is simpler if we ignore altogether the existence of the b quark (which was, indeed, unknown until about 1977) and the predicted t quark. In that case, U is a 2×2 matrix which can be parameterized by a single angle, the *Cabibbo angle* θ_C:

$$U = \begin{pmatrix} \cos\theta_C & \sin\theta_C \\ -\sin\theta_C & \cos\theta_C \end{pmatrix}. \tag{12.42}$$

The hadronic current (12.41) becomes

$$\mathcal{J}_h^\mu = (\bar{u}_L\gamma^\mu d_L + \bar{c}_L\gamma^\mu s_L)\cos\theta_C + (\bar{u}_L\gamma^\mu s_L - \bar{c}_L\gamma^\mu d_L)\sin\theta_C. \tag{12.43}$$

Consider, for example, the decay of a K^- meson, whose quark content is $(\bar{u}s)$, into a negative muon and an antineutrino ($K^- \to \mu^- + \bar{\nu}_\mu$). What happens, according to the GWS theory, is that the quark and antiquark annihilate to produce a virtual W^-, which subsequently decays to produce the leptons:

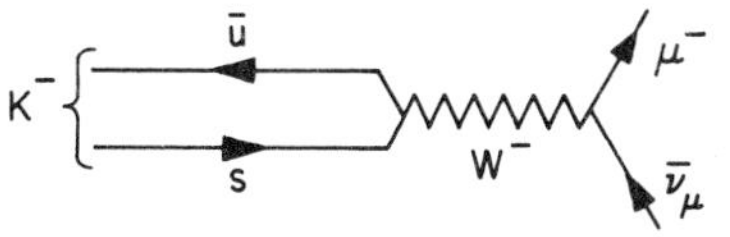

The field W_μ^+, which creates the W^-, couples to the hadronic current (12.43) in which the operator $\bar{u}\gamma^\mu s$ which annihilates the quarks has the coefficient $\sin\theta_C$. Thus, the $\bar{u}sW^-$ vertex has a factor of $\sin\theta_C$ and the decay rate a factor of $\sin^2\theta_C$. If there were no mixing, or, in other words, if d' and s' were identical with d and s, the decay could not take place. In terms of the Fermi theory, the K^- decay can be thought of as involving an effective Fermi constant $G_F \sin\theta_C$.

12.5. Colour and Quantum Chromodynamics

Although the GWS model as I have described it so far has a gauge-invariant action, it is not renormalizable. This is because of the occurrence of *anomalies*, which were mentioned in chapter 9. An example of part of a Feynman diagram whose divergence cannot be renormalized away is shown in figure 12.1. The theory will be renormalizable if the net contribution of all diagrams of this type is zero. Now, one such diagram can be formed with each charged fermion species circulating in the closed loop and, as it turns out, the condition for the divergences to cancel is that the sum of the charges of all these species is zero. In the standard model, this is true if two conditions are met. The first is that the fermion species fall into a number of complete

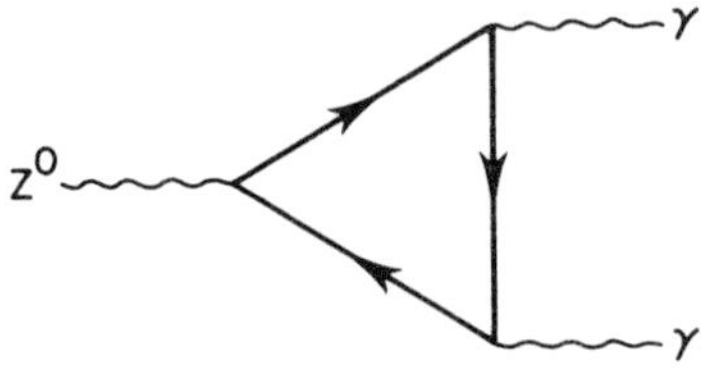

Figure 12.1 Example of a diagram which causes anomalies. Diagrams which contain this as a subdiagram cannot, in general, have their infinities removed by renormalization.

families or *generations*, each family comprising a neutrino, a negative lepton and a pair of quarks with charges of $\frac{2}{3}$ and $-\frac{1}{3}$. Provided that the expected top quark does exist, there are evidently three such families known, namely (ν_e, e, u, d), (ν_μ, μ, c, s) and (ν_τ, τ, t, b). The second condition is that each quark flavour should count as *three* species. In fact, it is believed that each flavour does indeed correspond to three distinct species, all having exactly the same mass and electroweak properties, but distinguished by a property called *colour*. There is no universal agreement on the three colours used to label these species, but the primary colours red, green and blue are commonly used. The earliest reason for this hypothesis was that some baryons appeared to consist of three identical quarks in a symmetric state, which is at variance with the fermionic nature of the quarks. This no longer presents a problem if the three quarks, while having the same flavour, are of different colours. Direct evidence for the existence of three colours comes from measuring the lifetime of the π^0. This particle is an antisymmetric combination of u$\bar{\text{u}}$ and d$\bar{\text{d}}$ bound states which decays to two photons via a Feynman diagram similar to figure 12.1. The integral turns out to be finite, but it is proportional to the number of quark species involved and gives the correct value only when allowance is made for three colours of quarks.

The existence of three quark colours provides the basis of the current theory of *strong interactions*, known as *quantum chromodynamics* or QCD. Here, I can do no more than outline some of its essential features. The three colours of a given quark flavour are taken to form a basic triplet

$$u = \begin{pmatrix} u_r \\ u_g \\ u_b \end{pmatrix} \qquad d = \begin{pmatrix} d_r \\ d_g \\ d_b \end{pmatrix} \quad \text{etc.} \tag{12.44}$$

The set of unitary transformations $u \to \exp[\mathrm{i}\boldsymbol{\alpha}(x)\cdot\boldsymbol{\lambda}]u$ which rearrange the three colours amongst themselves constitutes the *colour gauge group* SU(3). This group has eight generators. That is, there are eight linearly independent Hermitian λ matrices, analogous to the Pauli matrices of

SU(2). Consequently, when this group is used to construct a gauge theory, there are eight independent gauge fields and eight associated gauge bosons. These are called *gluons*, since they are held to form the 'glue' which binds quarks into the observed hadrons.

Unlike the electroweak theory, QCD contains no Higgs fields and so the gluons are massless. It might therefore appear that the colour forces should, like electromagnetic forces, have a long range and be easily detectable in the laboratory. It is believed, however, that QCD possesses a property known as *confinement*. The potential energy of two quarks increases linearly with the distance between them, rather than falling off. Thus, if we try to separate, say, the quark and antiquark in a pion, the increase in potential energy eventually favours the formation of a new quark–antiquark pair and we obtain not two widely separated quarks but two widely separated mesons. Only bound states which have no net colour (i.e. colour singlets) have a finite energy and this, in outline, explains why isolated quarks and gluons are never observed. The very different properties of QCD and QED can be traced to the non-Abelian nature of SU(3). As we saw in chapter 8, this implies that the gluons themselves carry a colour 'charge' and thus interact directly with each other, in contrast to photons which are electrically neutral.

While few theorists doubt the validity of this picture, it has not, as far as I know, been possible to give a definitive proof. The difficulty is that perturbation theory cannot be used. Perturbation theory, after all, assumes that the field operators in the theory can, to a first approximation, be interpreted as creation and annihilation operators for observable free particles, and in QCD this is not true. It has proved fruitful to consider an approximate theory in which spacetime is replaced by a discrete four-dimensional lattice of points, quite analogous to the lattice models of statistical mechanics. For such *lattice gauge theories*, the confinement property can be proved, but the proof does not necessarily remain valid when the lattice spacing is taken to zero. Using large computers, lattice gauge theories can also be used to investigate numerically properties such as the masses of hadrons formed from quarks and gluons. Such calculations have yielded encouraging results, but they do not, at the present time, appear capable of giving precise, reliable information. This is mainly because, in practice, the lattices which can be used are too small to be a good approximation to continuous spacetime.

The confinement of quarks (or, more accurately, of colour) is a large-distance of low-energy phenomenon. At high energies, QCD has the complementary property of *asymptotic freedom*. This means that the running coupling constant $\alpha_s(-q^2)$, the strong interaction equivalent of the energy-dependent fine structure constant (9.92), becomes very small at high energies. In fact, the result analogous to (9.94) is

$$\alpha_s(-q^2) = \alpha_s(\mu^2)\left[1 + (11 - \tfrac{2}{3}n_f)\frac{\alpha_s(\mu^2)}{4\pi}\ln\left(\frac{-q^2}{\mu^2}\right)\right]^{-1} \quad (12.45)$$

where n_f is the number of quark flavours. This high-energy behaviour is the opposite of that found in QED and clearly depends on the sign of the quantity $(11 - \frac{2}{3}n_f)$. The term $-\frac{2}{3}n_f$ arises from quark–gluon interactions, which have the same charge screening effect as electron–photon interactions in QED. The positive term, 11, comes from the self-interactions of gluons, and we see that it is this self-interaction which leads to asymptotic freedom, provided that there are no more than 16 flavours. Because of this property, perturbation theory can be used successfully to study high-energy processes for which $\alpha_s(-q^2)$ is sufficiently small. An example is the *deep inelastic scattering* of electrons by nucleons. The electron, of course, undergoes only electroweak interactions, and these processes can be understood in terms of interactions of a virtual photon with a single quark or gluon. A striking feature of the experimental data is the formation of *jets* of hadronic particles. These are interpreted as signalling the ejection from the nucleon of individual quarks or gluons, which subsequently acquire, through the creation of quark–antiquark pairs, the partners needed to form a shower of colourless hadrons (see figure 12.2). Detailed calculations of these and other processes are in quite good agreement with experimental data. To the extent that the success of perturbative QCD demonstrates the validity of asymptotic freedom, it places a limit, namely 16, on the number of quarks flavour (or a limit of eight on the number of families) which we might expect to discover. A much more stringent limit of four families can, however, be derived from cosmological considerations, which I shall mention in the next chapter. Unfortunately, it does not appear likely that either theoretical calculations or experiments could be devised which would provide tests of perturbative QCD comparable in accuracy with the tests of QED discussed in chapter 9.

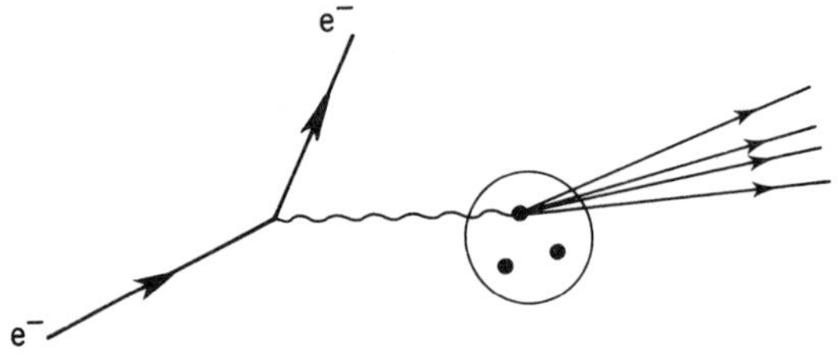

Figure 12.2 Schematic illustration of the formation of a quark jet. All the visible products of such an event are colourless hadrons. Some of them emerge as a roughly collimated jet, whose total momentum is that imparted to a single quark which was struck by the virtual photon.

It should be emphasized that QCD describes the strong interactions which bind quarks inside the observed hadrons. The forces which act between these hadrons, for example those which bind protons and neutrons in atomic nuclei or account for the low-energy scattering of protons and neutrons, should also have their origins in QCD, but they cannot be attributed to exchange of gluons. They are better understood as 'residual' strong forces. A fair analogy is provided by the van der Waals forces between gas molecules at large separations. Because each molecule is electrically neutral, there is no $1/r^2$ Coulomb force. However, some molecules possess permanent electric dipole moments and all molecules have fluctuating dipole moments which can polarize their neighbours. The resulting dipole–dipole interactions between molecules can be described as a residual electrostatic interaction resulting from incomplete cancellation of the principal Coulomb forces between their constituent particles.

Figure 12.3 illustrates, in terms of the flow of quarks, how the force between a proton and neutron can be attributed to the exchange of, for example, a neutral pion. The fundamental origin of the force is the QCD interaction which binds quarks in all three hadrons and causes the creation and annihilation of quark–antiquark pairs. However, their net effect at low energies or relatively large distances can be modelled by treating the pion as a fundamental spin-0 particle. This leads to a *one-particle exchange potential* of the form (9.85), with M being the pion mass, approximately equal to 135 MeV. The range of such a force is $\hbar/Mc \approx 1.5 \times 10^{-15}$ m which is indeed typical of the separation of two nucleons in a nucleus. This model is of rather restricted applicability. To improve on it, account must be taken of other mesons which might be exchanged and of the internal structure of these particles.

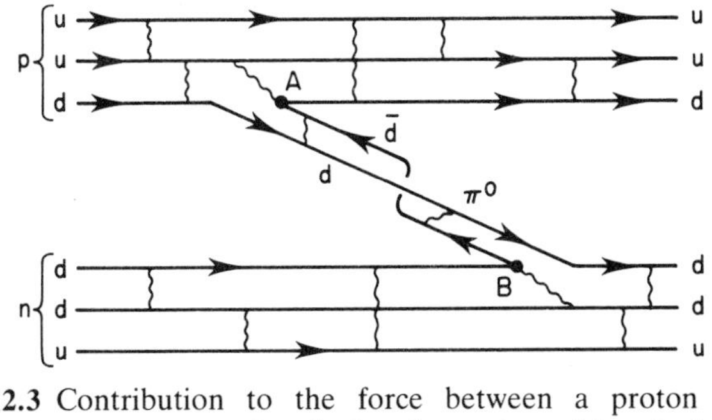

Figure 12.3 Contribution to the force between a proton and a neutron due to exchange of a π^0. Quarks are bound into hadrons by exchange of gluons. At A a gluon annihilates to form a d$\bar{d}$ pair and at B a d$\bar{d}$ pair annihilates to form a gluon. The net effect is the exchange of a π^0. Backward-pointing arrows denote a forward-moving antiquark.

12.6 Grand Unified Theories

The gauge theory whose construction I have outlined so far constitutes the *standard model* of particle physics. Within the uncertainties involved in actually calculating quantities which can be directly compared with experimental data, it appears to be consistent with all known phenomena. From a theoretical point of view, it is nevertheless held to be unsatisfactory because it contains a large number of parameters which simply have to be adjusted to values determined by experiment. I shall give just two examples of the improvements that might be sought.

The first concerns the question of *charge quantization*. We saw in chapter 8 that the numbers λ_i (in (8.17), for example) which express the charges of different particles as multiples of the fundamental charge e could have any values. There is no explanation for the fact that they are observed to have integer or, in the case of quarks, simple fractional values. In the GWS electroweak theory, the charges of particles belonging to an SU(2) doublet must differ by one unit, but the hypercharge of each multiplet, which gives the actual charges through the Gell-Mann–Nishijima formula, is assigned simply to fit the observed facts.

The second unsatisfactory feature is that the standard model involves three independent gauge coupling constants, namely the g and g' of the electroweak theory and a third, g_s, for QCD. This is because the gauge symmetry group is SU(3) $\times$ SU(2) $\times$ U(1), which means that the SU(3) transformations which rearrange colours, the SU(2) weak isospin rotations and the U(1) phase transformations all act independently of each other. It is, of course, satisfying that the strong, weak and electromagnetic interactions, which have at first sight very different physical effects, can all be described in essentially the same terms as gauge theories. Moreover, the weak and electromagnetic interactions are intimately related in the GWS theory. Indeed, the relative weakness of the weak interactions, as measured by the Fermi constant G_F, is seen from (12.36) to be due to the relatively large masses of the gauge bosons rather than to the size of the coupling constant g, which is actually greater than e. This and the different ranges of the two interactions are seen to be consequences of spontaneous symmetry breaking. That having been said, we still need three coupling constants to account for the three forces. In the view of most theorists, it would be much more satisfactory if we could account for all three forces using a single coupling constant, with all differences arising from spontaneous symmetry breaking. In particular, we would like to be able to *predict* the value of the Weinberg angle which, according to (12.26), just measures the ratio of g and g'.

Considerations such as these have led to the invention of *grand unified theories*, whose principal feature is that the fundamental gauge group should be *simple*. This means that it cannot be expressed as the product of several independent groups, which immediately implies the existence of only a single gauge coupling constant. The earliest and simplest of these theories was invented by H Georgi and S Glashow (1974), who took the gauge group to be SU(5). The 15 fermions of a single family fit into two SU(5) multiplets, of which the simpler is

$$\begin{pmatrix} \nu_{\mathrm{L}} \\ e_{\mathrm{L}} \\ d^{\mathrm{c}}_{\mathrm{rR}} \\ d^{\mathrm{c}}_{\mathrm{gR}} \\ d^{\mathrm{c}}_{\mathrm{bR}} \end{pmatrix}. \tag{12.46}$$

In this notation, $d^{\mathrm{c}}_{\mathrm{rR}}$, for example, denotes the charge conjugate of the right-handed component of the field operator for a red down quark. The charge conjugate of a right-handed component is left handed (see exercise 7.8), so all the field operators are in fact left handed. In terms of particles, the electron and its neutrino are grouped with the anti-down quark, whose charge is $+\frac{1}{3}$.

The gauge transformations which act on this multiplet are of the form $\exp[\mathrm{i}\boldsymbol{\alpha}(x)\cdot\boldsymbol{\xi}]$, where the matrices ξ^i, of which there are 24, are the SU(5) analogues of the Pauli matrices. The standard model is included in the SU(5) model, because some of these correspond to the gauge transformations of SU(3) × SU(2) × U(1). Thus, three of the ξ^i can be written as

$$\begin{pmatrix} \begin{pmatrix} & \tau^i & \end{pmatrix} & \begin{pmatrix} 0 & 0 & 0 \\ 0 & 0 & 0 \end{pmatrix} \\ \begin{pmatrix} 0 & 0 \\ 0 & 0 \\ 0 & 0 \end{pmatrix} & \begin{pmatrix} 1 & 0 & 0 \\ 0 & 1 & 0 \\ 0 & 0 & 1 \end{pmatrix} \end{pmatrix} \tag{12.47}$$

where τ^i are the SU(2) Pauli matrices. These generate the weak isospin transformations of the electron–neutrino doublet, leaving the right-handed quarks unchanged. A further eight are

$$\begin{pmatrix} \begin{pmatrix} 1 & 0 \\ 0 & 1 \end{pmatrix} & \begin{pmatrix} 0 & 0 & 0 \\ 0 & 0 & 0 \end{pmatrix} \\ \begin{pmatrix} 0 & 0 \\ 0 & 0 \\ 0 & 0 \end{pmatrix} & \begin{pmatrix} & & \\ & \lambda^i & \\ & & \end{pmatrix} \end{pmatrix} \tag{12.48}$$

λ^i being the SU(3) matrices which generate colour transformations of the quarks without affecting the leptons. Charge quantization is now automatic because there is a linear combination of SU(5) generators which can be written as

$$\xi_Q = \mathrm{diag}\,(0, -1, \tfrac{1}{3}, \tfrac{1}{3}, \tfrac{1}{3}).$$

This clearly generates the desired electromagnetic U(1) transformations *without* calling upon an overall phase transformation and therefore without involving an additional gauge coupling constant.

As readers should already have guessed, the SU(5) theory requires a total of 24 gauge bosons, of which 12 can be identified as the photon, $W^{\pm}$, Z^0 and gluons of the standard model. The remainder, called collectively the X bosons, correspond to new interactions, for which there is no experimental evidence. The fact that these new effects are not observed can be accounted for by spontaneous symmetry breaking. If the X bosons acquire a very large mass, then their effects will be very weak, because of relations like (12.36). One effect which should in principle be observable is *proton decay*. In the standard model, the currents which couple to the weak gauge fields contain only terms of the form $\bar{q}\gamma^\mu q$ or $\bar{l}\gamma^\mu l$, where q and l generically denote quarks and leptons. It follows that a quark can be transformed into a different quark by emitting a weak gauge boson, but not into a lepton. Consequently, a baryon can decay only into a lighter baryon, together with a virtual weak boson, which subsequently produces a lepton–antilepton pair, as in the beta decay of a free neutron. The proton, being the lightest baryon, cannot decay at all. The reason for this is that leptons and quarks are contained in separate SU(2) multiplets. Each SU(5) multiplet, however, contains both quarks and leptons. Therefore, the currents which couple to X gauge fields contain terms of the form $\bar{q}\gamma^\mu l$ and $\bar{l}\gamma^\mu q$, which permit the transformation of a quark into a lepton by the emission of an X boson. Moreover, the second, more complicated SU(5) multiplet, which I have not shown explicitly, contains both left-handed quark components and the charge conjugates of right-handed quark components, and this permits the transformation of a quark into an antiquark. Because of this, proton decay becomes possible, and figure 12.4 shows one mechanism whereby it can decay into a π^0 and a positron.

In order to recover the standard model, the SU(5) symmetry must be broken in two stages, which involve two sets of Higgs fields. The first stage leaves the SU(3) × SU(2) × U(1) symmetry intact, with massless weak gauge bosons, but gives large masses M_X to the X's. At this stage, the gauge couplings g, g' and g_s are given in terms of the SU(5) coupling g_5 by $g = g_s = g_5$ and $g' = \surd(3/5)g_5$. The last of these is determined by the particular combination of generators which constitutes ξ_Q in (12.49). This gives a value for the Weinberg angle corresponding to $\sin^2\theta_W = 3/8 = 0.375$, which is rather larger than the measured value. The second stage of symmetry breaking is that of the standard model which provides the masses of the weak gauge bosons.

At this point, it is necessary to take account of the running coupling constants which result from renormalization. At energies greater than

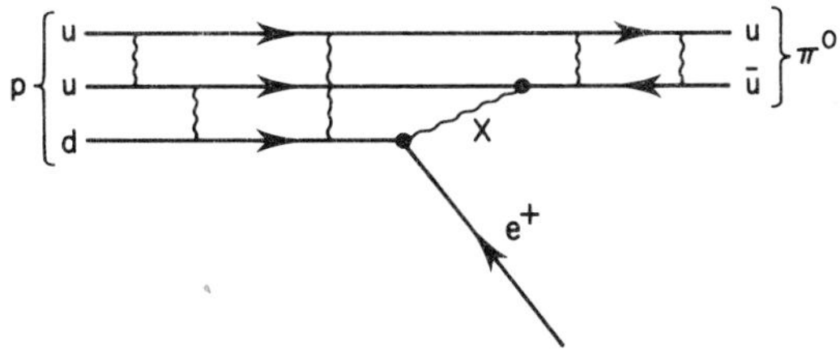

Figure 12.4 A contribution to the decay of a proton, producing a positron and a π^0. The X boson has a charge of $-4e/3$.

M_X, all gauge bosons contribute equally to this renormalization, and there is a single running coupling constant $g_5(\mu)$. At lower energies, because of the broken symmetry, the three couplings $g_3 = g_s$, $g_2 = g$ and $g_1 = \sqrt{(5/3)}g'$ depend differently on energy. The equality $g_1 = g_2 = g_3 = g_5$ holds at an energy of M_X while the couplings of the standard model corresponds to an energy of M_W. This is, of course, a somewhat imprecise argument since, for example, the W and Z masses are not exactly equal. Calculations of the running coupling constants give the result sketched in figure 12.5: they become equal at an energy which is taken as an estimate of M_X and is found to be approximately 4×10^{14} GeV. The prediction for the Weinberg angle at energies comparable with M_W is $\sin^2\theta_W \approx 0.21$, which agrees well with the measured value.

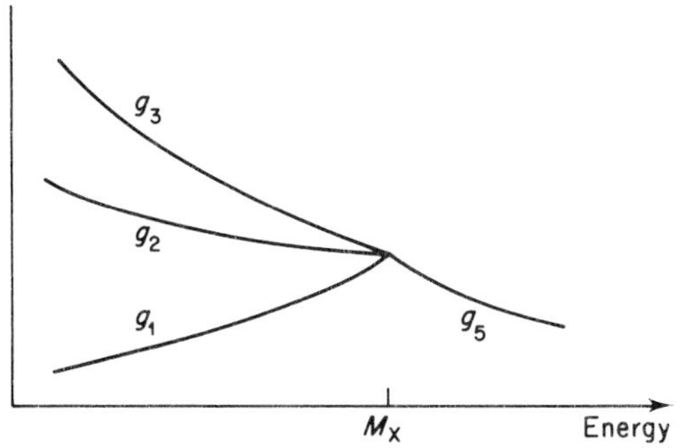

Figure 12.5 Energy dependence of the running coupling constants in the SU(5) grand unified theory. The spontaneously broken symmetry which gives the U(1), SU(2) and SU(3) couplings at low energy is restored at an energy approximately equal to the typical X boson mass.

By comparison with W and Z masses or with the few hundred GeV which can be produced in contemporary accelerators, 4×10^{14} GeV is an enormous energy, which accounts for our failure to observe most processes involving X bosons. The exception is proton decay, which is predicted to occur with a lifetime of about 10^{30} years. Large though this lifetime is, it should be possible to observe a few events in a sufficiently large volume of matter, and several well publicized experiments have been undertaken, usually deep underground to avoid the intrusion of cosmic radiation. At the time of writing, no decays have been observed. Although theoretical uncertainties in the estimate of the lifetime are quite large, most experts are agreed that these experiments rule out the SU(5) model as a theory of the real world. It is, however, possible to devise numerous other grand unified theories which would predict longer lifetimes for the proton.

Obviously, the value of grand unified theories lies much more in their aesthetic theoretical appeal in providing a completely unified description of the three interactions and suggesting an explanation of charge quantization, than in their utility for interpreting hard experimental data. (It is worth mentioning that they also suggest an explanation for the fact that the universe contains much more matter than antimatter, but I must ask readers to consult the literature for details of this.) Even their aesthetic appeal has its limitations. One is the awkwardness with which the fermions are fitted into multiplets of the unifying gauge group. It is hard, for example, to see any good physical reasons for treating the particles in (12.46) on a different footing from the others. Another is that, although there is only one fundamental gauge coupling constant, there are many other undetermined parameters, such as the masses and coupling constants associated with Higgs fields. Thus, the price of obtaining a prediction for one more measurable quantity, the Weinberg angle, is the introduction of further quantities which cannot even be measured. According, at least, to the SU(5) theory, it would be necessary to conduct experiments at inconceivably high energies in order to find phenomena other than proton decay which cannot be fitted into the standard model. This situation is, of course, unsatisfactory and it is not clear whether any real improvement can be hoped for.

12.7 Gravity and Supersymmetry

When introducing the idea of gauge theories in chapter 8, I emphasized that they are in some respects quite analogous to the relativistic theory of gravity. Many theorists hold the view that it should be possible to construct a fully unified theory of fundamental interactions in which gravity appears on much the same footing as the other three. Amongst other things, this involves promoting the classical theory of general relativity (or perhaps some variant of it) to a quantum field theory. I have mentioned previously that, if the metric tensor is expanded as

$g_{\mu\nu} = \eta_{\mu\nu} + h_{\mu\nu}$, then $h_{\mu\nu}$ can be interpreted as the wavefunction or field operator for a massless spin-2 particle, the *graviton*, which would serve as the gauge boson for gravity. ($\eta_{\mu\nu}$ denotes the metric of Minkowski spacetime, but it could, more generally, be expanded about a curved 'background' spacetime.) The difficulty with this is that the gravitational action, the last term in (4.14), does not lead to a renormalizable quantum field theory. Expressed in energy units, its coupling constant is $G/\hbar c^5 = (1.22 \times 10^{19}\ \text{GeV})^{-2}$, where G is Newton's constant. This constant has a negative dimension, so the dimensional criterion for renormalizability is not fulfilled. At a qualitative level, we can imagine that Einstein gravity is, like the Fermi theory of weak interactions, an effective low-energy theory, valid for processes which involve single-particle energies substantially lower than the so-called *Planck energy* of 1.22×10^{19} GeV. For almost all practical purposes, this is quite sufficient, but there is nevertheless a strong theoretical desire to construct a meaningful quantum-mechanical theory of gravity.

We have seen that the prospects of removing ultraviolet divergences from a quantum field theory are improved when the theory possesses symmetries, such as gauge invariance, which can lead to cancellations between potentially infinite integrals. An idea which has been vigorously pursued in recent years is that the existence of a new symmetry could result in cancellations which remove the infinities of quantum gravity. The symmetry in question is called *supersymmetry* and it has the superficially unlikely property of relating fermions and bosons within a single multiplet. (It has been suggested that supersymmetry might also solve a problem, known as the *gauge hierarchy problem*, which occurs in grand unified theories without gravity. This problem concerns the large difference between the masses of the weak gauge bosons and the X bosons. Mass renormalization involves interactions between all particle species in the theory, and it is difficult to see why the masses of similar species should not turn out to have comparable values. Indeed, the large mass differences can be achieved in grand unified theories only by 'fine tuning' various adjustable parameters. On the other hand, Goldstone bosons and photons are exactly massless by virtue of symmetries, despite their interactions with massive particles, and it is possible that supersymmetry might act in an analogous way to 'protect' the hierarchy of other gauge boson masses.)

The technical details of supersymmetry are well beyond the scope of this book, but I shall try to indicate how such symmetry transformations can be constructed. Consider first an isospin rotation through an infinitesimal angle α about the t_1 axis. To first order in α, the effect on an isospin doublet is

$$\begin{pmatrix} \phi_1 \\ \phi_2 \end{pmatrix} \to \begin{pmatrix} \phi_1 \\ \phi_2 \end{pmatrix} + \mathrm{i}\alpha\tfrac{1}{2}\begin{pmatrix} 0 & 1 \\ 1 & 0 \end{pmatrix}\begin{pmatrix} \phi_1 \\ \phi_2 \end{pmatrix}. \qquad (12.50)$$

The first-order changes in the two fields are

$$\delta\phi_1 = \tfrac{1}{2}\mathrm{i}\alpha\phi_2 \quad \text{and} \quad \delta\phi_2 = \tfrac{1}{2}\mathrm{i}\alpha\phi_1 \tag{12.51}$$

and we want to find a modification of these equations which will work when, say, one of the fields is a scalar and the other a spinor. If the two fields are to be, in some sense, interchangeable, they must represent the same number of physical degrees of freedom. One way of achieving this is to consider a complex scalar field ϕ representing a massless spin-0 particle and its antiparticle, and a left-handed spinor ψ representing a massless left-handed spin-$\frac{1}{2}$ particle and its right-handed antiparticle. Since the spinor represents only two physical degrees of freedom, it is possible to treat it as a two-component matrix rather than as a four-component Dirac spinor. In this two-component formalism, the 4×4 γ matrices are replaced by four 2×2 matrices, σ^μ, of which σ^0 is the unit matrix and the other three are the Pauli matrices (see exercise 12.3). The small change in ϕ must be a linear combination of the two components of ψ with, in general, complex coefficients

$$\delta\phi = \mathrm{i}\alpha^i\psi_i \tag{12.52}$$

where a sum over $i = 1, 2$ is implied. We must now take account of two important differences between the scalar and spinor fields. First, ϕ is a commuting quantity while ψ is an anticommuting Grassmann quantity. Therefore the 'angle' α must be a Grassmann variable. Second, in terms of the dimensional analysis of §9.6, ϕ has dimension $D = 1$, while ψ has $D = \frac{3}{2}$. Therefore, α must have $D = \frac{1}{2}$. Now, by analogy with (12.51), $\delta\psi$ must be proportional to α (actually, it is found, to α^*) and to ϕ. This will certainly give an anticommuting quantity but, to get the dimensions right, $\delta\psi$ must also be proportional to another quantity with dimension $D = 1$. The only available quantity is the derivative ∂_μ and, to maintain Lorentz covariance, this must be contracted with σ^μ. The transformation which is found to work is

$$\delta\psi_i = \alpha^{*j}\sigma^\mu_{ij}\partial_\mu\phi. \tag{12.53}$$

(In the interests of accuracy, I should mention that, in order to construct a mathematically consistent symmetry transformation, it is necessary to add to the (ϕ, ψ) multiplet an 'auxiliary field', but this, like the gauge degrees of freedom encountered in chapter 9, has no physical significance.)

Working along these lines, it is possible to devise field theories which are invariant under supersymmetry transformations, and in which all the fields fall into supersymmetry multiplets of fermions and bosons. It is found that infinities do indeed cancel. In fact, some supersymmetric theories are actually *finite*, in the sense that *all* their infinities cancel, even without renormalization. The drawback is that the observed particles of the standard model alone cannot be fitted into supersymmetry multiplets. What can be done is to invent *supersymmetric extensions* of the standard model, in which each observed particle is supplied with a supersymmetric partner: a spin-$\frac{1}{2}$ 'photino' for the photon, scalar 'selectron', 'sneutrino' and 'squarks' for the known fermions, and so on. There is absolutely no hint of experimental evidence for the existence of any such particles, but this does not appear to dampen the enthusiasm of many theorists for supersymmetry! If supersymmetry does exist in nature at some level of description, then it must be broken to such an extent that the masses of these particles are too large, or their couplings to ordinary matter too weak at laboratory energies, for them to have been detected.

The supersymmetry transformation described above is *global* because the 'rotation angle' α is constant. The importance of supersymmetry for theories of gravity emerges when we try to convert it to a *local* transformation where, as in a gauge theory, α varies in space and time. The crucial point is the derivative in (12.53). As we saw, for example, in (3.24), this derivative is the generator of spacetime translations which are, therefore, involved in the supersymmetry transformations. A local supersymmetry transformation must involve position-dependent translations, which are examples of general coordinate transformations. Therefore, a locally supersymmetry theory must be generally covariant and its gauge fields must include the affine connection. A quantum field theory with local supersymmetry therefore includes a quantum theory of gravity. Such theories, called *supergravity* theories, have been devised. They are generally rather complicated, and their exact physical implications are hard to establish. As far as I know, none of these theories can be put forward with any confidence as giving a realistic account of our world.

The most recent supersymmetric theories are the *superstring* theories, in which the fundamental entities are taken to be one-dimensional 'strings' rather than point particles. The length of a string would be something like the *Planck length* $(G\hbar/c^3)^{1/2} = 1.61 \times 10^{-35}$ m, and observed particles would correspond, roughly, to various quantized states of vibration of a fundamental string. There is some hope that this idea leads to a unified, finite theory of gravity and the other forces. Whether any such theory can be made to reproduce the standard model at low energies or to yield new, experimentally testable predictions is, however, not at all clear.

Exercises

12.1 Let $\psi = (\psi_1, \psi_0, \psi_{-1})^{\mathrm{T}}$ be a triplet of scalar fields with weak isospin 1. Show that the matrices which generate isospin rotations of this triplet are

$$t_1 = \frac{1}{\sqrt{2}}\begin{pmatrix} 0 & 1 & 0 \\ 1 & 0 & 1 \\ 0 & 1 & 0 \end{pmatrix} \qquad t_2 = \frac{-\mathrm{i}}{\sqrt{2}}\begin{pmatrix} 0 & 1 & 0 \\ -1 & 0 & 1 \\ 0 & -1 & 0 \end{pmatrix}$$

$$t_3 = \begin{pmatrix} 1 & 0 & 0 \\ 0 & 0 & 0 \\ 0 & 0 & -1 \end{pmatrix}.$$

Why do these matrices differ from the generators shown in equations (A2.2) and (A2.4) of appendix 2?

12.2 Consider an extended version of the GWS model where, in addition to the Higgs field (12.18), there is a triplet Higgs field, such as the ψ of the previous exercise, whose vacuum expectation value is $(0, 0, w)^{\mathrm{T}}$. What weak hypercharge must be assigned to ψ? Show that the value of the measurable parameter $\rho = M_{\mathrm{W}}^2/M_{\mathrm{Z}}^2 \cos^2\theta_{\mathrm{W}}$, which is found experimentally to be very close to 1, is given by

$$\rho = \frac{1 + 2w^2/v^2}{1 + 4w^2/v^2}.$$

Aside from the value of this parameter, why could an electroweak theory involving massive fermions not be constructed using ψ alone?

12.3 Verify that the *chiral representation* of the Dirac γ matrices, given by

$$\gamma^0 = \begin{pmatrix} 0 & -I \\ -I & 0 \end{pmatrix} \qquad \gamma^i = \begin{pmatrix} 0 & \sigma^i \\ -\sigma^i & 0 \end{pmatrix}$$

satisfies the Clifford algebra (7.24) and find the matrix γ^5 in this representation. Show that right- and left-handed spinors have the form $(\chi_{\mathrm{R}}, 0)^{\mathrm{T}}$ and $(0, \chi_{\mathrm{L}})^{\mathrm{T}}$ respectively, where, χ_{R} and χ_{L} are two-component spinors, and that the Dirac equation for massless particles can be written in the form $\sigma^\mu \partial_\mu \chi = 0$, where $\sigma^0 = 1$ for a right-handed spinor and $\sigma^0 = -1$ for a left-handed spinor.

13

The Early Universe

In this final chapter, I shall discuss an area of investigation which illustrates many of the theoretical ideas developed in the rest of the book, namely cosmology and the early history of the universe. Only in the last 50 years or so has it been possible to treat cosmology as a matter for serious scientific enquiry rather than philosophical speculation. Since we cannot (presumably) create a new universe in the laboratory, any theory concerning the history of our own universe must remain to some extent speculative. If, however, it is accepted that our knowledge of physics as established in the laboratory and by astronomical observations continues to be valid in the distant past, then a remarkable amount can be said with a fair degree of confidence. For example, the present age of the universe is known to within a factor of 2: it cannot be much less than 10 billion years (1 billion = 10^9) nor much greater than 20. Our established knowledge of physics can, of course, be applied with confidence only when conditions in the universe were such that a confident extrapolation can be made from conditions which can be created in the laboratory. This has been true ever since the universe was about one millisecond old. In the first millisecond, however, events moved extremely rapidly.

As we shall see, the temperature of the matter in the universe increases, without any known limit, as we progress backwards in time, and our reasoning about what the sequence of events may have been becomes increasingly speculative as we encounter energies at which our confidence in the standard model of particle physics begins to falter. Conversely, it is potentially fruitful to speculate about early events on the basis of theoretical models, such as grand unified theories, which cannot be rigorously tested in the laboratory. The reason is that the very early cosmological events implied by these models may have consequences for the present constitution of the universe which can be checked by astronomers. This opens the enticing possibility of using the early universe as a high-energy physics laboratory in which energies are available which could not conceivably be produced by man. Some fragments of information have already been obtained in this way. Clearly, however, the reliability of such information is no greater than the reliability with which the detailed consequences of theoretical models can be worked out. At present, there is, in my view at least, little cause for complacency in this respect.

I shall begin by outlining the standard *big bang* model of the history of the universe.

13.1 The Robertson–Walker Metric

Modern cosmology is based upon the description of spacetime geometry given by general relativity. As we saw in the last chapter, there is a widespread belief that general relativity is inadequate as a fundamental theory of geometry, to the extent that it is non-quantum-mechanical. If this is so, then there is a limit to the validity of the standard cosmological model which I shall mention in due course. For the moment, let us assume that general relativity is good enough. We need to write down the metric tensor of the universe. Obviously, it is impossible to do this in any detail. Fortunately, astronomical evidence shows that the overall structure of that part of the universe which can be observed is very simple. If the distribution of matter is averaged over distances which are large enough to encompass many clusters of galaxies, it is found to be *isotropic*, which means that it looks the same in all directions, and *homogeneous*, which means that it would look the same from any vantage point. The best evidence for isotropy actually comes from measurements of cosmic microwave radiation, which we shall have cause to discuss later on. Our first basic assumption, then, is that the universe is isotropic and homogeneous. This assumption is sometimes dignified as the *cosmological principle*. It can be seen as embodying the philosophical prejudice that our own location in the universe has nothing whatever to distinguish it from any other location. Its only scientific value, however, is that it is in reasonable accord with observations and that it makes further progress possible.

From the assumption of homogeneity and isotropy, it can be shown to follow that there is a coordinate system in which the line element (2.7) has the form (with $c = 1$)

$$\mathrm{d}\tau^2 = \mathrm{d}t^2 - a^2(t)\left(\frac{1}{1-kr^2}\,\mathrm{d}r^2 + r^2\,\mathrm{d}\theta^2 + r^2\sin^2\theta\,\mathrm{d}\phi^2\right). \quad (13.1)$$

This is called the Robertson–Walker line element. The second term, in which k is a constant, measures distances in a spatial section of the spacetime, which exists at an instant t of *cosmic time*. The physical distance in this space between two points separated by fixed coordinate intervals $\mathrm{d}r$, $\mathrm{d}\theta$ and $\mathrm{d}\phi$ varies with time in proportion to the function $a(t)$, called the *scale factor*, which depends only on time. As in the Schwarzschild line element (4.26), the coordinate r does not provide a linear measure of distance. However, t does measure a genuine time. The proper time τ measured by any observer whose spatial coordinates r, θ and ϕ are fixed is clearly the same as t. Moreover, such an observer is moving through the spacetime along a geodesic and is therefore in free fall, which would not be the case in the Schwarzschild spacetime. (It would be a good exercise for readers to verify this point by deriving the geodesic equations, using the method suggested by (4.25).) The sequence of spatial sections corresponding to successive instants of time can be thought of as a three-dimensional space which expands or contracts uniformly with time according to the variation of $a(t)$. The surfaces of constant r, θ and ϕ expand or contract in the same way, like a grid of lines painted on the surface of an inflating balloon, and these coordinates are said to be *comoving*.

The constant k in (13.1) may be positive, negative or zero. If it is non-zero, then we can make the change of variables $r \to r/|k|^{1/2}$ and $a(t) \to a(t)|k|^{1/2}$, so that the magnitude of k disappears. We can therefore always choose the coordinates so that k has one of the three values 1, 0 or -1. If $k = 0$, then the spatial part of (13.1) is just the line element of a three-dimensional Euclidean space and the universe is *flat*. To understand the spatial geometry when $k = 1$, consider the two-dimensional surface $\theta = \pi/2$. The three-dimensional space can be thought of as the volume of revolution of this surface. The surface is in fact the surface of a Euclidean sphere of radius $a(t)$, as sketched in figure 13.1. In terms of the angles α and ϕ, the element of length $\mathrm{d}s$ on this surface is clearly given by $\mathrm{d}s^2 = a^2(\mathrm{d}\alpha^2 + \sin^2\alpha\,\mathrm{d}\phi^2)$, and this reproduces the spatial part of the Robertson–Walker line element when r is identified as $\sin\alpha$, as shown. It will be seen that the coordinates r and ϕ cover only half of the sphere, with $\alpha < \pi/2$ or $r < 1$, and that the singularity at $r = 1$ is only a coordinate singularity. The spherical surface obviously is isotropic and homogeneous, and the origin $r = 0$ could be placed anywhere on it. At a given instant of time, the volumes inside and outside the spherical surface in figure 13.1 have nothing to do with the Robertson–Walker *space* and serve only as a means of visualizing the surface. On the other hand, the sequence of spatial sections which are obtained as $a(t)$ varies with time can be envisaged as a set of concentric spherical surfaces which fill all or part of this volume. Each spatial section can be described as having a (spatially) constant positive radius of curvature $a(t)$.

Consider now a sphere drawn in the Robertson–Walker space at fixed coordinate radius r. Its physical radius is

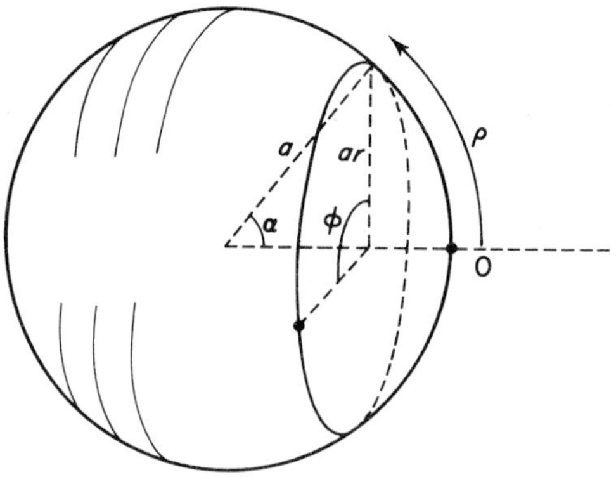

Figure 13.1 The surface of a sphere of radius a represents two of the spatial dimensions of a closed Robertson–Walker universe. The volumes inside and outside the surface are not part of the space. Relative to the origin at O, the coordinates r and ϕ can be visualized as shown. The physical distance from O to a point on the circle of radius ar in the figure is ρ. In the full three-dimensional spatial section, ρ would be the physical radius of a sphere centred at O.

$$\rho(r) = a \int_0^r \frac{dr}{(1 - r^2)^{1/2}} = a \sin^{-1}(r) = a\alpha. \tag{13.2}$$

The circumference of a great circle drawn on this sphere, say the equator $\theta = \pi/2$, is

$$c(r) = a \int_0^{2\pi} r\, d\theta = 2\pi a r = 2\pi a \sin(\rho/a) \tag{13.3}$$

which is always smaller than $2\pi\rho$, as is evident from figure 13.1. This circumference has a maximum value of $2\pi a$ at $\rho = \pi a$ and decreases to zero at $\rho = 2\pi a$. Thus, for $k = 1$, the spatial section of the Robertson–Walker universe is a three-dimensional spherical surface and is said to be *closed*.

For $k = -1$, the spatial section has a constant negative radius of curvature and is more difficult to imagine pictorially. The radius and circumference of a sphere are

$$\rho(r) = a \sinh^{-1}(r) \quad \text{and} \quad c(r) = 2\pi a \sinh(\rho/a). \tag{13.4}$$

The circumference is always greater than $2\pi\rho$ and both can be arbitrarily large. This universe has an infinite spatial extent and is said to be *open*.

We shall need to know the Ricci tensor which appears in the field equations (4.15). The metric tensor, whose components appear in (13.1), is diagonal, with $g_{00} = 1$ and spatial components given by

$$g_{ij} = -a^2 \begin{pmatrix} (1-kr^2)^{-1} & 0 & 0 \\ 0 & r^2 & 0 \\ 0 & 0 & r^2\sin^2\theta \end{pmatrix}. \tag{13.5}$$

We find that the Ricci tensor is also diagonal, given by

$$R_{00} = -3\frac{\ddot{a}}{a} \quad \text{and} \quad R_{ij} = -\left(\frac{\ddot{a}}{a} + 2\frac{\dot{a}^2}{a^2} + 2\frac{k}{a^2}\right)g_{ij} \tag{13.6}$$

where the overdots stand for $\partial/\partial t$. The Ricci scalar is

$$R = g^{\mu\nu}R_{\mu\nu} = -6\left(\frac{\ddot{a}}{a} + \frac{\dot{a}^2}{a^2} + \frac{k}{a^2}\right) \tag{13.7}$$

and the Einstein curvature tensor $G_{\mu\nu} = R_{\mu\nu} - (R/2)g_{\mu\nu}$ is diagonal with components given by

$$G_{00} = 3\left(\frac{\dot{a}^2}{a^2} + \frac{k}{a^2}\right) \quad \text{and} \quad G_{ij} = \left(2\frac{\ddot{a}}{a} + \frac{\dot{a}^2}{a^2} + \frac{k}{a^2}\right)g_{ij}. \tag{13.8}$$

If the metric of our universe is approximately of the Robertson–Walker form, and if the scale factor does change with time, then a simple consequence is *Hubble's law*. Assume that our galaxy and those we observe are comoving, so that their spatial coordinates are fixed. Then the physical distance between two galaxies separated by a coordinate distance d_0 is $d = a(t)d_0$. Their relative velocity is therefore

$$v = \frac{\mathrm{d}}{\mathrm{d}t}\,d(t) = \frac{\dot{a}(t)}{a(t)}\,d(t). \tag{13.9}$$

This velocity is proportional to the distance between the galaxies, with the proportionality factor

$$H(t) = \dot{a}(t)/a(t). \tag{13.10}$$

It is, of course, unlikely that galaxies will be exactly comoving. Nevertheless, it was discovered by E Hubble in 1929 that distant galaxies are, on average, receding from us with velocities proportional to their distances from us. The velocity of recession can be measured as a redshift of spectral lines, and the distance in terms of the apparent luminosity of an object whose absolute luminosity is known. When the redshift $z = (\lambda_o/\lambda_e) - 1$, where λ_o is the observed wavelength and λ_e the emitted wavelength, is small, it can be interpreted as a non-relativistic Doppler shift. More generally, however, careful account must be taken of the change in $a(t)$ between the moments of emission and reception. The relation between luminosity distance d_L and redshift can be written as a power series (see exercise 13.1)

$$d_L = H_0^{-1}[z + \tfrac{1}{2}(1 - q_0)z^2 + \ldots] \tag{13.11}$$

where *Hubble's constant* H_0 is the present value of $H(t)$ and q_0 is the present value of the *deceleration parameter*

$$q = -a\ddot{a}/\dot{a}^2. \tag{13.12}$$

The values of H_0 and q_0 are not known with very high precision. Hubble's constant is usually quoted as

$$H_0 = h \times 100\,\text{km}\,\text{s}^{-1}\,\text{Mpc}^{-1} = h(9.78 \times 10^9\ \text{years})^{-1} \tag{13.13}$$

and the dimensionless number h deduced from observations is between 0.5 and 1. (Clearly, H has dimensions $(\text{time})^{-1}$, but the units in which it is traditionally measured are recessional velocity ($\text{km}\,\text{s}^{-1}$) per unit distance to a galaxy, measured in megaparsecs, with $1\,\text{Mpc} = 3.086 \times 10^{22}$ m.) The value of q_0 is probably less than 2, but not much more can be said about it with any confidence.

For many purposes, including the derivation of (13.11), it is necessary to understand the behaviour of light waves in the Robertson–Walker universe. It will be sufficient to consider the case of a wave emitted by a comoving atom, say at $r = r_e$ and $\theta = \phi = 0$, and received by a comoving observer at $r = 0$. The light ray moves along a null geodesic whose equation, according to (13.1), is $\mathrm{d}t = -a(t)\,\mathrm{d}r/(1 - kr^2)^{1/2}$, the negative square root corresponding to a ray moving towards the origin. If a wave crest is emitted at time t_e and received at time t_o, then

$$\int_{t_e}^{t_o} \frac{\mathrm{d}t}{a(t)} = \int_0^{r_e} \frac{\mathrm{d}r}{(1 - kr^2)^{1/2}} = d_0 \tag{13.14}$$

where d_0 is independent of both t_e and t_o. If the following crest is emitted at time $t_e + \Delta t_e$ and received at time $t_o + \Delta t_o$, then

$$\int_{t_e+\Delta t_e}^{t_o+\Delta t_o} \frac{\mathrm{d}t}{a(t)} = d_0 + \frac{\Delta t_o}{a(t_o)} - \frac{\Delta t_e}{a(t_e)} = d_0 \tag{13.15}$$

and so the observed frequency and wavelength are related to those of the emitted wave by

$$\frac{\nu_o}{\nu_e} = \frac{a(t_e)}{a(t_o)} \quad \text{or} \quad \frac{\lambda_o}{\lambda_e} = \frac{a(t_o)}{a(t_e)}. \tag{13.16}$$

As seen by a comoving observer, therefore, the physical wavelength of a photon changes in proportion to the scale factor. In exercise 13.2, readers are invited to investigate this effect in terms of a covariant wave equation.

13.2 The Friedmann–Lemaître Models

The Robertson–Walker metric on its own tells us nothing about the time dependence of the scale factor. To investigate this, we have to study the field equations (4.15), which involve the stress tensor for whatever matter is present. From the form of the metric tensor and the Einstein curvature tensor (13.8), it is clear that the stress tensor must be diagonal, with elements

$$T_{00} = \rho(t) \quad \text{and} \quad T_{ij} = -p(t)g_{ij} \tag{13.17}$$

where $\rho(t)$ and $p(t)$ are functions of time only. This is the only form of stress tensor which is consistent with the assumptions of isotropy and homogeneity. In a sufficiently small region, the metric must be approximately that of Minkowski spacetime and we can choose new spatial coordinates in which g_{ij} is diagonal, with each diagonal component equal to -1. Then, by comparing (13.17) with (3.43), we can identify ρ as the energy density and p as the pressure, provided that the matter behaves as a fluid in thermal equilibrium. The field equations now provide two independent equations relating $a(t)$, $\rho(t)$, $p(t)$ and the cosmological constant Λ, which are

$$3\left(\frac{\dot{a}^2}{a^2} + \frac{k}{a^2}\right) = \kappa\rho + \Lambda \tag{13.18}$$

and

$$2\frac{\ddot{a}}{a} + \frac{\dot{a}^2}{a^2} + \frac{k}{a^2} = -\kappa p + \Lambda. \tag{13.19}$$

We saw in (4.22) that the quantity Λ/κ, which appears in (13.18) as an additional energy density, cannot be greater than the average density of matter in the universe, which is extremely small compared with the densities of everyday materials. In general relativity, Λ is a fundamental constant, independent of the properties or distribution of any matter the universe may happen to contain. To decide how large or small Λ is in a meaningful way, we must compare it with another fundamental quantity. The only constants at our disposal are G, $\hbar$ and c, and from these we can construct a quantity with the dimensions of a mass density for comparison with (4.22). It is $c^5/G^2\hbar \approx 5 \times 10^{93}\,\mathrm{g\,cm^{-3}}$. Thus, a dimensionless measure of the size of Λ is a number of the order of 10^{-122}. The staggering smallness of this number leads many theorists to suppose that the cosmological constant must be identically zero. Whether this number is really significant is hard to say. Since its derivation involves $\hbar$, any detailed understanding must require a reliable understanding of the relation between spacetime geometry and quantum mechanics, which we do not have. A number of speculative arguments to the effect that Λ should be zero, or at least very small, have been put forward, but none of them is conclusive. The standard cosmological model assumes that Λ can be neglected and this is what I shall do. The cosmological models based on the Robertson–Walker metric and Einstein's field equations with $\Lambda = 0$ are known as *Friedmann–Lemaître models*.

It is convenient to rewrite (13.18) and (13.19) with $\Lambda = 0$ as

$$\dot{a}^2 + k = \tfrac{1}{3}\kappa\rho a^2 \tag{13.20}$$

$$\ddot{a} = -\tfrac{1}{6}\kappa(\rho + 3p)a. \tag{13.21}$$

The first of these is sometimes referred to as the *Friedmann equation*. By differentiating it, we may easily show that

$$\frac{d}{dt}(\rho a^3) = -p\frac{d}{dt}(a^3). \tag{13.22}$$

This equation is equivalent to $\nabla_\nu T^{\mu\nu} = 0$ and is usually said to express the conservation of energy. The physical volume V occupied by a given amount of matter is proportional to a^3, so if the internal energy of this matter is U, then (13.22) asserts that $dU/dt = -p\,dV/dt$.

To draw detailed conclusions from (13.20) and (13.21), we need information about ρ and p. Some general conclusions can be obtained without very detailed information, however. First, suppose that $k = 0$, so that the universe is flat. Then (13.20) gives a relation between the density and the Hubble parameter (13.10):

$$\rho(t) = \rho_c(t) = \frac{3}{\kappa}\frac{\dot{a}^2(t)}{a^2(t)} = \frac{3}{\kappa}H^2(t). \tag{13.23}$$

The quantity $\rho_c(t)$ is called the *critical density*. When k is not necessarily equal to zero, it is convenient to measure the density as a fraction of the critical density, defining

$$\Omega(t) = \rho(t)/\rho_c(t). \tag{13.24}$$

Equation (13.20) becomes

$$\dot{a}^2(\Omega - 1) = k \tag{13.25}$$

and we see that in a closed universe, with $k = +1$, the density always exceeds the critical density ($\Omega > 1$), while in an open universe it is always less than the critical density ($\Omega < 1$).

For an ordinary fluid in thermal equilibrium, the density and pressure are both positive. Even if the matter is not in thermal equilibrium, the quantity $\rho + 3p$ in (13.21) is almost invariably positive, which is a special case of the *strong energy condition* discussed by Hawking and Ellis (1973). This being so, (13.21) shows that $\ddot{a}$ is always negative, and therefore $\dot{a}(t)$ always decreases with time. (Readers may like to note the somewhat counter-intuitive result that a positive pressure acts to slow down rather than accelerate the expansion.) Since the universe is now observed to be expanding, the expansion rate increases as we look further back in time. It follows (see figure 13.2) that at some time in the past the scale factor a was equal to zero and that the time which has elapsed since then is less than $1/H_0$. When the scale factor is zero, the universe is infinitely compressed (although, if it is open or flat, its spatial extent is still infinite). This is a highly singular state, containing matter at an infinite density. From a mathematical point of view, the metric becomes ill-defined, and the instant of time at which this occurs should be excluded from our spacetime manifold. Physically, we have no way of knowing what might happen in the extreme conditions prevailing near this singularity. From either point of view, the singularity marks the earliest time at which our universe can meaningfully be said to have existed. If we set $t = 0$ at the initial singularity, then the estimates of H_0 given above yield an upper bound to the present age of the universe t_0 of

$$t_0 < 2 \times 10^{10} \text{ years.} \tag{13.26}$$

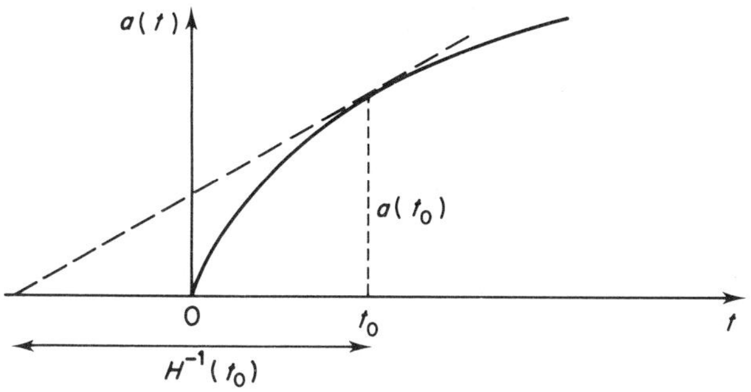

Figure 13.2 Scale factor of a Robertson–Walker universe in which the expansion rate always decreases. The age t_0 of the universe is less than $H^{-1}(t_0)$.

These conclusions are based on assumptions which could turn out to be false. The first was that the universe is homogeneous and isotropic, which is certainly not exactly true. We might wonder whether the occurrence of an initial singularity is a consequence of the high degree of symmetry, which might be avoided if allowance were made for anisotropies and inhomogeneities. It seems (as discussed, for example, by Hawking and Ellis (1973)) that this is not so and that under quite general conditions an initial singularity must have occurred. On the other hand, the behaviour of the metric in the neighbourhood of the singularity may be much more complicated in an anisotropic and/or inhomogeneous universe than in the Friedmann–Lemaître models (see, for example, Misner *et al* (1973)). Another assumption was the strong energy condition $(\rho + 3p) > 0$. If this is not true, then there need not be an initial singularity because, going backwards in time, $\ddot{a}$ might become positive, allowing a to pass through a minimum and then increase. For ordinary matter, the strong energy condition holds. Later on, in connection with the inflationary universe, we shall encounter a situation in which the condition may cease to hold for a brief period of time, but this does not in itself avoid the initial singularity. Finally, the entire argument is based on a classical spacetime geometry. If, as is generally believed, this geometry is ultimately subject to quantum-mechanical laws, then we may expect these laws to become important when the universe is sufficiently small. Since we have no reliable quantum theory of gravity, it is not possible to be certain about when quantum effects will be important. A rough estimate can be obtained by requiring that the energy density should exceed the characteristic value of $c^5/G^2\hbar$. At high densities, as we shall see shortly, the curvature term k/a^2 in (13.20) is negligible, even though a is very small. Using this equation (and dimensional analysis to convert to laboratory units), we find that quantum gravity effects are likely to be important when

$$H^{-1} \lesssim (G\hbar c^{-5})^{1/2} = 5 \times 10^{-44}\ \text{s}. \tag{13.27}$$

Since H^{-1} is a rough measure of the age of the universe, this time, called the *Planck time*, is the time at which we expect that quantum gravity effects ceased to be important.

At the present time, it appears that the matter in the universe is fairly well described as a uniform comoving distribution which exerts no pressure, known to cosmologists as *dust*. As a first approximation, it is instructive to suppose that this always has been and always will be true. The solution of (13.22) is

$$\rho(t) = M/a^3(t) \tag{13.28}$$

where M is a constant equal to the mass contained in a comoving region of physical volume a^3. With $p = 0$, the deceleration parameter can be expressed as

$$q(t) = \frac{\kappa\rho(t)}{6H^2(t)} = \tfrac{1}{2}\Omega(t). \tag{13.29}$$

The variation of the scale factor with time can now be found by solving (13.20). For $k = 0$, the solution is

$$a(t) = (\tfrac{3}{4}\kappa M)^{1/3} t^{2/3}. \tag{13.30}$$

For $k = \pm 1$, it can be written in parametric form in terms of an angle θ:

$$a = \tfrac{1}{3}\kappa M \sin^2\theta \qquad t = \tfrac{1}{3}\kappa M(\theta - \tfrac{1}{2}\sin 2\theta) \quad \text{for } k = 1 \tag{13.31}$$

$$a = \tfrac{1}{3}\kappa M \sinh^2\theta \qquad t = \tfrac{1}{3}\kappa M(\tfrac{1}{2}\sinh 2\theta - \theta) \quad \text{for } k = -1. \tag{13.32}$$

These solutions are sketched in figure 13.3, and we see that both the open and flat universes continue to expand for ever, while the expansion of the closed universe eventually comes to a halt and this universe recollapses to a final singularity. The situation is quite analogous to that of a projectile launched from the Earth's surface, the flat universe corresponding to an initial velocity equal to the escape velocity.

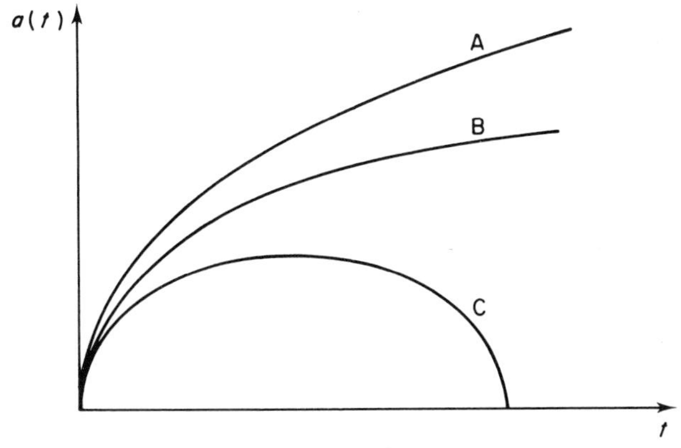

Figure 13.3 Variation of the scale factor with time in Friedmann–Robertson–Walker models: A, open universe, $k = -1$; B, flat universe, $k = 0$; C, closed universe, $k = 1$.

From the above solutions, it is possible to derive a relation between the age of the universe t, the Hubble parameter H and the density ratio Ω of the form

$$t = H^{-1} f(\Omega). \tag{13.33}$$

Since the open and closed universes correspond to $\Omega < 1$ and $\Omega > 1$ respectively, the function $f(\Omega)$ has different forms in these two ranges:

$$\begin{aligned} f(\Omega) &= \frac{1}{1-\Omega} - \frac{\Omega}{2}(1-\Omega)^{-3/2}\cosh^{-1}\left(\frac{2}{\Omega}-1\right) && \text{for } \Omega \leqslant 1 \\ &= \frac{\Omega}{2}(\Omega-1)^{-3/2}\cos^{-1}\left(\frac{2}{\Omega}-1\right) - \frac{1}{\Omega-1} && \text{for } \Omega \geqslant 1. \end{aligned} \tag{13.34}$$

At $\Omega = 1$, both expressions reduce to $f(1) = \frac{2}{3}$, and $f(\Omega)$ is in fact a perfectly smooth function, plotted in figure 13.4.

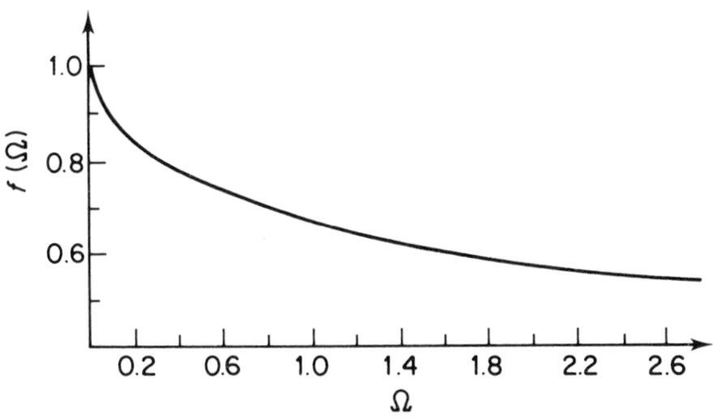

Figure 13.4 The function $f(\Omega)$ given in equation (13.34).

13.3. Matter, Radiation and the Age of the Universe

From (13.33) and (13.34), we can determine the present age of the universe, provided that (i) we can assume that $p = 0$; (ii) we have an estimate of H_0; and (iii) we have an estimate of Ω_0, the present density as a fraction of the present critical density. The assumption that $p = 0$ is, for this purpose, perfectly safe. The period during which this has been true is called the *matter-dominated era* and, as we shall discover shortly, it began when the universe was about one-millionth of its present age. The value of H_0 is, as we have seen, uncertain to something like a factor of 2, so the error in assuming that $p = 0$ is negligible by comparison. The value of Ω_0 is also rather uncertain. Direct observations reveal, of course, only *luminous* matter, namely that contained in stars whose radiation we can detect. There are, however, a number of reasons for believing that there is a considerable amount of additional matter, called *dark matter* or *missing matter*.

The masses of distant galaxies are estimated by means of the *virial theorem*, which asserts, roughly, that the mass of a gravitationally bound system, such as a solar system, a galaxy or a cluster of galaxies, is given by

$$M \approx D\langle v^2\rangle/G \tag{13.35}$$

where D is the characteristic size of the system and $\langle v^2\rangle$ a mean square velocity relative to the centre of mass. This is obviously true, for example, for a star of mass M with a small planet in a circular orbit of radius D with orbital velocity v. The masses of galaxies inferred in this way may be of the order of 10 times the mass which can be accounted for by visible stars. The *galactic halos* which contain this extra mass probably extend well beyond the visible part of the galaxy. For large clusters of galaxies, the inferred total mass may be several hundred times the mass of luminous matter. Since the critical density (13.23) is proportional to H^2, it might be thought that estimates of Ω_0 should depend on the value assumed for H_0. Actually, this is not true. The reason is that large distances are measured in terms of redshifts, by using the relation (13.11), and each distance measured in this way is thus proportional to H_0^{-1}. The *velocity dispersion* $\langle v^2\rangle$ in (13.35) is estimated from distributions of redshifts around the mean for the object concerned and is independent of distance, so the estimate of mass is proportional to a distance estimate. Estimates of density are therefore proportional to $(\text{distance})^{-2}$ or to H_0^2, and estimates of Ω_0 are independent of the value assumed for H_0. The value of Ω_0 which includes dark matter inferred from the virial theorem to exist in galactic halos and in the intergalactic medium in clusters is roughly in the range $0.1 < \Omega_0 < 0.3$.

An indirect estimate of Ω_0 can be made via (13.29) from measurements of the deceleration parameter. These measurements suggest a value of $q_0 \approx 1$, though with quite large uncertainties. This would imply a value for Ω_0 of about 2, which is one reason for thinking that there may be more dark matter than is inferred from the virial theorem. This is not unreasonable, since observations can infer only the presence of dark matter which is in the neighbourhood of luminous matter, and there are large regions of the observable universe in which luminous matter is scarce. Evidently, we cannot yet be sure whether the universe is open or closed. There is a theoretical prediction, arising from the *inflationary scenario* which I shall describe later, to the effect that Ω should be almost exactly equal to 1, and this would also require additional dark matter. What the dark matter might consist of has been the subject of intense and sometimes exotic speculation, but this has so far proved inconclusive and I cannot enter into it here.

From (13.13) and (13.33), our estimate of the present age of the universe is

$$t_0 = 9.78 \times 10^9 h^{-1} f(\Omega_0) \text{ years.} \tag{13.36}$$

If we take the limits on observed parameters as $0.5 < h < 1$ and $0.1 < \Omega_0 < 2$ then, referring to figure 13.4, we find that the age of the universe is within the limits

$$6 \times 10^9 \text{ years} < t_0 < 1.8 \times 10^{10} \text{ years.} \tag{13.37}$$

It is possible, of course, to place lower bounds on the age of the universe by estimating the age of objects it contains. Radio dating of terrestrial, lunar and meteoric rocks puts the age of the oldest material at about 4.5×10^9 years. It is believed, however, that this material is not primordial, but was formed in the cores of ancient stars, so the universe should be rather older than this. Estimates of the age of the oldest stars in our galaxy, those in globular clusters, suggest ages of about 1.6×10^{10} years. Thus, these data are consistent with (13.37), with a preference for the higher end of the range.

In addition to matter, the universe contains radiation. The most important component from a cosmological point of view is the *cosmic microwave background* which has, to a good approximation, a black-body spectrum corresponding to a temperature of 2.7 K. This radiation, first observed by A A Penzias and R W Wilson (1965), is found to be isotropic to about one part in 10^4, except for a dipolar anisotropy which can be attributed to the motion of the Earth relative to comoving coordinates. This microwave background provides a vital clue to the early history of the universe. Because of its high degree of isotropy, it cannot have originated in observed galaxies, and it is generally held to be a relic of an early period in which the content of the universe was a hot, dense plasma of particles and radiation. Because the universe must have been expanding very rapidly during this early phase, the standard cosmological model is often referred to as the *hot big bang* model. The importance of the microwave background for our present discussion is twofold. First, its isotropy provides the best evidence for the isotropy of the observable universe upon which the Robertson–Walker metric depends. Second, black-body radiation exerts a pressure as well as contributing to the energy density, so we can use it to estimate the duration of the matter-dominated era, during which the approximation $p = 0$ holds good.

From (10.86) with $g = 2$ for photons, we find for the energy density ρ_{rad} or the equivalent mass density ρ_{rad}/c^2 of the microwave background

$$\rho_{\text{rad}}(t_0) = 4.02 \times 10^{-14}\,\text{J}\,\text{m}^{-3} \quad \text{or} \quad \rho_{\text{rad}}(t_0)/c^2 = 4.47 \times 10^{-34}\,\text{g}\,\text{cm}^{-3}. \tag{13.38}$$

The overall density is given by (13.24) as

$$\rho(t_0) = h^2\Omega_0 \times 1.88 \times 10^{-29}\,\text{g}\,\text{cm}^{-3} \tag{13.39}$$

which we can take as about $1 \times 10^{-29}\,\text{g}\,\text{cm}^{-3}$. At the present time, therefore, the contribution of the radiation to the energy density is negligible and its pressure, which is $\frac{1}{3}$ of its energy density, is also negligible in (13.21). However, we saw in (13.16) that the frequency or energy of a photon is proportional to $1/a(t)$. Therefore, if we assume a constant number of photons in a given comoving region, the energy density of radiation is proportional to $1/a^4$, whereas that of non-relativistic matter is proportional to $1/a^3$ as in (13.28). Thus $\rho_{\text{rad}}(t)/\rho(t) \propto 1/a(t)$ and the radiation becomes more important at earlier times. To see how long the universe has been matter dominated, we can estimate the time t_{m} at which the densities of matter and radiation are approximately equal, which is also the time at which the radiation pressure becomes significant in (13.21). The condition is

$$\frac{a(t_{\text{m}})}{a(t_0)} = \frac{\rho_{\text{rad}}(t_0)}{\rho(t_0)} \approx 5 \times 10^{-5}. \tag{13.40}$$

Since Ω is fairly close to 1, it is sufficient to use (13.30) to get

$$t_{\text{m}}/t_0 \approx (5 \times 10^{-5})^{3/2} \approx 4 \times 10^{-7}. \tag{13.41}$$

This result shows that the time which has elapsed since the universe became matter dominated is about 1 million times that which had elapsed previously. For various reasons, it is not a very accurate estimate. Quite apart from the uncertainties in h and Ω_0, the relation $a(t) \propto t^{2/3}$ is not valid before t_{m}. A better approximation is, as we shall see, $a(t) \propto t^{1/2}$. This does not, however, invalidate the conclusion that t_{m} is only a tiny fraction of t_0 and, therefore, that the zero-pressure model can be used to estimate t_0 to the accuracy permitted by other uncertainties.

The foregoing argument assumed that the pressure was entirely due to radiation, so it would be as well to check that the pressure exerted by the matter itself is also negligible. Since the energy density and pressure of the radiation are proportional to $1/a^4$ and also, according to (10.88), to T^4, it follows that the temperature T is proportional to $1/a$ and therefore equal to about 5×10^4 K at t_m. At this temperature, the matter consisted (as it turns out) mainly of ionized hydrogen and helium. The equivalent energy $k_B T$ is about 4.3 eV, so the kinetic energy even of the electrons was much less than their rest energy of 511 keV. Thus, the kinetic energy density and pressure of the matter was negligible compared with the density of its rest energy.

13.4 The Fairly Early Universe

Processes occurring in the early universe at temperatures below about 10^{12} K (at which $k_B T$ is approximately equal to the mass of a muon, 106 MeV) can be investigated quite thoroughly on the basis of well established physics. This temperature probably occurred when the universe was about 10^{-4} s old, and I shall refer to the period between then and t_m, when the universe became matter dominated, as the 'fairly early' universe. The fairly early history of the universe has been carefully documented by, for example, Peebles (1971) and Weinberg (1972), and I shall largely follow Weinberg's account.

A few gross features are easily deduced. First of all, the curvature of space was unimportant. If space is curved, then $|k| = 1$ in (13.20). Let us define the ratio of k to the right-hand side of this equation as

$$K(t) = 3k/\kappa\rho(t)a^2(t). \tag{13.42}$$

From (13.23)–(13.25), we find that $K = (\Omega - 1)/\Omega$, and the limits on the present value of Ω imply for the present value K_0 of K that $|K_0| < 10$. During the matter-dominated era in which (13.28) is true, $K(t)$ is proportional to $a(t)$. Therefore, according to (13.40), $|K(t_m)| < 5 \times 10^{-4}$. At earlier times, $K(t)$ was even smaller, so k has a negligible effect in (13.20) and may be taken as zero. For much of the time, the content of the early universe was *radiation dominated*, in the sense that its contents behaved like black-body radiation with $p = \rho/3$. When this is true, (13.20) and (13.21) with $k = 0$ give

$$\frac{d^2}{dt^2}(a^2) = 2(a\ddot{a} + \dot{a}^2) = 0 \tag{13.43}$$

whose solution is of the form

$$a(t) = (At + B)^{1/2}. \tag{13.44}$$

The assumption usually made is that ρ can be evaluated as the sum of densities of several species of particles, each behaving as an ideal gas in thermal equilibrium, and p as the sum of their pressures. Consider a comoving region, whose volume V is proportional to a^3, and suppose that the particles it contains can be divided into groups such that the particles in each group interact with each other but not with those in other groups. The idea is that, within each group, the interactions are strong enough for the temperature and relative numbers of particles to be determined by the condition of thermal equilibrium, but sufficiently weak that the interaction energy does not contribute significantly to the energy density and pressure. According to the fundamental relation (10.30) of equilibrium thermodynamics, we have for the ith group of particles

$$\frac{\mathrm{d}U_i}{\mathrm{d}t} + p_i \frac{\mathrm{d}V}{\mathrm{d}t} = T_i \frac{\mathrm{d}S_i}{\mathrm{d}t} + \sum_j{}^{(i)} \mu_j \frac{\mathrm{d}N_j}{\mathrm{d}t} \tag{13.45}$$

where the sum is over particle species belonging to the ith group. If we sum this over all groups of particles to get the total energy density and pressure, then (13.22) gives

$$\sum_i T_i \frac{\mathrm{d}S_i}{\mathrm{d}t} = -\sum_j \mu_j \frac{\mathrm{d}N_j}{\mathrm{d}t} \tag{13.46}$$

where the sum on i is over all groups of mutually interacting particles and the sum on j is over all particle species.

It will sometimes be important to know the numbers of particles present as well as their contributions to the energy density and pressure. Using the basic distribution functions (10.60) and denoting by q the magnitude of a particle's 3-momentum, we find that the number of particles of a given species per unit physical volume with momentum in the range q to $q + \mathrm{d}q$ is

$$n(q)\mathrm{d}q = \frac{g}{2\pi^2\hbar^3}\{\exp[\beta(\varepsilon(q) - \mu)] \pm 1\}^{-1} q^2\,\mathrm{d}q \tag{13.47}$$

where $\varepsilon(q) = c(q^2 + m^2c^2)^{1/2}$. If $\mu = 0$ and the particles are highly relativistic, so that their mass can be neglected, the total number per unit volume is

$$n = \frac{g}{2\pi^2}\left(\frac{k_{\mathrm{B}}T}{c\hbar}\right)^3 \int_0^\infty (\mathrm{e}^x \pm 1)^{-1} x^2\,\mathrm{d}x = \frac{\zeta(3)}{\pi^2}\begin{pmatrix}3/4\\1\end{pmatrix} g\left(\frac{k_{\mathrm{B}}T}{c\hbar}\right)^3 \tag{13.48}$$

where the upper values refer to fermions, the lower ones to bosons and ζ is the Riemann zeta function with $\zeta(3) = 1.202\ldots$. At present, the microwave background contains some 400 photons per cm^3. If we take the present density of matter as $10^{-29}\,\mathrm{g\,cm^{-3}}$ and assume that it is primarily composed of nucleons, each with a mass of about 1.7×10^{-24} g, then the ratio of the number of nucleons to the number of photons, conventionally denoted by η is

$$\eta = n_{\mathrm{N}}/n_{\gamma} \approx 10^{-8}. \tag{13.49}$$

If, as is widely thought, much of the dark matter is not nucleonic, then the actual value may be 10 or a hundred times smaller. At any rate η is small and, since T is almost always proportional to $1/a$, it is constant in time, until we reach temperatures of the order of 10^{13} K at which nucleon–antinucleon pairs could be copiously produced in collisions.

We can now work out what conditions must have been like at temperatures a little below 10^{12} K. The nucleons which still exist today were present but their numbers, energy density and pressure were negligible compared with those of the black-body photons. The typical energy of a photon was such that electron–positron pairs could be copiously produced and these pairs can also annihilate to photons. Under the assumption of thermal equilibrium, the balance between these processes leads to an energy spectrum for electrons and positrons of the form (13.47), and I shall discuss shortly the conditions under which this assumption is likely to be valid. Likewise, the electron- and muon-type neutrinos and antineutrinos could be produced and annihilated by weak interaction processes and also had a thermal spectrum. The particles present in substantial numbers were therefore γ, e^-, ν_{e}, ν_μ, together with the antileptons, and there were also a few nucleons. All known heavier particles which will have been present at higher temperatures undergo rapid decays or particle–antiparticle annihilation, whose final products are the particles I have listed.

Under conditions of thermal equilibrium, the abundant particles have energy spectra of the form (13.47), so we need to know their chemical potentials. As we saw in chapter 10, the equilibrium density operator (10.55) can contain only operators associated with conserved quantities. For the particle species of interest, there are four conserved quantities, namely electric charge Q (measured in units of e), electron number E, muon number M and baryon number B. The values of these numbers for the various particles are

	e	ν_{e}	ν_μ	p	n
Q	−1	0	0	1	0
E	1	1	0	0	0
M	0	0	1	0	0
B	0	0	0	1	1

(13.50)

with opposite values for their antiparticles. These conservation laws are embodied in the standard GWS model. For example, any interaction vertex which creates an electron also either creates a positron or antielectron neutrino or annihilates an electron neutrino, so electron number is conserved. In grand unified theories which allow processes like proton decay, the lepton and baryon numbers are not separately conserved. However, processes which violate these conservation laws will occur at significant rates only when collision energies are greater than the X boson masses of about 10^{14} GeV or at temperatures above 10^{27} K. In (10.56), we can introduce independent chemical potentials μ_Q, μ_E, μ_M and μ_B for each conserved quantity and then, using (13.50), express Q, E, M and B in terms of particle numbers:

$$\mu_Q\hat{Q} + \mu_E\hat{E} + \mu_M\hat{M} + \mu_B\hat{B} = \mu_Q[\hat{N}_{e^+} - \hat{N}_{e^-} + \hat{N}_p] + \mu_E[\hat{N}_{e^-} + \hat{N}_{\nu_e} - \hat{N}_{e^+} - \hat{N}_{\bar{\nu}_e}] + \mu_M[\hat{N}_{\nu_\mu} - \hat{N}_{\bar{\nu}_\mu}] + \mu_B[\hat{N}_p + \hat{N}_n]. \tag{13.51}$$

From this, we can read off the chemical potentials for the particle species themselves. For example

$$\mu_{e^+} = \mu_Q - \mu_E = -\mu_{e^-}. \tag{13.52}$$

As in (11.3), we now adjust the chemical potentials to give the correct mean numbers of particles. Consider the total electric charge. All the evidence is that this is now exactly zero so, since charge is conserved, it must have been zero in the early universe as well. Adding up the charges of all the particle species, we have

$$Q = N_{e^+} - N_{e^-} + N_p = N(\mu_{e^+}) - N(\mu_{e^-}) + N_p = 0 \tag{13.53}$$

where $N(\mu)$ is the integral of (13.47) with the electron mass and the appropriate chemical potential. Under the conditions we are considering, the numbers of electrons and positrons are comparable with the number of photons and thus, according to (13.49), very much greater than the number of protons. To a good approximation, therefore, the numbers of electrons and positrons must be equal, and their chemical potentials must also be equal. So, in view of (13.52), these chemical potentials must be zero. I shall follow the usual assumption that the chemical potentials of the neutrinos are also zero, which appears reasonable, though there is no firm evidence for it. As in (13.52), we find that the chemical potentials of a neutrino and its antiparticle are equal and opposite. Large neutrino chemical potentials lead to a condition called *degeneracy*, similar to that which characterizes electrons in metals. The consequences of neutrino degeneracy can be investigated, and the main effect is to increase the contribution of neutrinos to the total energy density. This in turn affects several predictions of the standard cosmological model, notably those for nucleosynthesis which I shall discuss below. These effects serve to place limits on the size of the chemical potentials, and interested readers will find some discussion of them in Weinberg (1972).

To continue the story of the fairly early universe, it is necessary to understand the conditions under which thermal equilibrium can be maintained. Readers will recall from chapter 10 that the ensemble averages of statistical mechanics correspond to long time averages for a single system. In order for the scattering processes which maintain the balance of particle numbers to be effective, it must be possible for a reasonable number of these events to occur before any great change takes place in the environment. To obtain a criterion for this, consider the mean free path λ of a particle between scattering events. Under laboratory conditions, we have $\lambda = 1/n\sigma$, where n is the number of particles per unit volume and σ the scattering cross-section. In the expanding universe, consider a particle with velocity v relative to comoving coordinates, attempting to collide with a comoving target particle a distance λ away. The expansion is carrying the target particle away with a velocity given by Hubble's law as $H\lambda$. A rough criterion for scattering to take place at a reasonable rate is that v should be considerably greater than $H\lambda$. Another way of putting this is that the mean time between collisions under laboratory conditions, $\tau = \lambda/v$, should be less than the characteristic expansion time H^{-1}:

$$\tau H = H/n\sigma v \ll 1. \tag{13.54}$$

Let us apply this to the weak interactions which are supposed to maintain the thermal distribution of neutrinos. The energies we are considering are much smaller than the masses of the weak gauge bosons, so the Fermi theory (with the addition of neutral currents) is adequate. Scattering cross-sections are proportional to G_F^2 where, as we saw in chapter 12, $G_F/(\hbar c)^3 = 1.16\ \text{GeV}^{-2}$. Since $k_B T$ is much greater than the electron rest energy, it is the only relevant quantity with the dimensions of energy, and dimensional analysis shows that the cross-sections must be given by

$$\sigma \approx G_F^2 (k_B T)^2 (\hbar c)^{-4}. \tag{13.55}$$

If we take $H = (\kappa\rho/3)^{1/2}$, ρ and n to be given by the thermal distributions for a few species of particles and, for neutrinos and highly relativistic electrons, $v = c$, we obtain the estimate

$$\tau H \approx (10^{10}/T)^3 \tag{13.56}$$

when T is measured in degrees kelvin. As our story starts, just below 10^{12} K, this is small enough for thermal equilibrium to become established, if it had not already been, and maintained. As the temperature fell to around 10^{10} K, however, the rate of neutrino scattering became very small, so that, in effect, the neutrinos ceased to interact with the other particles or, as the jargon has it, became *decoupled*. The thermal distributions of neutrinos did not disappear, however. Their temperature simply continued to fall as $1/a$ and they are, presumably, here to this day, though it would be extremely difficult to detect them. Their present temperature is, as we are about to see, rather less than that of the microwave background and their contribution to the energy density correspondingly smaller. The cross-section for electromagnetic scattering of electrons, positrons and photons is greater than the weak cross-sections, and these particles continued to interact.

The rest energy of an electron corresponds to a temperature of about 5.9×10^9 K. As the temperature dropped below this value, electron–positron pairs could no longer be produced by collisions. The electrons and positrons which had been present annihilated rapidly, the extra photons they produced heating the black-body radiation. Since the neutrinos had ceased to interact, their temperature was unaffected, so the temperature of the photons was now greater than that of the neutrinos, and has remained so since. We can work out the ratio of the photon and neutrino temperatures from (13.46). The right-hand side of this equation is zero, as may be seen in the following way. The chemical potentials of the electrons, positrons and photons are zero. The only other particles present in significant numbers are the neutrinos and, since these have ceased to interact, the number of them in a comoving volume is constant. So, regardless of their chemical potentials, neutrinos do not contribute to the right-hand side of (13.46). On integrating (13.47) for a neutrino species, with $m = 0$, we find that the total number in a comoving volume proportional to a^3 can be expressed as $(aT)^3 f(\mu/T)$, where f is the function determined by the integral and multiplying constants. Since this number is constant, and T is proportional to $1/a$, the ratio μ/T is constant. The neutrino entropy in the comoving volume can, as readers may easily check, be expressed in the same form but with a different function f, so it too is constant and makes no contribution to the left-hand side of (13.46). Thus, the left-hand side of (13.46) has significant contributions only from electrons, positrons and photons which, since they still interact, have the same temperature. What (13.46) tells us, therefore, is that the total entropy of electrons, positrons and photons in a comoving volume is constant, irrespective of the neutrino chemical potentials.

While the electron–positron annihilation is taking place, the electron mass is comparable with $k_B T$, and the integral for the entropy cannot be computed analytically. For our present purpose, however, this is not necessary. We consider a time 'before' the annihilation when the electrons were relativistic, and a time 'after' when they had vanished. In each case, we can use (10.87) for the electron–positron–photon entropy. The multiplicity factor g is given by (10.83) as

$$g_{\text{before}} = 4 \times \tfrac{7}{8} + 2 = \tfrac{11}{2} \quad \text{and} \quad g_{\text{after}} = 2 \tag{13.57}$$

since the electron, positron and photon each have two polarizations. Conservation of this entropy implies

$$g_{\text{before}}(aT)^3_{\text{before}} = g_{\text{after}}(aT)^3_{\text{after}} \tag{13.58}$$

where T is the photon temperature. For the neutrino temperature T_ν, on the other hand, we have

$$(aT_\nu)_{\text{after}} = (aT_\nu)_{\text{before}} = (aT)_{\text{before}} \tag{13.59}$$

and so, after the annihilation,

$$T_\nu = (4/11)^{1/3}T = 0.714T. \tag{13.60}$$

The present neutrino temperature is therefore about 1.9 K.

As far as the abundant species of particles are concerned, there were no further significant events until the universe became matter dominated. The state of the nucleons did indeed undergo important changes, which are discussed in the next section, but these had no significant effect upon the energy density, pressure or expansion rate. We can now estimate the periods of time which elapsed between the various events I have described so far. Consider a period during which the multiplicity factor g^* for the total number of abundant species is constant. (Note that g^* is different from the g given in (13.57), which counts only those particles interacting efficiently with photons.) Since we have set $k = 0$, we may use equations (13.44), (13.10), (13.23) and (10.86) to express a time difference $t_1 - t_2$ in terms of the temperatures T_1 and T_2 which prevailed at those times. The result is

$$\begin{aligned} t_1 - t_2 &= \left(\frac{3c^3}{64\pi G\sigma}\right)^{1/2} g^{*-1/2}(T_1^{-2} - T_2^{-2}) \\ &= 3.26 \times 10^{20} g^{*-1/2}(T_1^{-2} - T_2^{-2}) \end{aligned} \tag{13.61}$$

where the times are in seconds and temperatures in degrees kelvin.

In order to make use of this result, we need to know the value of g^*, which means that we need to know all the species of particles which were present. We have seen that the electron- and muon-type neutrinos were decoupled at temperatures below about 10^{10} K but still contributed to the energy density and pressure. However, we also saw in chapter 12 that a further neutrino species, the tau-type neutrinos, are known to exist. These and perhaps further, as yet unknown, species of neutrinos or other light particles will also have been present. Whatever these species are, we know from laboratory experiments that they do not interact strongly at the temperatures we have considered, so they do not affect our calculations up to this point. They will, however, affect any calculations which require us to know periods of time rather than merely temperatures, and this is one point where theoretical models of particle physics have cosmological consequences which can be confronted with observations. Each additional species has, presumably, a thermal energy spectrum similar to that of the neutrinos. As we have seen, however, the temperature of the electron and muon neutrinos was changed relative to that of photons by the electron–positron annihilation. Depending on the temperature at which a given species decoupled, its temperature may have been similarly affected by earlier particle–antiparticle annihilation processes, of which we have no definitive understanding.

These matters can be dealt with in detail only on the basis of some definite model of particle physics and, in general, some additional assumptions about the sequence of events in the very early universe, at temperatures above 10^{12} K. For the sake of argument, I shall suppose that there are N_ν species of neutrinos, all at the same temperature. In that case, the value of g^* prior to electron–positron annihilation is

$$g^* = \frac{11}{2} + \frac{7}{4} N_\nu \quad \text{for } 10^{12}\,\text{K} > T > 6 \times 10^9\,\text{K}. \tag{13.62}$$

After the annihilation, we can take account of the different neutrino temperature (13.60) by including an appropriate factor in g^*:

$$g^* = 2 + \frac{7}{4}\left(\frac{4}{11}\right)^{4/3} N_\nu \quad \text{for } T < 6 \times 10^9\,\text{K}. \tag{13.63}$$

Let us calculate some representative time intervals, taking $N_\nu = 3$ to include just the known neutrinos. The time taken for the temperature to fall from 10^{12} to 10^{11} K was 9.8×10^{-3} s. The further time to reach 10^{10} K was, obviously, a hundred times this, 0.98 s. Near their annihilation temperature, the electrons and positrons are non-relativistic, so our equations based on black-body radiation are not valid, and a numerical calculation using the correct distribution is needed. It is a fair approximation, however, to imagine that the annihilation occurred instantaneously, using (13.62) just above and (13.63) just below 6×10^9. With this approximation, the time to get from 10^{10} to 6×10^9 K was 1.77 s, and the further time to get to 10^9 K was 4.9 h. According to (13.40), the universe became matter dominated at a temperature of $2.7/5 \times 10^{-5} = 5.4 \times 10^4$ K. If we neglect the density of nuclear matter up to that point, then it happened about 2000 years after the events we have been considering. This temperature and time depend on the values assumed for h and Ω_0, and a time of, say, 10 000 years to matter domination would be entirely reasonable.

To estimate the time from the initial singularity to our starting point at 10^{12} K, we would need to know what happened during that time. If we assume that (13.61) remains valid, then the value of g^* obviously increases with temperature. Thus it is reasonable to guess that this time is no greater than what we obtain by using (13.62) and setting the initial temperature to infinity, namely about 10^{-4} s. Clearly, using the figures given above, we might as well say that the temperature was 10^{10} K at 0.98 s after the initial singularity, and so on.

13.5 Nucleosynthesis

Although protons and neutrons made a negligible contribution to the overall composition of matter in the early universe, they were nevertheless able to take part in interactions which had important consequences. There is a narrow range of temperatures around 10^9 K at which nuclear reactions could take place which fused protons and neutrons into larger nuclei. These reactions have been well studied in the laboratory, and it is possible to work out quite accurately the relative numbers in which various light nuclei would have been formed. The process is called *nucleosynthesis* and it is important for at least two reasons. One is that the predicted abundances can be compared with matter observed in the present universe. After allowance is made for later reactions occurring in the cores of stars, good agreement is found between the predicted and observed abundances, and this provides an important test of the standard big bang model. The second reason is that the predicted abundances depend on the value of N_ν which is used in (13.62), and the successful prediction of nuclear abundances thus places a constraint on this number. Taken at face value, N_ν is the number of families of fermions in the standard model of particle physics, which cannot yet be determined from laboratory experiments (see, however, the discussion at the end of this section). It turns out that hydrogen and helium-4 are by far the most abundant nuclear species, and I shall give a simplified account of the calculation of their relative abundances. Interested readers will find more details and further references in, for example, Peebles (1971), Weinberg (1972), Dolgov and Zeldovich (1981), Barrow (1983), the papers by G Steigman in the conference proceedings edited by Baier and Satz (1985), and in Bernstein *et al* (1989).

The relative abundances of nuclei obviously depend on the relative numbers of protons and neutrons and, to estimate their ratio, we must begin the story of nucleosynthesis at a temperature of about 10^{11} K. Although the total number of nucleons cannot change at this temperature, lepton–nucleon scattering can easily interconvert protons and neutrons by weak interaction processes such as ($e^- + p \leftrightarrow n + \nu_e$). The energy absorbed or released by these conversions is the neutron–proton mass difference $\Delta m = m_n - m_p = 1.29\,\text{MeV}$. As long as the weak interactions are effective in maintaining thermal equilibrium, the ratio of the numbers of protons and neutrons can adequately be determined from classical statistical mechanics and is given by

$$n_n/n_p = \exp(-\Delta m/k_B T). \tag{13.64}$$

At about the time the neutrinos cease to interact with electrons, the interconversions of protons and neutrons also cease, and the ratio becomes frozen.

For good accuracy, it is necessary to determine the ratio precisely, and this requires a detailed analysis of the reaction rates, which I am not going to reproduce here. It is easy to see, however, that the ratio depends on the value of g^* at the temperature T_f where the freeze occurs. Consider, for example, neutron–neutrino scattering, for which the cross-section is roughly the same as (13.55). As readers may convince themselves, the number of scattering events per unit time per unit volume is $\sigma n_\nu n_n c$, where n_ν and n_n are the number densities of neutrinos and neutrons respectively. The number of events per unit time per neutron is therefore $\sigma n_\nu c$. The mean time between scattering events for a particular neutron is the reciprocal of this quantity and, roughly speaking, the freeze occurs when this time equals the expansion time H^{-1}. To estimate T_f, we use (13.48) with $g = 1$ for n_ν, (13.23) for H, and estimate ρ using (10.86) with g equal to the g^* given in (13.62) for all the abundant species present at temperatures near 10^{10} K. The result is

$$T_f \approx 2.6 \times 10^{10} g^{*1/6}\ \text{K}. \tag{13.65}$$

Inserting this value into (13.64) gives a good indication of how the neutron–proton ratio depends on g^* and hence on N_ν, but the number 2.6×10^{10} is merely a guess. The results of a more careful analysis, in so far as they can be approximated by an equation of the form (13.64), indicate that this number should be replaced by something like 6.4×10^9.

At the prevailing nucleon densities, the probability of more than two particles colliding simultaneously is negligible, so nuclei can be built up only by two-particle collisions. The first nucleus which can be formed is deuterium, consisting of one proton and one neutron. Now, deuterium has a binding energy of only about 2.2 MeV and, at temperatures near 10^{10} K, there are many photons capable of dissociating it. Deuterium nuclei remain intact in sufficient numbers for further reactions to proceed only when the temperature has fallen to a value which is estimated at about 8×10^8 K. This value depends somewhat on the actual numbers of nucleons present, which in turn must be deduced from uncertain observations of the present matter density. Studies of the reactions which may then ensue show that almost all the available neutrons are used to form helium-4, the excess protons remaining single. Only very small quantities of heavier nuclei such as lithium-7 emerge, together with small amounts of deuterium and tritium.

The relative abundance of hydrogen (protons) and helium is thus essentially determined by the neutron–proton ratio at 8×10^8 K, and I shall now estimate it, taking N_ν to be 3. At the temperature T_f, which is 9.5×10^9 K, the ratio n_n/n_p is given by (13.64) to be 0.206, and the fraction $X_n = n_n/(n_p + n_n)$ is 0.171. The time which elapses as the temperature falls from T_f to 8×10^8 K is found from (13.61) to be 274 s. During this time, a few neutrons decay, each one to a proton plus leptons, with a mean lifetime of 917 s. Thus, when nucleosynthesis begins, we have

$$X_n = \frac{n_n}{n_n + n_p} = 0.171 \exp(-274/917) = 0.127. \qquad (13.66)$$

Since each helium-4 nucleus contains two neutrons and two protons and has almost exactly four times the mass of a proton, the fraction by weight of helium, $M_{He}/(M_{He} + M_H)$, is, as readers may check, just twice this number, or about 25.3%. I emphasize that, while this calculation illustrates the essential argument, a much more thorough analysis is needed to obtain reliable results. This analysis does predict a helium abundance of around 25%, but there are three main sources of uncertainty. One is the lifetime of the neutron which is, in fact, known only with an accuracy of about 1%. Another is the total density of nucleons, which affects the exact time at which nucleosynthesis starts. What matters is the nucleon–photon ratio η (13.49), which was the same during nucleosynthesis as it is now. The third is the number N_ν of light-particle species. What is affected by this number is the expansion rate which in turn affects the value of the freezing temperature T_f and the length of time between freezing and the onset of nucleosynthesis during which neutrons can decay. For $N_\nu = 2$ or 4, my calculation gives helium abundances of 23.9% or 26.7%, respectively, compared with 25.3% for $N_\nu = 3$, and more careful calculations give about the same variation.

Observations indicate that even today most of the matter in the universe is hydrogen and helium in the ratio of about three to one by weight, while the abundances of heavier nuclei are consistent with their having been produced mainly in stars. Attempts to deduce precise values for the primordial abundances are, as readers may imagine, fraught with difficulty. Nevertheless, experts are able to convince themselves that the primordial helium abundance can be determined with fair accuracy. Steigman, in the papers referred to above, quotes a value of $(24 \pm 2)\%$. Allowing for the various uncertainties, it appears that the maximum value of N_ν consistent with observations is 4. Taken at face value, this means that the number of fermion families in the standard model of particle physics cannot be greater than four. It should be borne in mind, however, that more neutrinos or other light particles could exist if, for some reason, their temperature in the fairly early universe was too small for them to contribute significantly to g^*. As we saw in chapter 12, the fact that asymptotic freedom appears to hold in QCD places a limit of eight on the number of families, but the cosmological limit is the most stringent which is so far available. Actually, the true number of families may be known by the time this book is published, from measurements of the decay rate of the weak vector boson Z^0. This particle can decay into a neutrino–antineutrino pair of any species, and its decay rate depends on the number of species. At the time of writing, an experiment to measure this rate with sufficient accuracy is imminent at CERN. (See note on p 337.)

13.6 Recombination and the Horizon Problem

By the time of nucleosynthesis, almost all the electrons and positrons which had once been present had annihilated. Assuming, however, that the universe is electrically neutral, there must have been a small residual number of electrons to balance the charge of the protons. When the temperature fell to a small enough value, T_r, these electrons will have combined with the positive nuclei to form neutral atoms. This process is called *recombination*. To estimate T_r with reasonable accuracy, it is sufficient to consider a universe filled entirely with hydrogen. Near T_r, the fraction x of ionized atoms is determined by thermal equilibrium, maintained by atomic collisions, and this is described by the *Saha equation* (exercise 10.8). This equation involves the number density of protons, which can be expressed in terms of the density of photons and the nucleon–photon ratio η. Taking the ionization energy as 13.6 eV, we obtain

$$x^2/(1 - x) = 1.19 \times 10^{14}\eta^{-1}T^{-3/2}\exp(-1.578 \times 10^5/T). \quad (13.67)$$

The numerical solution of this equation is easy. If we take η to be about 10^{-9}, we find that x falls abruptly from near 1 to near zero at a temperature T_r of about 4000 K. It also falls to zero at about 9000 K, but at this and higher temperatures there are enough photons with energy greater than the ionization energy to keep all the matter ionized, and the Saha equation no longer applies.

While electromagnetic radiation interacts strongly with charged particles, it interacts hardly at all with a neutral gas of hydrogen and helium, which is almost completely transparent. It follows that the microwave background radiation we observe today was last scattered at the time of recombination and has travelled freely toward us ever since. This leads to a conundrum known as the *horizon problem*, which I shall now explain. The path of a light ray is found by setting $d\tau = 0$ in (13.1) where, for simplicity, I shall take $k = 0$. As measured by comoving coordinates, the distance it travels between times t_1 and t_2 is

$$L = \int_{t_1}^{t_2} \frac{\mathrm{d}t}{a(t)}. \tag{13.68}$$

Recombination occurred, as readers may work out, somewhat after the universe became matter dominated. For simplicity again, however, I shall assume that $a(t)$ was proportional to $t^{1/2}$ right up to recombination, since this will not greatly affect our conclusion. The coordinate distance d which a non-interacting light ray could have travelled between the initial singularity and the time t_r of recombination is

$$d = 2t_r/a(t_r). \tag{13.69}$$

Of course, light rays did interact strongly. The point is that no signal of any kind could have travelled a distance greater than d, and so any causal influences could have acted only within a 'causally connected' region whose diameter was no greater than d, which is called the *causal horizon*.

Since recombination, the universe has been matter dominated and, to a reasonable approximation, we can use (13.30) to write

$$\frac{a(t)}{a(t_r)} = \left(\frac{t}{t_r}\right)^{2/3}. \tag{13.70}$$

Then the coordinate distance D which a photon we now detect has travelled towards us since recombination is

$$D = \frac{3t_r}{a(t_r)}\left(\frac{a(t_0)}{a(t_r)}\right)^{1/2}. \tag{13.71}$$

The angle subtended at the Earth by one causally connected region is the ratio

$$\frac{d}{D} = \frac{2}{3}\left(\frac{a(t_r)}{a(t_0)}\right)^{1/2} = \frac{2}{3}\left(\frac{T_0}{T_r}\right)^{1/2} \approx 0.017\,\mathrm{rad} \approx 1^\circ. \tag{13.72}$$

What is puzzling about this is that the observed radiation is completely isotropic. Thus, at the time of recombination, very many regions which could never have communicated with each other were, to at least one part in 10^4, at the same temperature.

13.7 The Flatness Problem

Cosmologists speak of a second puzzle concerning the standard model, which is called the *flatness problem*. During the whole history of the universe, the scale factor $a(t)$ has been roughly proportional to a power of t, say t^x, with x equal to either $\frac{1}{2}$ or $\frac{2}{3}$. To make matters simple, suppose that x was always $\frac{1}{2}$. Crudely, we can then use (13.25) to compare the present energy density with that at earlier times:

$$\Omega(t) - 1 \approx \left(\frac{\dot{a}(t_0)}{\dot{a}(t)}\right)^2 (\Omega_0 - 1) \approx \left(\frac{t}{t_0}\right)(\Omega_0 - 1). \tag{13.73}$$

It will be recalled that the value $\Omega = 1$ corresponds to a flat universe, and it seems most unlikely that Ω_0 could differ from this value by more than a factor of 100. When the universe was, say, 1 second old, Ω must have been equal to 1 with an accuracy of at least one part in 10^{15}, and this seems to represent a degree of fine tuning which would not be expected to occur without some good reason.

Whether this should be regarded as a genuine puzzle is to some extent a matter of philosophical taste. Even though (13.73) is not exactly correct, it is obvious that, whatever the value of Ω_0, $\Omega(t)$ will be arbitrarily close to 1 if we choose a sufficiently early time. It is worth reflecting, however, that all the events which determined the overall constitution of the universe took place within the first few seconds, if we are content to regard nucleosynthesis as a relatively minor rearrangement of the particles which already existed. Thus, all the relevant timescales which naturally arise from physics are of the order of a second or less and, unless Ω is for some reason exactly equal to 1, we might have expected some appreciable variation by that time. It is sometimes said, indeed, that the only truly fundamental timescale is the Planck time (13.27), at which $\Omega - 1$ was less than 10^{-60}, and that we might have expected some appreciable difference of Ω from 1 by then. At any rate, if Ω is exactly equal to 1, then we would certainly like to know why. If it is not, then, since $|\Omega - 1|$ grows at least as fast as $t^{2/3}$, we may reasonably wonder why the difference is still fairly small after at least 10 billion years.

The horizon and flatness problems do not make the standard cosmological model incompatible with observations, but they do show that the model requires very special initial conditions. Any explanation of these initial conditions must be sought in the very early universe, at temperatures well above 10^{12} K.

13.8 The Very Early Universe

As we attempt to look into the very early universe, by which I mean roughly the first 10^{-4} s, we soon encounter energies of a few hundred GeV at which the standard model of particle physics has been only incompletely tested in the laboratory. (Readers may like to bear in mind that an energy of 1 GeV corresponds to a temperature of 1.16×10^{13} K.) At still higher energies, the standard model may well be quite inadequate. It is widely thought that grand unified theories and/or the various supersymmetric theories should come into play, but there is no firm experimental foundation for any of these. Little of what is said about the very early universe can therefore be taken as reliably established and much of it is purely conjectural. As I said at the beginning of this chapter, however, it is possible in principle to work out the consequences of these theoretical conjectures and confront them with observations.

It seems clear that a prominent role must have been played by *phase transitions* of various kinds. The first of these we encounter, moving backwards in time, is the *quark–hadron* or *deconfinement* transition. The idea is that, at sufficiently high temperature and density, quarks and gluons cease to be bound in identifiable hadronic particles but exist instead in a relatively weakly interacting plasma along with the photons and leptons. Approximate calculations based on the lattice version of QCD suggest that this change takes place at a sharp phase transition which, at the fairly low density of nucleons present in the early universe, would have occurred at a temperature around 10^{12}–10^{13} K. Experimental studies of heavy-ion collisions, which produce, for a short time, large densities of nuclear matter at high energy, provide some evidence for this kind of effect. Deconfinement is related to the property of *asymptotic freedom* which means, as readers will recall from chapter 12, that the effective strength of the strong interactions decreases at high energy. Were it not for asymptotic freedom, indeed, very little could be said at all about the first millisecond. Most of what we believe about the fairly early universe is based on treating radiation and matter as nearly ideal gases. If the 'strong' interactions continued to be strong at nucleon densities approaching those in nuclei, then the difficulty of applying statistical mechanics to such a strongly interacting fluid would become prohibitive. If the idea of asymptotic freedom is correct, then we do not encounter such densities until the temperature is high enough, and the strong interaction weak enough, for the ideal gas approximation to be adequate.

If the gauge theories of fundamental interactions are correct, then we may expect phase transitions to occur at which their symmetries cease to be spontaneously broken. The possibility of symmetry restoration at high temperatures was first recognized by D A Kirzhnits and A D Linde (1972). These phase transitions are quite analogous to the superconducting transition, with critical temperatures given, very roughly, by the masses of the relevant gauge bosons. What happens is not that the particle's mass becomes negligible in comparison with its kinetic energy, but rather that its effective mass, as it moves through the hot plasma, is actually equal to zero.

To indicate how this happens, I shall consider a single scalar field ϕ, which could be one of the Higgs fields in a gauge theory. For simplicity, I shall take it to be real, with a finite temperature action analogous to (10.71) given by

$$S_\beta(\phi) = \int_0^\beta \mathrm{d}\tau \int \mathrm{d}^3x \left[\frac{1}{2}\left(\frac{\partial\phi}{\partial\tau}\right)^2 + \frac{1}{2}\nabla\phi\cdot\nabla\phi + \frac{\lambda}{4!}(\phi^2 - v^2)^2\right]. \tag{13.74}$$

Up to loop corrections in perturbation theory, the vacuum expectation value of ϕ is one of the two values $\pm v$, which are the two minima of the potential term in (13.74). A high-temperature state is, however, not a vacuum state, and we need to estimate the expectation value in this state. To that end, we introduce a source J for the field and, as in (10.75), define a thermodynamic potential density $\Omega(\beta, J)$ by

$$\exp[-\beta V\Omega(\beta, J)] = Z_{\mathrm{gr}} = \int \mathcal{D}\phi \exp\left[-S_\beta + J\int \mathrm{d}\tau\mathrm{d}^3x\phi(\boldsymbol{x}, \tau)\right]. \tag{13.75}$$

The expectation value of ϕ should be independent of $\boldsymbol{x}$ and τ and is given by

$$\bar{\phi} \equiv \langle\phi\rangle = -\left.\frac{\partial\Omega}{\partial J}\right|_\beta. \tag{13.76}$$

Consequently, the thermodynamic relation analogous to (10.32) is

$$\mathrm{d}\Omega = -s\,\mathrm{d}T - \bar{\phi}\,\mathrm{d}J \tag{13.77}$$

where s is the entropy density. If we now make a Legendre transformation to a free energy density defined by

$$F(\beta, \bar{\phi}) = \Omega + J\bar{\phi} \tag{13.78}$$

then we have

$$\mathrm{d}F = -s\,\mathrm{d}T + \bar{\phi}\,\mathrm{d}J \tag{13.79}$$

and consequently

$$\left.\frac{\partial F}{\partial\bar{\phi}}\right|_\beta = J. \tag{13.80}$$

Thus, when J is zero, the expectation value we require is a minimum of F which, as we shall see, is equal to the potential in (13.74) plus a temperature-dependent correction.

A satisfactory calculation of F is slightly complicated, but I shall present a simple calculation which captures the main result. (Astute readers will notice the weak points in the argument and may like to think how they can be overcome.) The calculation is essentially first-order perturbation theory. We write ϕ as $\bar{\phi} + \psi$ and expand S_β to quadratic order in ψ, leaving out the interaction terms:

$$\begin{aligned} S_\beta(\phi) = {} & \beta V\left[\frac{\lambda}{4!}(\bar{\phi}^2 - v^2)^2 - J\bar{\phi}\right] \\ & + \int_0^\beta \mathrm{d}\tau \int \mathrm{d}^3x \left[\frac{1}{2}\left(\frac{\partial\psi}{\partial\tau}\right)^2 + \frac{1}{2}\nabla\psi\cdot\nabla\psi + \frac{1}{2}m^2(\bar{\phi})\psi^2 \right. \\ & \left. + \left(\frac{\lambda}{6}\bar{\phi}(\bar{\phi}^2 - v^2) - J\right)\psi\right] \end{aligned} \tag{13.81}$$

where

$$m^2(\bar{\phi}) = \frac{\lambda}{6}(3\bar{\phi}^2 - v^2). \tag{13.82}$$

We now estimate Ω by substituting this into (13.75) and carrying out the integral. The largest contribution to the integral should come from values of ϕ near $\bar{\phi}$ which, to a first approximation, is $\pm v$. The coefficient of ψ in (13.81) vanishes there, so we ignore the final term. The functional integral is now identical to that which led to (10.79), except that we have only a single particle species. So, after a renormalization which removes the vacuum energy, our approximation to the free energy is

$$F(\beta, \bar{\phi}) = \frac{\lambda}{4}(\bar{\phi}^2 - v^2)^2 + \frac{1}{2\pi^2\beta^4}\int_0^\infty \mathrm{d}x\, x^2 \ln\{1 - \exp[-(x^2 + \beta^2 m^2(\bar{\phi}))^{1/2}]\}. \tag{13.83}$$

The import of this result becomes clearer if we make a high-temperature expansion, the first few terms of which are

$$F(\beta, \bar{\phi}) = \left[\frac{\lambda}{24}v^4 - \frac{\pi^2}{90}(k_{\mathrm{B}}T)^4\right] + \frac{\lambda}{12}\left[\frac{1}{4}(k_{\mathrm{B}}T)^2 - v^2\right]\bar{\phi}^2 + \frac{\lambda}{4!}\bar{\phi}^4 + \ldots. \tag{13.84}$$

This is similar to a Ginzburg–Landau expansion, and we see that the critical temperature at which symmetry is restored is given by $k_{\mathrm{B}}T_{\mathrm{c}} = 2v$. When ϕ is a Higgs field, this critical temperature is related to gauge boson masses at zero temperature by equations similar to (12.23), so unless the gauge coupling constant is very large or very small, these masses give a fair indication of T_{c}. The expectation value of ϕ is clearly given by $\pm v[1 - (T/T_{\mathrm{c}})^2]^{1/2}$.

If a phase transition of this kind leads to restoration of the SU(3) × SU(2) × U(1) symmetry of the standard model, then this occurred at a temperature between 10^{15} and 10^{16} K, at a time of about 10^{-12} s. It does not appear that this would have had any great effect on the expansion rate. In the case of a grand unified theory, the transition would occur at a temperature of some 10^{27} K, about 10^{-35} s after the initial singularity. According to what is called the *inflationary scenario*, the effect of this may have been spectacular. The idea of inflation was proposed by A Guth (1981) as a possible solution to the horizon and flatness problems. According to Guth, the universe may, at a very early time, have undergone a short period of very much more rapid expansion than is envisaged in the standard cosmological model.

To see how this might come about, consider a period during which the temperature is falling towards the critical temperature for a symmetry-breaking phase transition, involving a scalar field with an action similar to (13.74). The expectation value of ϕ is zero. Below T_c, the state of thermal equilibrium is one in which the expectation value is non-zero, but the field will require some period of time to adjust to this new state. During this time, equilibrium statistical mechanics is not valid. As the universe continues to expand, the thermal energy, measured by the expectation values of the derivative terms in the action, falls. The expectation value of the whole action which, presumably, we should use in place of the matter terms in (4.14) to derive Einstein's field equations, is dominated by the potential energy term. Now, (13.74) is the finite temperature action. What we actually want to use in (4.14) is the true action, whose expectation value will be the spacetime integral of $-\lambda v^4/4!$. We see that this is equivalent to a cosmological constant $\Lambda = \kappa\lambda v^4/4!$. The fact that $\Lambda = 0$ in our present vacuum is taken into account by taking the potential term of (13.74) to be zero when $\phi = v$. The low-temperature state with $\phi = 0$ is sometimes referred to as a *false vacuum*.

In the false vacuum state, the rate of expansion of the universe is governed by (13.18) and (13.19), in which we must set ρ and p to zero. Equivalently, we can keep the original cosmological constant equal to zero and take $\rho = -p = \Lambda(v)/\kappa$, where $\Lambda(v)$ is the effective cosmological constant we have just derived. The two equations are equivalent to a single equation

$$\dot{a}^2 + k = \tfrac{1}{3}\Lambda a^2. \tag{13.85}$$

Suppose that this equation first becomes approximately true at a time t_i when the scale factor is a_i. For a flat universe, with $k = 0$, the solution is

$$a(t) = a_i \exp\left[(\tfrac{1}{3}\Lambda)^{1/2}(t - t_i)\right]. \tag{13.86}$$

The cosmological model in which this is always true is called the *de Sitter* model. It is not a good model of our universe because it contains no ordinary matter. The exponential expansion is much faster than the $t^{1/2}$ expansion envisaged in the standard model. If this period of inflation lasts long enough, $a(t)$ can expand by a very large factor. The inflationary period comes to an end as the broken symmetry state of the field theory becomes established, and the potential energy of the false vacuum is converted into kinetic energy of particles and radiation. A feature of this process which runs counter to normal physical intuition is that the effective cosmological constant corresponds roughly to a constant *energy density*, so the total potential energy of the universe increases in proportion to a^3. There is therefore enough energy available to create a new plasma of particles and radiation at a temperature comparable with the critical temperature T_c at which inflation started.

From that point onwards, the history of the universe is that described by the standard model.

The behaviour of the scale factor in the inflationary and standard models is sketched in figure 13.5. It should be clear that inflation can solve the horizon problem. During the period before inflation, two small regions from which we now receive background radiation were much closer together than is allowed for in the standard model and could, after all, have communicated with each other. Conditions in these two regions would then naturally be very similar. To see how inflation can solve the flatness problem, we must solve (13.85) with $k = \pm 1$. The solution is

$$a(t) = a_i \cosh\left[\left(\frac{\Lambda}{3}\right)^{1/2}(t - t_i)\right] + \left(a_i^2 - \frac{3k}{\Lambda}\right)^{1/2} \sinh\left[\left(\frac{\Lambda}{3}\right)^{1/2}(t - t_i)\right]. \tag{13.87}$$

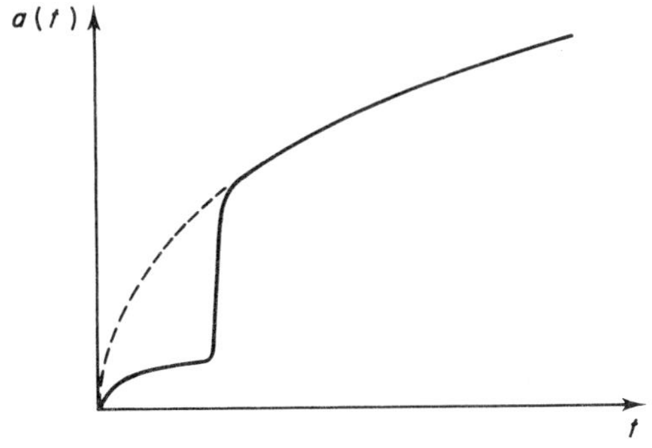

Figure 13.5 Schematic comparison of scale factors in the standard model (broken curve) and some versions of the inflationary model (full curve). The amount of inflationary expansion would be far greater than is actually shown.

For large values of their argument, both $\cosh\theta$ and $\sinh\theta$ are approximately equal to $\exp(\theta)/2$, so if inflation lasts long enough we again have exponential expansion. During this expansion, the Hubble parameter $H = \dot{a}/a$ is just a constant, equal to $(\Lambda/3)^{1/2}$. If a itself becomes very large, then k/a^2 becomes negligible compared with H^2, and the universe is very close to being flat. Intuitively, we may imagine, for example, a closed universe as a balloon inflated to a very large size. What we observe is only a very small part of its surface, which will appear almost flat. At the end of inflation, the potential energy density in Λ is converted into an equivalent energy density in particles and radiation, equal to the critical density $3H^2/\kappa$. In a flat universe, as we have seen, the energy density continues to be equal to the current critical density. Thus, if inflation lasted long enough for the universe to become extremely flat, the present value of Ω should also be close to 1.

If inflation does take place in the way I have described, the crucial question is, how long does it actually last? During the inflationary period, the state of the quantum fields is neither a true vacuum state nor a state of thermal equilibrium. How this non-equilibrium state evolves with time and how the broken-symmetry state finally emerges are very complicated questions, and the answers certainly depend on details of the field theory model considered. The traditional procedure (which is open to a number of objections) is to assume that the non-equilibrium state can be characterized by a time-dependent expectation value of the field, $\phi(t)$, governed by an equation of motion

$$\ddot{\phi} + 3\frac{\dot{a}}{a}\dot{\phi} + \gamma\dot{\phi} = -\frac{\mathrm{d}V(\phi)}{\mathrm{d}\phi}. \tag{13.88}$$

This is equivalent to the equation of motion for a classical particle moving in the potential $V(\phi)$ and subject to frictional forces. The frictional term proportional to γ represents the loss of energy due to the creation of particles and radiation, and the derivation of the remaining terms is the subject of exercise 13.3. The potential $V(\phi)$ is an *effective potential*, obtained from the generating functional (9.59) in much the same way that we obtained the free energy (13.83). It is therefore the sum of the potential for ϕ which appears in the original action and correction terms which arise from both the self-interaction of ϕ and its interactions with other fields. Depending on the nature of the field theory considered, quite complicated potentials can arise in this way. Readers who followed closely the derivation of (13.83) will realize that the free energy or effective potential is really valid only near its minimum, although more sophisticated derivations may obscure this fact. This is one reason why the kind of analysis I am describing is open to question (see Lawrie (1988) for further discussion and references).

The effective potentials which have mainly been considered are of the general form shown in figure 13.6 and may or may not have a subsidiary minimum, corresponding to the false vacuum at $\phi = 0$. If the evolution of the state of the quantum fields is represented as the motion of a particle in this potential, then the particle starts at $\phi = 0$. It must end at rest in one of the absolute minima which correspond to our true vacuum. The actual state of the matter in the universe at the end of inflation is a hot plasma existing in this true vacuum. If there is a minimum at the origin, the particle must first escape, by quantum tunnelling, after which it is considered to obey the classical equation of motion (13.88). Inflation lasts while the particle is in a region of large potential energy, where it moves slowly. At the end of this period, the particle falls into a potential well and begins rapid oscillations about the minimum. At this point, the frictional terms become important, and the oscillations are damped until the particle eventually comes to rest. Some of the energy lost by friction through the $\dot{a}/a$ term is effectively lost entirely. It can be thought of as an increase in gravitational potential energy. The remainder goes to creating particles and radiation.

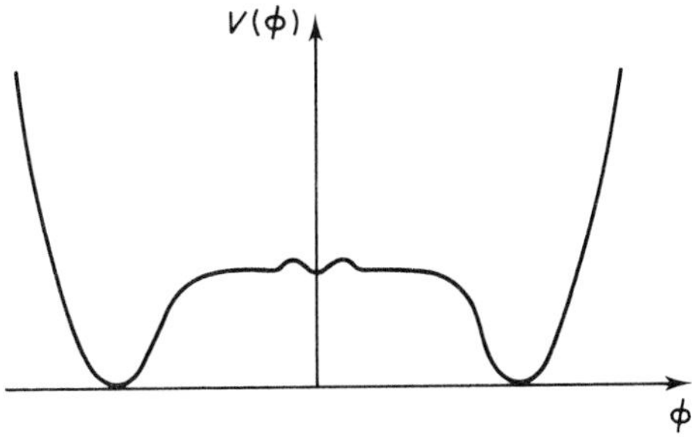

Figure 13.6 Qualitative form of the effective potential assumed in some versions of the inflationary model.

The results of detailed calculations along these lines depend very much on details of the field theory model which is used, but the amount of inflationary expansion can be enormous. If, as I have assumed, inflation is associated with the spontaneous breaking of grand unified symmetries, it can be estimated that, in order to solve the horizon and flatness problems, expansion by a factor of at least 10^{25} is necessary, and that this would occur in a time of some 10^{-32} s. It seems possible that the expansion could have taken place at an even more alarming rate than this.

Inflationary theories have their own drawbacks. In order to achieve enough expansion, and to make acceptable various other cosmological consequences which I have no space to explain in detail, it has been found necessary to adjust properties of the field theory very carefully. Since the idea was to avoid fine adjustment of the initial conditions in the standard model, it may seem that inflation merely converts the original problem into a different form. We would, of course, like inflation to appear as a natural consequence of our field theories of particle physics. In many inflationary models, this is not the case, since the field ϕ which is responsible for inflation is introduced solely for that purpose. My own opinion, for what it is worth, is that the methods of calculation which have so far been available for estimating the consequences of any given model are not sufficiently reliable for firm conclusions to be drawn.

Exercises

13.1. The *absolute luminosity* L of an astronomical object is the total power it radiates. Its *apparent luminosity* l is the power per unit area received by an observer. In Euclidean space, the apparent luminosity for an observer at a distance d is obviously $l = L/4\pi d^2$. In general, the *luminosity distance* of a source of known absolute luminosity is defined as $d_L = (L/4\pi l)^{1/2}$. Consider a comoving source and a comoving observer separated by a coordinate distance r in a Robertson–Walker spacetime. Radiation emitted at time t_e is received at time t_o. By considering both the rate at which photons are received and the redshift of each photon, show that

$$l = \frac{La^2(t_e)}{4\pi r^2 a^4(t_o)}.$$

The scale factor at time t can be expressed as a power series in $(t - t_o)$ as

$$a(t) = a(t_o)[1 + H_0(t - t_o) - \tfrac{1}{2}q_0 H_0^2(t - t_o)^2 + \ldots].$$

Use this expansion and (13.14) to express the redshift z and the coordinate distance r as power series in $(t_o - t_e)$ and hence express r as a power series in z. Show that the luminosity distance is given by (13.11).

13.2. A covariant action for a massless scalar field can be written as

$$S = \tfrac{1}{2}\int d^4x(-g)^{1/2} g^{\mu\nu}\partial_\mu\phi\partial_\nu\phi.$$

Considering a spatially flat Robertson–Walker spacetime, and using Cartesian spatial coordinates, derive the Euler–Lagrange equation. Show that it has plane-wave solutions of the form

$$\phi(\boldsymbol{x}, t) = [2\omega(t)]^{-1/2}\exp\left(i\int_{t_0}^{t}\omega(t')\,dt' - i\boldsymbol{k}\cdot\boldsymbol{x}\right)$$

where the time-dependent frequency satisfies the equation

$$\omega^2 + \frac{1}{2}\frac{\ddot{\omega}}{\omega} - \frac{3}{4}\frac{\dot{\omega}^2}{\omega^2} - \frac{k^2}{a^2} = 0.$$

By considering short time intervals during which a and ω are approximately constant, show that the wavelength of a massless particle is redshifted as in (13.16).

13.3. By adding to the action of the previous exercise a potential $V(\phi)$, and considering a spatially uniform field, deduce the equation of motion (13.88) with $\gamma = 0$.

13.4. For a radiation-dominated universe, show that the function $f(\Omega)$ in (13.33) is given by $f(\Omega) = 1/(1 + \sqrt{\Omega})$.

13.5. The discussion of §13.3 assumes a cosmological model in which $\rho(t) = \rho_{\text{matter}}(t) + \rho_{\text{rad}}(t) = M/a^3(t) + \Gamma/a^4(t)$, where M and Γ are constants, and $p(t) = p_{\text{rad}}(t) = \Gamma/3a^4(t)$. Verify that such a model is consistent with equations (13.18) and (13.19).

13.6. With a positive cosmological constant Λ, show that a static universe (the *Einstein universe*) with a, ρ and p all constant is possible provided that $\rho \leqslant 2\Lambda/\kappa$, and that this universe is closed. In the *Lemaître* universe, p is taken to be zero and the constant $M = \rho a^3$ is larger than the value required for a static universe. Show that (i) this model has an initial singularity with $a(t)$ initially proportional to $t^{2/3}$, (ii) the expansion slows down until $\dot{a}$ reaches a minimum when $a^3 = \kappa M/2\Lambda$, and (iii) after a sufficiently long time the expansion becomes exponential as in the de Sitter universe (13.86).

Note added in proof

Since the manuscript of this book was completed, measurements of the properties of the Z^0 have been reported by experimental groups at Fermilab (Abe *et al* (1989)), Stanford (Abrams *et al* (1989)) and, with greater accuracy, at CERN (Adeva *et al* (1989), Decamp *et al* (1989), Akrawy *et al* (1989) and Aarnio *et al* (1989)). Roughly summarized, these results are consistent with the existence of 3.2 ± 0.4 families of quarks and leptons. This interpretation *assumes* that the standard model, with an initially unspecified number of families, is correct. While the number of families should, of course, be an integer, some rival theories predict non-integer values for the outcome of the same calculation, which would not correspond directly to the number of particle species. Within the standard model, the existence of more than three families is still possible if, for some unknown reason, their neutrinos have masses greater than one-half that of the Z^0.

Some Snapshots of the Tour

Our tour having come to an end, readers may like to review some of its highlights with the aid of a few snapshots, which are provided on the following pages. The snapshots are intended to give an overall view of the logical structure of the principal theories we have visited and to summarize some of the important results.

Thank you for travelling with Unified Grand Tours; I hope you enjoyed your trip.

Snapshot of Geometry and Gravitation

Geometry

- Basic spacetime structure is *differentiable manifold* on which smooth curves can be drawn and which can support differentiable functions. But *parallelism, angle* and *length* are not defined.
- *Tensors* defined either as intrinsic geometrical objects or as sets of components with definite coordinate transformation laws:

$$T^{\mu'\ldots}{}_{\alpha'\ldots} = \left[\frac{\partial x^{\mu'}}{\partial x^{\mu}} \frac{\partial x^{\alpha}}{\partial x^{\alpha'}} \ldots\right] T^{\mu\ldots}{}_{\alpha\ldots}$$

  Typical (contravariant) vector is tangent vector $V^\mu = (\mathrm{d}/\mathrm{d}\lambda)x^\mu(\lambda)$. Typical one-form (covariant vector) is gradient $\omega_\mu = \partial f/\partial x^\mu$.
- *Affine connection:* Γ defines parallel transport yielding *covariant derivative*

$$\nabla_\mu T^{\alpha\ldots}{}_{\beta\ldots} = \partial_\mu T^{\alpha\ldots}{}_{\beta\ldots} + \Gamma^\alpha{}_{\lambda\mu} T^{\lambda\ldots}{}_{\beta\ldots} - \Gamma^\lambda{}_{\beta\mu} T^{\alpha\ldots}{}_{\lambda\ldots} + \ldots$$

- *Curvature* (for symmetric connection) defined by

$$[\nabla_\mu, \nabla_\nu] V^\alpha = R^\alpha{}_{\beta\mu\nu} V^\beta$$

  with the *Riemann tensor*

$$R^\alpha{}_{\beta\mu\nu} = \Gamma^\alpha{}_{\beta\nu,\mu} - \Gamma^\alpha{}_{\beta\mu,\nu} + \Gamma^\alpha{}_{\sigma\mu}\Gamma^\sigma{}_{\beta\nu} - \Gamma^\alpha{}_{\sigma\nu}\Gamma^\sigma{}_{\beta\mu}$$

- Ricci tensor is

$$R_{\mu\nu} = R^\alpha{}_{\mu\alpha\nu}.$$

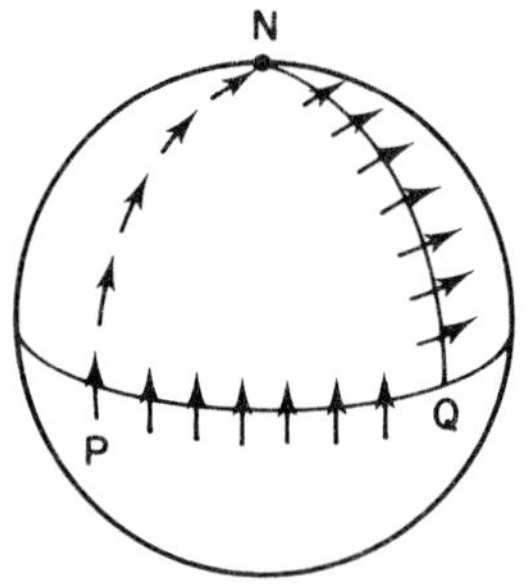

■ *Geodesics* are self-parallel curves (generalization of straight line)

$$\frac{d^2x^\mu}{d\lambda^2} + \Gamma^\mu{}_{\nu\sigma}\frac{dx^\nu}{d\lambda}\frac{dx^\sigma}{d\lambda} = 0$$

for affine parametrization.

■ *Metric tensor* $g_{\mu\nu}(x) = g_{\nu\mu}(x)$ defines:
(i) distance along a curve

$$ds^2 = g_{\mu\nu}(x)\,dx^\mu\,dx^\nu$$

(ii) scalar product of two vectors $g_{\mu\nu}V^\mu V^\nu$ (cf. dot product $\boldsymbol{u}\cdot\boldsymbol{v}$)
(iii) one-to-one correspondence between vectors and one-forms

$$V_\mu = g_{\mu\nu}V^\nu \qquad V^\mu = g^{\mu\nu}V_\nu \qquad g^{\mu\sigma}g_{\sigma\nu} = \delta^\mu{}_\nu$$

■ *Metric connection* preserves lengths and angles under parallel transport:

$$\nabla_\mu g_{\alpha\beta} = 0$$

determines connection coefficients as

$$\Gamma^\mu{}_{\nu\sigma} = \tfrac{1}{2}g^{\mu\lambda}(g_{\nu\lambda,\sigma} + g_{\sigma\lambda,\nu} - g_{\nu\sigma,\lambda})$$

■ Ricci curvature scalar

$$R = g^{\mu\nu}R_{\mu\nu}$$

■ *Vierbein* embeds local inertial coordinates y^a into large-scale coordinate system x^μ:

$$e^a{}_\mu = \frac{\partial y^a}{\partial x^\mu} \qquad e^\mu{}_a = \frac{\partial x^\mu}{\partial y^a} \qquad e^a{}_\mu e^\mu{}_b = \delta^a{}_b \qquad e^\mu{}_a e^a{}_\nu = \delta^\mu{}_\nu$$

$$e^a{}_\mu e^b{}_\nu \eta_{ab} = g_{\mu\nu} \qquad e^\mu{}_a e^\nu{}_b g_{\mu\nu} = \eta_{ab}$$

Gravitation and cosmology

- *Equivalence principle:* at any point P, can find coordinate system such that

$$g_{\mu\nu}(\mathrm{P}) = \eta_{\mu\nu} = \begin{pmatrix} 1 & 0 & 0 & 0 \\ 0 & -1 & 0 & 0 \\ 0 & 0 & -1 & 0 \\ 0 & 0 & 0 & -1 \end{pmatrix}$$

(Minkowski metric of special relativity) and $g_{\mu\nu,\sigma}(\mathrm{P}) = 0$ but $g_{\mu\nu,\sigma\tau}(\mathrm{P}) \neq 0$ in general.
- Presence of a *gravitational field* is indicated if no coordinates can be found such that $g_{\mu\nu} = \eta_{\mu\nu}$ everywhere.
- *Response of test particles to gravitational field:* In absence of non-gravitational forces, particle follows a geodesic path. Connection terms interpreted as gravitational forces. In *Newtonian approximation* of weak fields and slowly moving particles ($g_{\mu\nu} = \eta_{\mu\nu} + h_{\mu\nu}$)

$$\boxed{\frac{\mathrm{d}^2 x^i}{\mathrm{d}t^2} \approx -\frac{\partial V}{\partial x^i}}$$

$$\boxed{V = \tfrac{1}{2}c^2 h_{00}}$$

Equality of inertial and gravitational masses is implied.
- *Response of geometry to distribution of matter:* Einstein's *field equations* (follow, for example, from principle of least action)

$$\boxed{R_{\mu\nu} - (\tfrac{1}{2}R - \Lambda)g_{\mu\nu} = \kappa T_{\mu\nu}}$$

$$\boxed{\kappa = 8\pi G/c^4}$$

In Newtonian approximation, field equations give Poisson's equation

$$\boxed{\nabla^2 V \approx 4\pi G\rho/c^2}$$

ρ/c^2 is mass density.
- For a perfect fluid,

$$\boxed{T^{\mu\nu} = c^{-2}(\rho + p)u^\mu u^\nu - pg^{\mu\nu}}$$

where u^μ = 4-velocity of a fluid element and ρ and p are the energy density and pressure measured in its rest frame.

- In general, stress tensor is conserved:

$$\boxed{\nabla_\nu T^{\mu\nu} = 0}$$

- *Schwarzschild solution:* Valid in vacuum outside a spherically symmetric body of mass M:

$$\boxed{c^2\,\mathrm{d}\tau^2 = (1 - r_S/r)c^2\,\mathrm{d}t^2 - (1 - r_S/r)^{-1}\,\mathrm{d}r^2 - r^2(\mathrm{d}\theta^2 + \sin^2\theta\,\mathrm{d}\phi^2)}$$

Schwarzschild radius $r_S = 2GM/c^2$.

- *Robertson–Walker metric* (homogeneous, isotropic) with $c = 1$:

$$\boxed{\mathrm{d}\tau^2 = \mathrm{d}t^2 - a^2(t)\,[(1 - kr^2)^{-1}\,\mathrm{d}r^2 + r^2\,\mathrm{d}\theta^2 + r^2\sin^2\theta\,\mathrm{d}\phi^2]}$$

Cosmic time t is proper time for comoving observers. Spatial sections may be closed ($k = 1$), flat ($k = 0$) or open ($k = -1$). Stress tensor must have perfect fluid form.

- Field equations in Friedmann–Robertson–Walker models are:

$$\boxed{3(\dot{a}^2 + k)/a^2 = \kappa\rho + \Lambda}$$

$$\boxed{2\frac{\ddot{a}}{a} + \frac{\dot{a}^2}{a^2} + \frac{k}{a^2} = -\kappa p + \Lambda}$$

These imply energy conservation in the form

$$\boxed{\frac{\mathrm{d}}{\mathrm{d}t}(\rho a^3) = -p\,\frac{\mathrm{d}}{\mathrm{d}t}(a^3)}$$

- *Hubble's law*

$$\boxed{\text{recessional velocity} = H(t) \times \text{distance};\ H = \dot{a}/a}$$

- *Critical density* ($\Lambda = 0$)

$$\boxed{\rho_c(t) = 3H^2(t)/\kappa}$$

$$\boxed{\Omega(t) = \rho(t)/\rho_c(t)}$$

$\Omega > 1 \Rightarrow k = 1$, $\Omega = 1 \Rightarrow k = 0$, $\Omega < 1 \Rightarrow k = -1$.

- Matter-dominated universe ($p = 0$): $a(t) \sim t^{2/3}$.
- Radiation-dominated universe ($p = \rho/3$): $a(t) \sim t^{1/2}$.

Snapshot of Field Theory

Free fields

- *Klein–Gordon equation:*

$$E^2 - p^2 - m^2 = 0 \Rightarrow \boxed{(\Box + m^2)\phi = 0}$$

- *Dirac equation* (spin-$\frac{1}{2}$)

$$\boxed{(\mathrm{i}\gamma^\mu \partial_\mu - m)\psi = 0}$$

To reproduce Klein–Gordon equation:

$$\boxed{\{\gamma^\mu, \gamma^\nu\} = 2g^{\mu\nu}} \quad \textit{Clifford algebra.}$$

- To accommodate negative energies write plane-wave solutions as

$$\phi(x) = \int \mathrm{d}^3k[(2\pi)^3 2\omega_k]^{-1}\{a(\boldsymbol{k})\mathrm{e}^{-\mathrm{i}k\cdot x} + c^\dagger(\boldsymbol{k})\mathrm{e}^{\mathrm{i}k\cdot x}\}$$

where $a(\boldsymbol{k})$ annihilates particles and $c^\dagger(\boldsymbol{k})$ creates antiparticles.

Interacting fields

- *Asymptotic states:* Initial and final states of scattering/decay processes created by free fields, $\phi_{\text{in}}(x)$, $\phi_{\text{out}}(x)$. With adiabatic switching

$$\begin{aligned} \phi(x) &\to \sqrt{Z}\phi_{\text{in}}(x) \quad t \to -\infty \\ &\to \sqrt{Z}\phi_{\text{out}}(x) \quad t \to +\infty. \end{aligned}$$

- Amplitudes $\langle k'_{1\ldots}; \text{out} \mid k_{1\ldots}; \text{in}\rangle$ related to vacuum expectation values of interacting fields by *reduction formulae*

$$\boxed{\langle \ldots; \text{out} \mid \ldots; \text{in}\rangle = \int \mathrm{d}x_1 \ldots \langle 0 | T[\phi \ldots \phi^\dagger] | 0\rangle + \ldots}$$

- Time-ordered products

$$T[\phi(x_1) \ldots \phi(x_N)]\text{: latest-on-left ordering of fields.}$$

Reordering involves factor (-1) for each interchange of two fermionic fields.

- Vacuum expectation values of time-ordered products represented as path integrals:

$$\boxed{\langle 0|T[\phi(x_1)\ldots\phi^{\dagger}(x_N)|0\rangle = \int \mathcal{D}\phi\,\phi(x_1)\ldots\phi^{*}(x_N)\mathrm{e}^{\mathrm{i}S(\phi)}}$$

- *Perturbation theory:* Expansion in powers of coupling constants can be expressed in terms of *Feynman diagrams*, e.g. $\lambda\phi^4$ theory:

$\mathrm{i}\langle 0|T[\phi(x)\phi^{\dagger}(y)]|0\rangle =$ —— +  + 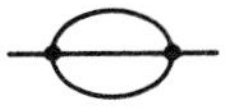 + ⋯

Unperturbed propagators:

$$\text{scalar: ——} = G_{\mathrm{F}}(x-y) = -\int \frac{\mathrm{d}^4k}{(2\pi)^4}\,\mathrm{e}^{-\mathrm{i}k\cdot(x-y)}\frac{1}{k^2-m^2+\mathrm{i}\varepsilon}$$

$$\text{spin-}\tfrac{1}{2}\text{: —>—} = S_{\mathrm{F}}(x-y) = \int \frac{\mathrm{d}^4k}{(2\pi)^4}\,\mathrm{e}^{-\mathrm{i}k\cdot(x-y)}\frac{(\not{k}+m)}{k^2-m^2+\mathrm{i}\varepsilon}.$$

- *Renormalization:* Re-expresses calculated results in terms of physically measurable masses etc, which are modified by interactions. In *renormalizable* theories, infinities in Feynman integrals are removed by renormalization.
- *Running coupling constants:* Net effect of interactions involves both coupling constants and energy–momentum-dependent Feynman integrals. Effective expansion parameter depends on energy. Also related to renormalization and characteristic length scales of physical processes.

Gauge fields

- Fundamental forces due to communication between different points of spacetime.
- Wavefunction or field exists in an *internal space*—collection of internal spaces at all spacetime points is a *fibre bundle*.
- Comparison of fields at different points requires a *gauge connection* to define parallel transport through fibre bundle.
- Gauge-covariant derivative

$$\mathrm{D}_\mu \psi = (\partial_\mu + \mathrm{i}gA_\mu)\psi.$$

- Curvature of fibre bundle (cf. Riemann tensor)

$$\begin{aligned} F_{\mu\nu} &= -(\mathrm{i}/g)[\mathrm{D}_\mu, \mathrm{D}_\nu] \\ &= \partial_\mu A_\nu - \partial_\nu A_\mu \text{ (Abelian case: Maxwell field strength tensor of electromagnetism)} \\ &= \partial_\mu A_\nu - \partial_\nu A_\mu + \mathrm{i}g[A_\mu, A_\nu] \text{ (non-Abelian).} \end{aligned}$$

- Theories should be covariant under *gauge transformations* (cf. general coordinate transformations)

$$\psi \to U\psi$$

$$A_\mu \to UA_\mu U^{-1} + (\mathrm{i}/g)(\partial_\mu U)U^{-1}.$$

In Abelian case (electromagnetism) $U = \mathrm{e}^{\mathrm{i}\theta}$ is phase transformation. Covariant derivative transforms as

$$\mathrm{D}_\mu \psi \to U\mathrm{D}_\mu \psi.$$

- Gauge-invariant action for gauge fields and fermions has the form

$$S = \int d^4x[-\tfrac{1}{4}F_{\mu\nu}F^{\mu\nu} + \bar{\psi}(i\gamma^\mu D_\mu - m)\psi].$$

- *Gauge boson masses:* The quantity $M^2 A_\mu A^\mu$ is not gauge invariant. Masses can be introduced in a gauge-invariant manner through a *Higgs scalar field:*

$$S_{\text{Higgs}} = \int d^4x[(D_\mu\phi)^\dagger(D_\mu\phi) - V(\phi^\dagger\phi)]$$

  which acquires a non-zero vacuum expectation value.
- When left- and right-handed chiral components of ψ have different gauge transformation laws, the quantity $m\bar{\psi}\psi = m\bar{\psi}_L\psi_R + m\bar{\psi}_R\psi_L$ is not gauge invariant either. In the standard electroweak theory, fermion masses can be generated in a gauge-invariant manner through Yukawa couplings to Higgs fields

$$\Delta\mathcal{L} = -f(\bar{l}_e\phi e_R + \bar{e}_R\phi^\dagger l_e).$$

Snapshot of Statistical Mechanics

- *Classical ergodic theory:* Ensemble average of dynamical quantity is

$$\boxed{\bar{A}(t) = \int \mathrm{d}X \rho(X, t) A(X)}$$

- Liouville equation for probability density in phase space

$$\boxed{\frac{\partial \rho}{\partial t} = \{H, \rho\}_P}$$

- To describe thermal equilibrium need $\partial\rho/\partial t = 0$, so ρ depends on X only through conserved quantities.
- *Closed isoenergetic system:* (E, N, V fixed) $\Rightarrow$ *microcanonical ensemble*

$$\boxed{\rho_{\text{micro}}(X, E) = \delta[H_N(X) - E]/\Sigma(E, N, V)}$$

$$\boxed{\Sigma(E, N, V) = \int \mathrm{d}X \delta[H_N(x) - E]}$$

Ergodic system $\leftrightarrow$ microcanonical average = long time average
$\leftrightarrow$ ρ_{micro} is unique time-independent distribution.
Relation to thermodynamics:

$$\boxed{\text{Entropy } S(E, N, V) = k_B \ln[\Sigma(E, N, V,)/h^{3N}N!]}$$

- *Closed isothermal system:* (T, N, V fixed) $\Rightarrow$ *canonical ensemble:* statistical independence of non-interacting systems $\Rightarrow$

$$\boxed{\rho_{\text{can}}(X, \beta) = \mathrm{e}^{-\beta H_N(X)} \Big/ \int \mathrm{d}X \mathrm{e}^{-\beta H_N(X)}}$$

$\beta = 1/k_B T$

$$\boxed{Z_{\text{can}}(\beta, N, V) = (h^{3N}N!)^{-1} \int \mathrm{d}X \mathrm{e}^{-\beta H_N(X)}}$$

Relation to thermodynamics:

$$\boxed{\text{Helmholtz free energy } F(\beta, N, V) = -k_B T \ln Z_{\text{can}}(\beta, N, V)}$$

- Relation between canonical and microcanonical ensembles:

$$Z_{\text{can}}(\beta, N, V) = \int \mathrm{d}E \mathrm{e}^{-\beta E} \Sigma(E, N, V)/h^{3N} N!$$

(Laplace transform).
In thermodynamic limit

$$F = E - TS$$

(Legendre transform).

- *Open isothermal system:* (T, μ, V fixed) $\Rightarrow$ *grand canonical ensemble:* Poisson distribution of particle numbers $\Rightarrow$

$$\boxed{\rho_{\text{gr}}(X, N) = \mathrm{e}^{\beta\mu N} \mathrm{e}^{-\beta H_N(X)}/h^{3N} N! Z_{\text{gr}}}$$

$$\boxed{Z_{\text{gr}}(\beta, \mu, V) = \sum_N \mathrm{e}^{\beta\mu N} Z_{\text{can}}(\beta, N, V)}$$

(Laplace transform). Relation to thermodynamics:

$$\boxed{\text{Grand potential } \Omega(\beta, \mu, V) = -k_B T \ln Z_{\text{gr}}(\beta, \mu, V)}$$

In thermodynamic limit

$$\Omega(\beta, \mu, V) = F - \mu N$$

(Legendre transform).

- *Quantum statistical mechanics*
  Expectation value of dynamical quantity is

$$\boxed{\bar{A}(t) = \mathrm{Tr}\,[\hat{A}\,\hat{\rho}(t)]}$$

- Quantum Liouville equation

$$\boxed{\frac{\mathrm{d}}{\mathrm{d}t}\,\hat{\rho} = \frac{\mathrm{i}}{\hbar}\,[\hat{\rho},\,\hat{H}]}$$

- Canonical ensemble

$$\hat{\rho}_{\text{can}} = \mathrm{e}^{-\beta\hat{H}_N}/Z_{\text{can}}$$

$$Z_{\text{can}}(\beta,\,N,\,V) = \mathrm{Tr}\,\mathrm{e}^{-\beta\hat{H}_N}.$$

- Grand canonical ensemble

$$Z_{\text{gr}}(\beta,\,\mu,\,V) = \sum_N \mathrm{e}^{\beta\mu N} Z_{\text{can}}(\beta,\,N,\,V)$$

  or, using second quantization,

$$Z_{\text{gr}}(\beta,\,\mu,\,V) = \mathrm{Tr}\,[\mathrm{e}^{\beta\mu\hat{N}-\beta\hat{H}}].$$

- Grand canonical ensemble for particles ≡ canonical ensemble for fields.

■ *Field theories at finite temperature*

$$\hat{\phi}(\boldsymbol{x}, \tau) = \mathrm{e}^{\tau \hat{H}} \hat{\phi}(\boldsymbol{x}) \mathrm{e}^{-\tau \hat{H}}$$
$$\hat{\phi}^{\dagger}(\boldsymbol{x}, \tau) = \mathrm{e}^{\tau \hat{H}} \hat{\phi}^{\dagger}(\boldsymbol{x}) \mathrm{e}^{-\tau \hat{H}}$$

τ is *imaginary time* $(0 \leqslant \tau \leqslant \beta)$.

■ Path integral representation of expectation values:

$$\boxed{\mathrm{Tr}\,\{\hat{\rho} T_{\tau}[\hat{\phi}(x_1) \ldots \hat{\phi}^{\dagger}(x_N)]\} = Z_{\mathrm{gr}}^{-1} \int \mathcal{D}\phi\, \phi(x_1) \ldots \phi^*(x_N) \mathrm{e}^{-S_{\beta}(\phi)}}$$

and, for example, in $\lambda\phi^4$ theory

$$\boxed{S_{\beta}(\phi) = \int_0^{\beta} \mathrm{d}\tau \int \mathrm{d}^d x \left(\frac{\partial \phi^*}{\partial \tau} \frac{\partial \phi}{\partial \tau} + \boldsymbol{\nabla}\phi^* \cdot \boldsymbol{\nabla}\phi + m^2 \phi^* \phi + \frac{\lambda}{4} (\phi^* \phi)^2 \right)}$$

equivalent to a theory in $(d + 1)$ Euclidean dimensions, of finite extent β in the extra dimension and with periodic boundary conditions (antiperiodic for fermions).

■ Imaginary time propagator is

$$\boxed{G_0(\boldsymbol{x} - \boldsymbol{x}', \tau - \tau') = \beta^{-1} \int \frac{\mathrm{d}^d k}{(2\pi)^d} \mathrm{e}^{\mathrm{i}\boldsymbol{k}\cdot(\boldsymbol{x}-\boldsymbol{x}')} \sum_n \mathrm{e}^{\mathrm{i}\omega_n(\tau-\tau')} \widetilde{G}_0(\boldsymbol{k}, n)}$$

$$\boxed{\widetilde{G}_0(\boldsymbol{k}, n) = [k^2 + \omega_n^2 + m^2]^{-1}}$$

$\omega_n = 2\pi n/\beta$ are *Matsubara frequencies.*

Appendix 1

Some Mathematical Notes

A1.1 Delta Functions and Functional Differentiation

The *Kronecker delta* symbol, written as δ_{ij}, δ^{ij} or $\delta^{i}{}_{j}$ according to context, is defined to equal 1 if $i = j$ and 0 otherwise. It is mainly useful when we are dealing with summations, say of a set of quantities $\{f_i\}$, and it obviously has the property

$$\sum_i \delta_{ij} f_i = f_j. \tag{A1.1}$$

The *Dirac delta function* is a generalization of the Kronecker δ which allows us to deal with integrals in the same way. The function (known in rigorous mathematics as a *distribution*) $\delta(x - x_0)$ is equal to zero unless $x = x_0$, when it is infinite. The infinite value becomes meaningful when the delta function appears inside an integral, and the defining property of $\delta(x - x_0)$ is that, for any sufficiently smooth function $f(x)$,

$$\begin{aligned} \int_a^b \delta(x - x_0) f(x)\,\mathrm{d}x &= f(x_0) \quad \text{if } a < x_0 < b \\ &= 0 \qquad\quad \text{otherwise.} \end{aligned} \tag{A1.2}$$

This can be understood in terms of the Riemann definition of the integral, in which we divide the interval $[a, b]$ into N segments of length $\Delta x = (b - a)/N$ and take the limit $N \to \infty$. If x_0 lies in the jth segment, then the integral (A1.2) can be represented as

$$\lim_{N\to\infty} \sum_i \Delta x[\delta_{ij}/\Delta x] f(x_i) = f(x_0) \tag{A1.3}$$

and so $\delta(x - x_0)$ is the limit as $\Delta x \to 0$ of $\delta_{ij}/\Delta x$.

Consider a function $F(\{f_i\})$ which depends on all the f. If we make a small change δf_i in each variable, the first-order change in F is given by

$$\delta F = \sum_i \delta f_i \frac{\partial F}{\partial f_i}. \tag{A1.4}$$

If F depends instead on a continuous function $f(x)$, then it is useful to generalize the partial derivative to a *functional derivative* defined by an analogous equation

$$\delta F = \int \mathrm{d}x \delta f(x) \frac{\delta F}{\delta f(x)}. \tag{A1.5}$$

Quite often, F will be defined in terms of integrals involving $f(x)$. The action (3.53) for electromagnetic fields is a case in point. In that case, F is not itself a function of x, but $\delta F/\delta f(x)$ is a function of x. However, we can also find the functional derivative of a quantity which is a function of x. For example, if F is $f(x)$ itself, then $\delta f(x)/\delta f(y) = \delta(x - y)$, because

$$\delta f(x) = \int \mathrm{d}y \delta f(y) \delta(x - y). \tag{A1.6}$$

The delta function in (A1.2) can be thought of as imposing the constraint that $x = x_0$. Sometimes, we may wish instead to impose the constraint $g(x) = g_0$, where g is some function. In (7.9), for example, we use $g(k^0) = (k^0)^2$ and $g_0 = \omega^2(\boldsymbol{k})$. We can do this by changing the integral over x to an integral over g:

$$\int \delta(g(x) - g_0) f(x)\,\mathrm{d}x = \int \mathrm{d}g \left(\frac{\mathrm{d}g}{\mathrm{d}x}\right)^{-1} \delta(g - g_0) f(x(g)) = \frac{f(x_0)}{g'(x_0)} \tag{A1.7}$$

where x_0 is the point at which $g(x_0) = g_0$. If there are several such points, then the integral is the sum of values of f/g' at these points. We can therefore write

$$\delta(g(x) - g_0) = \sum_i \left(\frac{\mathrm{d}g}{\mathrm{d}x}\right)^{-1} \delta(x - x_i) \tag{A1.8}$$

where x_i are the points at which $g(x_i) = g_0$.

In this book, I use $\delta(\boldsymbol{x} - \boldsymbol{x}')$ to stand for the product of delta functions $\delta(x - x')\delta(y - y')\delta(z - z')$.

A useful representation of the Dirac delta function is provided by the theory of Fourier transforms. If $f(x)$ is sufficiently well behaved, it can be expressed as

$$f(x) = \int_{-\infty}^{\infty} \mathrm{d}k \hat{f}(k) \mathrm{e}^{ikx} \tag{A1.9}$$

where

$$\hat{f}(k) = \frac{1}{2\pi} \int_{-\infty}^{\infty} \mathrm{d}x f(x) \mathrm{e}^{-ikx}. \tag{A1.10}$$

It is easily seen that, if $f(x)$ in (A1.2) is such that both these equations are valid, then the delta function can be represented by

$$\delta(x - x') = \frac{1}{2\pi} \int_{-\infty}^{\infty} \mathrm{d}k \mathrm{e}^{\pm ik(x-x')}. \tag{A1.11}$$

Under suitable conditions, other orthogonal functions may be used in place of the exponential.

The *Heaviside step function* $\theta(x - x_0)$ is defined to equal 0 for $x < x_0$ and 1 for $x > x_0$. It is not usually necessary to specify its value at $x = x_0$. A little thought will show that $d\theta(x - x_0)/dx = \delta(x - x_0)$.

A1.2 The Levi–Civita Tensor Density

The symbol $\varepsilon^{\mu\nu\sigma\tau}$, in which each index can take the values 0, 1, 2 or 3, is defined to be +1 when $(\mu, \nu, \sigma, \tau) = (0, 1, 2, 3)$ and to be antisymmetric under the interchange of *any* pair of indices: $\varepsilon^{\mu\nu\sigma\tau} = -\varepsilon^{\nu\mu\sigma\tau} = \varepsilon^{\nu\sigma\mu\tau}$, etc. It follows from this definition that $\varepsilon^{\mu\nu\sigma\tau}$ is +1 when (μ, ν, σ, τ) is an even permutation of (0, 1, 2, 3), −1 for an odd permutation, and zero otherwise. Clearly, any totally antisymmetric tensor will be proportional to ε. An ε symbol can be defined in any number of dimensions, d, by giving it d indices. The ε symbol can be made into a tensor-like quantity, called the *Levi–Civita tensor density*, by specifying its transformation properties. Suppose that its components have the values specified above in a particular coordinate system. In a new system let

$$\hat{\varepsilon}^{\mu'\nu'\sigma'\tau'} = \Lambda^{\mu'}{}_{\mu}\Lambda^{\nu'}{}_{\nu}\Lambda^{\sigma'}{}_{\sigma}\Lambda^{\tau'}{}_{\tau}\varepsilon^{\mu\nu\sigma\tau}. \qquad \text{(A1.12)}$$

Clearly, $\hat{\varepsilon}$ is also totally antisymmetric and therefore proportional to $\varepsilon^{\mu'\nu'\sigma'\tau'}$. Furthermore, we have

$$\hat{\varepsilon}^{0123} = \Lambda^{0}{}_{\mu}\Lambda^{1}{}_{\nu}\Lambda^{2}{}_{\sigma}\Lambda^{3}{}_{\tau}\varepsilon^{\mu\nu\sigma\tau} = \det|\Lambda^{\mu'}{}_{\mu}| \qquad \text{(A1.13)}$$

since the sum of products with alternating signs is just the rule for forming the determinant. If we define $\varepsilon^{\mu'\nu'\sigma'\tau'}$ to have the same values as $\varepsilon^{\mu\nu\sigma\tau}$, then the transformation law is

$$\varepsilon^{\mu'\nu'\sigma'\tau'} = (\det|\Lambda^{\mu'}{}_{\mu}|)^{-1}\Lambda^{\mu'}{}_{\mu}\Lambda^{\nu'}{}_{\nu}\Lambda^{\sigma'}{}_{\sigma}\Lambda^{\tau'}{}_{\tau}\varepsilon^{\mu\nu\sigma\tau}. \qquad \text{(A1.14)}$$

An object which transforms like a tensor, but with an extra factor of $(\det|\Lambda|)^n$, is called a *tensor density of weight n*, so ε is a tensor density of weight −1.

The metric determinant g can be written as

$$g = \det|g_{\mu\nu}| = \tfrac{1}{4!}\varepsilon^{\mu\nu\sigma\tau}\varepsilon^{\alpha\beta\gamma\delta}g_{\mu\alpha}g_{\nu\beta}g_{\sigma\gamma}g_{\tau\delta} \qquad \text{(A1.15)}$$

and is easily seen to be a scalar density of weight −2.

A1.3 The Covariant Divergence

The derivative of g can be expressed as

$$\partial_\lambda g = \tfrac{1}{3!}\varepsilon^{\mu\nu\sigma\tau}\varepsilon^{\alpha\beta\gamma\delta}g_{\mu\alpha}g_{\nu\beta}g_{\sigma\gamma}(\partial_\lambda g_{\tau\delta}) = g g^{\mu\nu}(\partial_\lambda g_{\mu\nu}). \qquad \text{(A1.16)}$$

Therefore, using the metric connection (2.49) we find

$$\partial_\lambda(-g)^{1/2} = \tfrac{1}{2}(-g)^{1/2}g^{\mu\nu}(\partial_\lambda g_{\mu\nu}) = (-g)^{1/2}\Gamma^\mu{}_{\mu\lambda}. \quad \text{(A1.17)}$$

Using this result, the covariant divergence of a vector field may be expressed as

$$V^\mu{}_{;\mu} = \frac{1}{(-g)^{1/2}}\partial_\mu((-g)^{1/2}V^\mu). \quad \text{(A1.18)}$$

A1.4 Gauss' Theorem

The integral of the divergence of a vector field over a region D is a scalar quantity, provided that we use the covariant volume element (4.12). By using the version of Gauss' theorem which applies in Euclidean space, we can write it as a surface integral

$$\int_{\mathrm{D}} \mathrm{d}^4x(-g)^{1/2}V^\mu{}_{;\mu} = \int_{\mathrm{D}} \mathrm{d}^4x\partial_\mu((-g)^{1/2}V^\mu) = \int_{\mathrm{S}} (-g)^{1/2}V^\mu\,\mathrm{d}S_\mu \quad \text{(A1.19)}$$

where S is the surface which bounds the region D.

A1.5 Surface Area and Volume of a *d*-dimensional Sphere

Let Ω_d be the surface area of a sphere of unit radius in d Euclidean dimensions. The surface area of a sphere of radius r is $\Omega_d r^{d-1}$ and we find by integrating that its volume is $\Omega_d r^d/d$. To evaluate Ω_d, let $r^2 = x_1^2 + \ldots + x_d^2$ and consider the integral

$$\int_{-\infty}^{\infty} \mathrm{d}^d x \mathrm{e}^{-r^2} = \left[\int_{-\infty}^{\infty} \mathrm{d}x\mathrm{e}^{-x^2}\right]^d = \pi^{d/2}. \quad \text{(A1.20)}$$

The solid angle subtended by the surface of the sphere at its centre is Ω_d, so if we change to polar coordinates and integrate over the $(d-1)$ angular variables which do not appear in the integrand, this integral is

$$\pi^{d/2} = \Omega_d\int_0^{\infty} \mathrm{d}r\,r^{d-1}\mathrm{e}^{-r^2} = \tfrac{1}{2}\Omega_d\int_0^{\infty} \mathrm{d}t\,t^{d/2-1}\mathrm{e}^{-t} = \tfrac{1}{2}\Omega_d\Gamma(d/2) \quad \text{(A1.21)}$$

where $\Gamma(p) = (p-1)!$ is Euler's gamma function. Thus, we have

$$\Omega_d = \frac{2\pi^{d/2}}{\Gamma(d/2)}. \quad \text{(A1.22)}$$

Since $\Gamma(\tfrac{1}{2}) = \pi^{1/2}$ and $\Gamma(p+1) = p\Gamma(p)$, we find, for example, that $\Omega_2 = 2\pi$, which is the circumference of a unit circle, and $\Omega_3 = 4\pi$, which is the surface area of a unit sphere in three dimensions. When carrying out spacetime integrals, we need to know that $\Omega_4 = 2\pi^2$.

Appendix 2

The Rotation Group and Angular Momentum

The mathematical framework which allows a systematic study of the consequences of symmetry in physics is *group theory*. This appendix gives a brief account of the simpler aspects of rotations in three-dimensional Euclidean space. It illustrates some basic ideas of group theory but does not attempt to develop the mathematics in a detailed manner. Readers who wish to pursue the subject at a deeper level will find numerous books on the subject and will also find chapters devoted to group theory in textbooks on elementary particle theory. A few of these are mentioned in the bibliography.

Let $\boldsymbol{r}$ be a position vector in three-dimensional Euclidean space with Cartesian components (x^1, x^2, x^3). In matrix notation, a rotation about the x^3 axis through an angle α leads to a new set of components

$$\begin{pmatrix} x^{1'} \\ x^{2'} \\ x^{3'} \end{pmatrix} = \begin{pmatrix} \cos\alpha & \sin\alpha & 0 \\ -\sin\alpha & \cos\alpha & 0 \\ 0 & 0 & 1 \end{pmatrix} \begin{pmatrix} x^1 \\ x^2 \\ x^3 \end{pmatrix} = R_3(\alpha) \begin{pmatrix} x^1 \\ x^2 \\ x^3 \end{pmatrix}. \tag{A2.1}$$

These components can be regarded either as the components of a new vector $\boldsymbol{r}'$, obtained by rotating $\boldsymbol{r}$ through an angle $-\alpha$ (the *active* point of view) or as the components of $\boldsymbol{r}$ relative to a new coordinate system obtained by rotating the old axes through an angle $+\alpha$ (the *passive* point of view). It proves useful to express this rotation in terms of a *generator* matrix

$$\mathscr{J}^3 = \begin{pmatrix} 0 & -\mathrm{i} & 0 \\ \mathrm{i} & 0 & 0 \\ 0 & 0 & 0 \end{pmatrix}, \tag{A2.2}$$

which has the property that $(\mathscr{J}^3)^3 = \mathscr{J}^3$. It is easy to show that $R^3(\alpha) = \exp(\mathrm{i}\alpha\mathscr{J}^3)$. More generally, for a rotation about the direction of the unit vector $\boldsymbol{n}$, the rotation matrix is

$$R(\boldsymbol{\alpha}) = \exp(\mathrm{i}\boldsymbol{\alpha}\cdot\boldsymbol{\mathcal{J}}) \tag{A2.3}$$

where $\boldsymbol{\alpha} = \alpha\boldsymbol{n}$ and the other two generators are

$$\mathcal{J}^1 = \begin{pmatrix} 0 & 0 & 0 \\ 0 & 0 & -\mathrm{i} \\ 0 & \mathrm{i} & 0 \end{pmatrix} \qquad \mathcal{J}^2 = \begin{pmatrix} 0 & 0 & \mathrm{i} \\ 0 & 0 & 0 \\ -\mathrm{i} & 0 & 0 \end{pmatrix}. \tag{A2.4}$$

The effect of rotations upon different geometrical objects may be expressed in various ways. Consider, for example, the object $\boldsymbol{w} = \boldsymbol{u} \times \boldsymbol{v}$. In elementary vector analysis, this cross-product is regarded as forming a new vector. In terms of the totally antisymmetric ε symbol discussed in appendix 1, its components are given by $w^i = \varepsilon^{ijk}u^j v^k$. Under a rotation, these components transform according to (A2.1). It is, however, more or less an accident of three dimensions that this object can be regarded as a vector. Thus, the component $w^1 = u^2v^3 - u^3v^2$ is really associated with the x^2–x^3 plane, and it is only in three dimensions that this plane has a unique normal, which is the x^1 axis. More generally, the cross product should be regarded as defining an antisymmetric rank 2 tensor, $w^{ij} = u^i v^j - u^j v^i$, whose components form the matrix

$$w = \begin{pmatrix} 0 & w^3 & -w^2 \\ -w^3 & 0 & w^1 \\ w^2 & -w^1 & 0 \end{pmatrix}. \tag{A2.5}$$

It is easy to check that, when the components w^i are transformed according to (A2.1), this matrix transforms as

$$w' = R(\boldsymbol{\alpha})wR^{-1}(\boldsymbol{\alpha}). \tag{A2.6}$$

Readers may like to check that this also agrees with the general tensor transformation law (2.19).

The set of all rotation matrices $R(\boldsymbol{\alpha})$ forms a *group*. Mathematically, a group G is a set of elements $\{g\}$, which satisfies the following three requirements: (i) There is a rule for multiplying any two elements to form another element of the group, $g_1g_2 = g_3$, and this rule is *associative*, which means that, for any three elements, $g_1(g_2g_3) = (g_1g_2)g_3$. (ii) There is an *identity* element I such that $gI = Ig = g$ for any element g. (iii) Every element g has a unique inverse element g^{-1}, such that $gg^{-1} = g^{-1}g = I$. In the case of the rotation matrices, the multiplication rule is just matrix multiplication, I is the unit matrix and the inverse of $R(\boldsymbol{\alpha})$ is $R(-\boldsymbol{\alpha})$. The rotation matrices depend upon three continuously variable parameters, the rotation angles α^1, α^2 and α^3. A group of this kind is called a *Lie group*. Each matrix $R(\boldsymbol{\alpha})$ is an orthogonal 3×3 matrix with unit determinant, and the rotation group as we have considered it so far is the set of all such matrices. It is called the *special orthogonal group of order 3*, or SO(3). A group whose elements all commute with each other, that is such that $gg' = g'g$ for any two elements g and g', is called an *Abelian* group. In general, two rotations about different axes do not commute, so the rotation group is non-Abelian. On the other hand, two rotations about the same axis do commute, so the subgroup consisting of all rotations about a particular axis is Abelian. The set of spacetime translations is another example of an Abelian group.

Rotations also affect functions of the coordinates, such as wavefunctions for quantum-mechanical particles. To see how this works, it is useful to observe that all the properties of rotations can be deduced by considering only infinitesimal transformations. A rotation through a finite angle can be built up as a sequence of infinitesimal rotations:

$$R(\boldsymbol{\alpha}) = \exp(\mathrm{i}\boldsymbol{\alpha}\cdot\boldsymbol{\mathcal{J}}) = \lim_{N\to\infty}\left(1 + \frac{1}{N}\mathrm{i}\boldsymbol{\alpha}\cdot\boldsymbol{\mathcal{J}}\right)^N . \tag{A2.7}$$

A scalar function $\psi(\boldsymbol{r})$ will be expressed in a new coordinate system by a new function $\psi'(\boldsymbol{r}')$, such that $\psi'(\boldsymbol{r}') = \psi(\boldsymbol{r})$ when $\boldsymbol{r}'$ and $\boldsymbol{r}$ are the new and old coordinates of the same point. If we rewrite this as $\psi'(\boldsymbol{r}') = \psi(R^{-1}\boldsymbol{r}')$, then $\boldsymbol{r}'$ is just a dummy variable and, for notational convenience, we can drop the prime. If α is infinitesimal in (A2.1), we can use a Taylor series to express the transformation as

$$\begin{aligned}\psi'(x^1, x^2, x^3) &= \psi(x^1 - \alpha x^2, x^2 + \alpha x^1, x^3)\\ &= [1 + \mathrm{i}\alpha\mathcal{J}^3]\psi(x^1, x^2, x^3)\end{aligned} \tag{A2.8}$$

where the generator is now the differential operator

$$\mathcal{J}^3 = -\mathrm{i}\left(x^1\frac{\partial}{\partial x^2} - x^2\frac{\partial}{\partial x^1}\right). \tag{A2.9}$$

This and the two other generators of the same kind can be expressed as $\mathcal{J}^i = -\mathrm{i}\varepsilon^{ijk}x^j\partial/\partial x^k$.

Whether we are dealing with matrices or differential operators, the three generators have the commutation relations

$$[\mathcal{J}^i, \mathcal{J}^j] = \mathrm{i}\varepsilon^{ijk}\mathcal{J}^k. \tag{A2.10}$$

This is called the *Lie algebra* of the rotation group and most of the important properties of rotations follow from it.

In quantum mechanics, the operators which represent the Cartesian components of angular momentum are identified in terms of the rotation generators as $J^i = \hbar\mathcal{J}^i$. Their commutation relations are therefore

$$[J^i, J^j] = \mathrm{i}\hbar\varepsilon^{ijk}J^k. \tag{A2.11}$$

If this identification is to work, it must be possible to regard the angular momentum components as representing the cross product $\boldsymbol{r}\times\boldsymbol{p}$. Readers will find it instructive to examine how the generators themselves can be regarded as transforming according to the rule (A2.6). If, as discussed in chapter 5, we want to use a maximal set of observable quantities which includes angular momentum, we need to know the eigenvalues and eigenstates of these operators. Since the members of a maximal set all commute with each other, and therefore have simultaneous eigenstates, we can include at most one of the angular momentum components. Conventionally, we use J^3, and the x^3 axis which is singled out in this way is sometimes referred to as the *spin quantization axis*. This axis is singled out only by the way in which we choose to describe the system and has no physical significance. On the other hand, if there *is* a preferred direction in space, such as the direction of a magnetic field applied to our system, then it is usually advantageous to choose the spin quantization axis in this direction. Although no two J commute with each other, there is another operator, namely $\boldsymbol{J}^2 = (J^1)^2 + (J^2)^2 + (J^3)^2$ which commutes with all of them. If the generators are matrices, then $\boldsymbol{J}^2$ is proportional to the unit matrix. In general, an operator which commutes with all the generators is called a *Casimir* operator. The rotation group has only one Casimir operator, but other groups may have several.

The eigenvalues and eigenvectors of $\boldsymbol{J}^2$ and J^3 can be found by the same method that we used in chapter 5 to find the energy levels of the harmonic oscillator. The raising and lowering operators are

$$J^{\pm} = J^1 \pm iJ^2 \tag{A2.12}$$

and their commutators with J^3 are

$$[J^{\pm}, J^3] = \mp\hbar J^{\pm}. \tag{A2.13}$$

It is found that the eigenvectors can be grouped into *multiplets* of vectors $|j, m\rangle$, for which the eigenvalue of $\boldsymbol{J}^2$ is $j(j+1)\hbar^2$ and the eigenvalue of J^3 is $m\hbar$. For a given value of j, the values of m run from $-j$ to j in integer steps. There are therefore $(2j+1)$ states in the multiplet corresponding to a given j, so j must be either an integer or a half-odd-integer. In the case of *orbital* angular momentum, which a particle may possess by virtue of its motion in space, the relevant operators are the differential operators acting on wavefunctions. The eigenfunctions are the *spherical harmonics*, which exist only for integer values of j.

In non-relativistic quantum mechanics, we do not know *a priori* whether the half-odd-integer values of j have any relevance to physics. At any rate, they cannot describe orbital angular momentum. As it turns out, they are relevant for describing the *intrinsic* angular momentum or *spin* of certain particles, the most common of which are electrons, protons and neutrons, for which (using s for spin in place of j) $s = \frac{1}{2}$. It is customary to describe spin in the non-relativistic theory by using a two-component wavefunction

$$\psi(\boldsymbol{x}) = \begin{pmatrix} \psi_+(\boldsymbol{x}) \\ \psi_-(\boldsymbol{x}) \end{pmatrix} \tag{A2.14}$$

so that $|\psi_+(\boldsymbol{x})|^2$ is the probability density for finding the particle near $\boldsymbol{x}$ with a spin component of $+\hbar/2$ along the quantization axis and similarly for ψ_-. The operator s^3 must be a diagonal 2×2 matrix with eigenvalues $\pm\hbar/2$, and in fact the operators for the three spin components are $s^i = (\hbar/2)\sigma^i$, where σ^i are the *Pauli matrices* shown in (7.26). Readers may readily verify that these matrices obey the commutation relations (A2.11). The somewhat deeper understanding of spin which arises from the relativistic theory is discussed in chapter 7.

The description of spin-$\frac{1}{2}$ particles requires us to enlarge our view of the rotation group. The matrix $U(\boldsymbol{\alpha})$ which rearranges the components of a spin-$\frac{1}{2}$ wavefunction under a rotation is, according to the general rule (A2.3),

$$U(\boldsymbol{\alpha}) = \exp(\mathrm{i}\boldsymbol{\alpha}\cdot\boldsymbol{s}/\hbar) = \exp(\mathrm{i}\tfrac{1}{2}\boldsymbol{\alpha}\cdot\boldsymbol{\sigma}). \tag{A2.15}$$

Now, the square of each σ^i is the unit 2×2 matrix and it is straightforward to show that

$$U(\boldsymbol{\alpha}) = \cos(\tfrac{1}{2}\alpha) + \mathrm{i}\sin(\tfrac{1}{2}\alpha)\boldsymbol{n}\cdot\boldsymbol{\sigma}. \tag{A2.16}$$

Evidently, for a rotation through an angle of 2π, we get $U = -1$, whereas the rotation of a vector through this angle using R obviously leaves the vector unchanged. For spin-$\frac{1}{2}$ wavefunctions, any rotation angle between 0 and 4π leads to a distinct transformation. A rotation through an angle of $\alpha + 2\pi$ leaves the spin pointing in the same direction as a rotation through α but changes the sign of the wavefunction. Each matrix $U(\boldsymbol{\alpha})$ is a unitary 2×2 matrix with unit determinant and the group of all such matrices is called the *special unitary group of order 2*, or SU(2).

We now have two groups, SO(3) and SU(2), both of which describe rotations in three dimensions. In order to understand the relationship between them, it is first necessary to appreciate that the group itself is the set of all distinct *transformations* or, even more abstractly, simply a set of elements which can be combined in the prescribed manner. Given a multiplet of $(2j + 1)$ objects, for one of the allowed values of j, a rotation rearranges these objects amongst themselves. These objects and the operators which act on them to effect the transformation may be of different kinds. When we give them a concrete form, say by arranging the objects into a column matrix which is transformed by an appropriate set of $(2j + 1) \times (2j + 1)$ matrices, or by identifying them as spherical harmonics which are transformed by the action of differential operators, we say that we have constructed a *representation* of the group. More economically, we can regard the eigenvalue j as defining *the* $(2j + 1)$-

dimensional representation of the group, whatever concrete form we use to represent its elements. Since all the transformations in the group *can* be represented by matrices, the names of the groups are derived from the smallest matrices which can be constructed to have the required properties. These matrices belong to the smallest representation of the group, $j = 1$ for SO(3) or $j = \frac{1}{2}$ for SU(2), which is called the *fundamental representation*.

Both SU(2) and SO(3) have the same Lie algebra (A2.10), which largely determines their properties. The distinction between the two groups lies in their representations and in the attitude we adopt towards the transformations. The Lie algebra allows both integer and half-odd-integer representations. All of these representations are included as representations of SU(2), which is called the *covering group*. Rotations through angles of α and $\alpha + 2\pi$ have different effects upon the half-odd-integer representations and must be regarded as distinct transformations in SU(2). However, these two transformations have the same effect upon integer representations. If we allow only the integer representations, then we are entitled to regard rotations through α and $\alpha + 2\pi$ as identical. When we make this choice, the reduced set of distinct transformations constitutes the group SO(3).

Appendix 3

Natural Units

When we deal with everyday physical situations, it is convenient to use the SI system of units, based upon the metre as a unit of length, the second as a unit of time and the kilogram as a unit of mass. For doing fundamental physics, it is usually much more convenient to use a system of units, known as *natural units*, in which the constants $\hbar$ and c are both equal to 1. Since three basic units need to be defined, this leaves us with one unit still to be chosen. In experiments which study the properties of fundamental particles, the quantity which is most easily controlled is the energy of a particle which has been accelerated by means of electromagnetic fields, so a convenient choice for the remaining unit is some multiple of the electron-volt. To be definite, let us choose the MeV (10^6 eV), which is approximately twice the rest energy of an electron. The conversion factors which allow us to change between SI and natural units are:

$$\begin{aligned} 1\,\mathrm{MeV} &= 1.602\,189 \times 10^{-13}\,\mathrm{J} \\ \hbar &= 1.054\,589 \times 10^{-34}\,\mathrm{J\,s} = 6.582\,17 \times 10^{-22}\,\mathrm{MeV\,s} \\ c &= 2.997\,925 \times 10^{8}\,\mathrm{m\,s^{-1}} \\ \hbar c &= 1.973\,29 \times 10^{-13}\,\mathrm{MeV\,m.} \end{aligned}$$

Thus, for example, if $t(s)$ is a time interval measured in seconds, then $t(\mathrm{MeV}^{-1}) = t(\mathrm{s})/\hbar$ is the equivalent interval in natural units, where the unit of time is MeV^{-1}. Some useful conversions are:

$$\begin{aligned} &\text{time:} & t(\mathrm{s}) &= 6.582\,17 \times 10^{-22} t(\mathrm{MeV}^{-1}) \\ &\text{distance:} & l(\mathrm{m}) &= 1.973\,29 \times 10^{-13} l(\mathrm{MeV}^{-1}) \\ &\text{mass:} & m(\mathrm{kg}) &= 1.783\,48 \times 10^{-30} m(\mathrm{MeV}). \end{aligned}$$

From a theoretical point of view, the use of natural units is more than a matter of convenience: it embodies much of our understanding of the way the world is. If, for example, we measure the speed of sound in a particular material, it makes sense to ask why this speed has the particular value we measure. We can set about calculating it in terms of the density and elastic modulus of the material and these in turn depend on the masses of its constituent atoms and the forces which act on them. However, it makes no sense to ask the same question of the speed of light. According to the theories of relativity, the metrical structure of space and time implies that time intervals and distances are really things of the same kind, and there is no fundamental reason for measuring them in different units. The reason for the appearance of a fundamental 'velocity' c is just that we traditionally measure these two quantities relative to two different standards. The value $c = 1$ is, in every sense of the word, the *natural* value. The number $2.99\ldots \times 10^8$ does not represent the value of any genuine physical quantity. It is properly thought of as being merely a conversion factor which relates our procedures for calibrating rulers and clocks. There is, of course, a good reason for our using different standards for measurements of time and distance intervals. It is that our conscious experiences of these quantities are of quite different kinds. In the equations of theoretical physics, this obvious difference is represented by nothing more than the minus signs in (2.8). This leaves, in my view, deep unresolved questions about the relationship between the universe as described by physics and the actual perceptions of sentient beings such as physicists.

In a somewhat similar way, quantum theory tells us that the notions of energy and momentum are essentially equivalent to those of frequency and inverse wavelength. At an elementary level, this equivalence is manifest in the de Broglie relations (5.1) and (5.2). More fundamentally, it arises from the canonical commutation relations and the role of the energy and momentum operators as generators of spacetime translations. The real significance of Planck's constant is not that, for example, the magnitude of the right-hand side of (5.37) is $1.054\ldots \times 10^{-34}$ J s, but simply that it is *not zero*. The fact that this commutator is a non-zero constant means that there is a fundamental relationship between momenta and intervals of distance, and there is therefore no fundamental reason for measuring them in independent units. Thus, the *natural* way of measuring momentum is as an inverse length, and the constant $\hbar$ is a conversion factor which translates an inverse length into our traditional units of mass $\times$ velocity. Even though a momentum is not something we perceive directly, it is fair to say that the notion of momentum as an inverse distance does not correspond in an obvious way to our ordinary experience of the behaviour of physical objects. As with time and distance, therefore, there is a good reason for our traditional momentum units. The fact that momentum does not ordinarily appear to us as an inverse wavelength is, in my view, one of the deep unresolved mysteries of the interpretation of quantum theory. Whether this mystery is also bound up with the place of sentient beings in the physical world, I am not sure.

From a theoretical point of view, the SI system of units treats electromagnetic quantities in a curious way. In my opinion, this creates deep mysteries where none actually exist! In a vacuum, the electrostatic potential energy of, say, two electrons treated as classical particles a distance r apart, is

$$V(r) = e^2/4\pi\varepsilon_0 r$$

where the quantity $\varepsilon_0 = 8.854\,187\ldots \times 10^{-12}\ \mathrm{F\,m^{-1}}$ is called the *permittivity of free space*. The physical content of this is that the potential energy is proportional to $1/r$, with a constant of proportionality equal to $e^2/4\pi\varepsilon_0$. This quantity, which measures the strength of electrical forces, clearly has the dimensions of (energy $\times$ distance) and, in natural units, is equal to the *fine structure constant* $\alpha = e^2/4\pi\varepsilon_0\hbar c = 7.297\ldots \times 10^{-3} \approx 1/137$, which is dimensionless. The factor of 4π in the denominator has a geometrical significance (see equation (9.84)), being the surface area of a unit sphere, but the constant ε_0 is merely a conversion factor which relates the SI unit of electric charge, the Coulomb, to the units of energy and distance. It cannot be emphasized too strongly that ε_0 *does not* refer to any physical property of the vacuum. Similarly, magnetic forces involve a quantity μ_0, called the *permeability of free space*, whose value is *defined* to be $4\pi \times 10^{-7}\ \mathrm{H\,m^{-1}}$. Since its value is defined, μ_0 also cannot refer to any physical property of the vacuum and it too is no more than a conversion factor. The product $\varepsilon_0\mu_0$ is equal to $1/c^2$ which, as we have seen, is also a conversion factor in the relativistic view of the world. If, when dealing with electromagnetism in SI units, we were to measure all charges in units of $e/\sqrt{\varepsilon_0}$, then only the constant c would ever appear. The reason why c appears is that the magnetic field generated by a moving charge is obtained by a Lorentz transformation of the electric field in its rest frame and, if the velocity of the charge is v, depends on v/c.

There is, therefore, no real need for an independent unit of electric charge. Classically, the strength of electromagnetic forces involving an SI charge q is measured in purely mechanical units by q^2/ε_0. Quantum-mechanically, the strength of electromagnetic forces between fundamental particles is measured by the dimensionless number α, although a proper characterization of this strength requires the running coupling constant discussed in chapter 9.

It might be wondered whether some third fundamental constant, in addition to $\hbar$ and c, should be used to define a system of natural units in which no arbitrary choice of a third unit would be called for. One possibility would be to take the mass of some fundamental particle as a basic unit. The trouble here is that there are many particles to choose from. At present, we do not properly understand the origin of particle masses, and there is no good reason for regarding, say, the electron or muon as especially fundamental. It is quite possible to imagine a universe in which, although $\hbar$ and c had the same fundamental significance as in ours, there were no electrons or muons. The only serious candidate for a third truly fundamental constant is Newton's gravitational constant G. By using $\hbar$, c and G, we can construct three fundamental units of mass, length and time, which are the *Planck units*

$$\text{Planck time:} \quad (G\hbar c^{-5})^{1/2} = 5.389 \times 10^{-44}\ \text{s}$$

$$\text{Planck length:} \quad (G\hbar c^{-3})^{1/2} = 1.615 \times 10^{-35}\ \text{m}$$

$$\text{Planck mass:} \quad (\hbar c/G)^{1/2} = 2.176 \times 10^{-8}\ \text{kg}.$$

Unfortunately, it is not quite clear whether G has the same fundamental status as $\hbar$ and c. It appears, indeed, to be more like e, in that it measures the strength of gravitational forces. Whereas $\hbar$ and c are merely conversion factors, it seems possible to imagine that electromagnetism and gravity could have been either weaker or stronger than they actually are, and that in that sense e and G measure genuine physical properties of our particular world. In any system of units with $\hbar = c = 1$, e is properly measured by the dimensionless fine structure constant and cannot provide a third basic unit. G, on the other hand, cannot be combined with $\hbar$ and c to form a dimensionless measure of the strength of gravity. This fact, as discussed in chapter 12, is symptomatic of the difficulties we experience in trying to reconcile gravity with quantum mechanics, and might be an indication that G is not as fundamental as it appears.

Appendix 4

Scattering Cross-sections and Particle Decay Rates

When analysing the results of a high-energy scattering experiment, we typically consider an initial state containing two particles, with 4-momenta k_1 and k_2, and wish to know the probability of obtaining a given final state containing, say, N particles with momenta $k'_1, \ldots, k'_N$. Actually, the probability of a final state with *exactly* these momenta is generally zero, and we ask for the probability that the first final-state particle has its 3-momentum in the range $\mathrm{d}^3k'_1$ near $\boldsymbol{k}'_1$ and so on. These probabilities are conventionally expressed in terms of *cross-sections*, which can be understood picturesquely in the following way. We consider an incident particle, number 1, heading in the general direction of a stationary target particle, number 2. In the plane containing the target particle and perpendicular to the momentum of the incident particle, we draw an annulus surrounding the target particle of area $\mathrm{d}\sigma$, and imagine that any incident particle which passes through this annulus will give rise to the specified final state. The greater the probability of this event, the larger is the cross-section $\mathrm{d}\sigma$. This is not what actually happens—the picture simply gives a way of quantifying the probability in the following way. Suppose we have a beam of incident particles with a flux j equal to the number of particles crossing unit cross-sectional area per unit time and a target containing n particles per unit volume. The number of scattering events per unit time per unit volume of the target is given by

$$\text{number of events/unit time/unit volume} = jn\,\mathrm{d}\sigma. \qquad \text{(A4.1)}$$

Regardless of our simple picture, this defines the differential cross-section $\mathrm{d}\sigma$.

The quantities we can attempt to calculate theoretically are S-matrix elements of the form $\langle k'_1, \ldots, k'_N; \text{out}|k_1, k_2; \text{in}\rangle$.

Since energy and momentum are conserved, this matrix element is proportional to $\delta(P_\mathrm{f} - P_\mathrm{i})$ where P_i and P_f are the total 4-momenta of the initial and final states. To be specific, we write it as

$$\langle k_1', \ldots, k_N'; \mathrm{out}|k_1, k_2; \mathrm{in}\rangle = (2\pi)^4\delta(P_\mathrm{f} - P_\mathrm{i})T_\mathrm{fi}. \tag{A4.2}$$

According to (5.9), the probability we want is proportional to the square magnitude of this quantity, which involves the square of the δ function. In one of these two δ functions, we are entitled to set the argument to zero. This gives an infinite value, which can be interpreted in the following way. Using the representation (A1.11), but remembering that the argument of our function is a 4-momentum, we have

$$(2\pi)^4\delta(0) = \int \mathrm{d}^3x\,\mathrm{d}t. \tag{A4.3}$$

If we now imagine observing a large target volume V for a long time T, we can interpret this spacetime integral as the product VT.

We must now consider the normalization of our state vectors, which is defined by (7.18). The probability (5.9) is correct when the vectors are normalized to 1. Consider, therefore, a single-particle state $|\Psi\rangle$ such that $\langle\Psi|\Psi\rangle = 1$. For a particle in this state, the probability of finding it to have a 3-momentum in the range d^3k near $\boldsymbol{k}$ must be of the form

$$|\langle k|\Psi\rangle|^2 g(\boldsymbol{k})\,\mathrm{d}^3k = \langle\Psi|k\rangle\langle k|\Psi\rangle g(\boldsymbol{k})\,\mathrm{d}^3k \tag{A4.4}$$

where $g(\boldsymbol{k})$ is a function which accounts for the normalization of the momentum eigenstates. On integrating over the momentum, we must get a total probability of 1. We must therefore choose $g(\boldsymbol{k})$ so that

$$\int \mathrm{d}^3k\, g(\boldsymbol{k})|k\rangle\langle k| = \hat{I} \tag{A4.5}$$

where $\hat{I}$ is the identity operator (see exercise 5.4 and equation (9.20)). It is a simple exercise, using (7.18), to show that $g(\boldsymbol{k}) = [(2\pi)^3 2\omega(\boldsymbol{k})]^{-1}$.

To calculate the number of scattering events per unit time per unit volume which give rise to final states in our specified range, we take the squared magnitude of (A4.2), divide by VT, which equals $(2\pi)^4\delta(0)$ and multiply by the factor

$$\mathrm{d}\rho_\mathrm{f} = C_\mathrm{f}\prod_{i=1}^{N}\frac{\mathrm{d}^3k_i'}{(2\pi)^3 2\omega(\boldsymbol{k}_i')}. \tag{A4.6}$$

The number C_f is included to account for any sets of identical particles in the final state: for any set of n identical particles, C_f contains a factor of $1/n!$, because rearrangements of these particles do not count as distinct states. The quantity we arrive at in this way is the scattering rate per unit volume defined in (A4.1):

$$jn\,\mathrm{d}\sigma = (2\pi)^4\delta(P_\mathrm{f} - P_\mathrm{i})|T_\mathrm{fi}|^2\,\mathrm{d}\rho_\mathrm{f} \tag{A4.7}$$

provided that j and n are identified in accordance with the normalization of our state vectors. As we saw in chapter 7, this normalization implies that there are $2\omega(\boldsymbol{k})$ particles per unit volume. The target particles with mass m_2 are at rest, so $n = 2m_2$, and if the incident particles are travelling with a speed v, then their flux is $j = 2\omega(\boldsymbol{k}_1)v$. Thus, our result for the cross-section is

$$\mathrm{d}\sigma = \frac{1}{4Q}(2\pi)^4\delta(P_\mathrm{f} - P_\mathrm{i})|T_\mathrm{fi}|^2\,\mathrm{d}\rho_\mathrm{f} \tag{A4.8}$$

where $Q = \omega(\boldsymbol{k}_1)m_2 v$. This expression for Q is valid in the rest frame of the target particles, where the initial 4-momenta are $k_1 = (\omega(\boldsymbol{k}_1), \boldsymbol{k}_1)$ and $k_2 = (m_2, \boldsymbol{0})$. It is easily shown that $v = |\boldsymbol{k}_1|/\omega(\boldsymbol{k}_1)$ and that

$$Q = [(k_1 \cdot k_2)^2 - m_1^2 m_2^2]^{1/2}. \tag{A4.9}$$

This is a Lorentz scalar, expressed in terms of the 4-vector momenta, and is therefore valid in any frame.

In the same way, we can consider an initial state containing a single unstable particle and work out the probability per unit time, $\mathrm{d}\Gamma$, for it to decay into a final state specified as above. The result, valid in the rest frame of the decaying particle, is

$$\mathrm{d}\Gamma = \frac{1}{2m}(2\pi)^4\delta(P_\mathrm{f} - P_\mathrm{i})|T_\mathrm{fi}|^2\,\mathrm{d}\rho_\mathrm{f} \tag{A4.10}$$

where m is the mass of the decaying particle. If we integrate over all final state momenta and sum over all modes of decay, we get the total decay probability per unit time Γ, and the lifetime of the particle is $1/\Gamma$.

Bibliography

I give here a list of textbooks and review articles on the various subjects covered in this book. Some, marked (B), will be most useful for preliminary background reading. A few, marked (A), are considerably more advanced than this one. The others, while treating their specialized subjects in more detail than I do, should be readily understood by anyone who has mastered the contents of this book.

General Theoretical Physics

Longair M S 1984 *Theoretical Concepts in Physics* (Cambridge: Cambridge University Press) (B)

Classical Mechanics and Classical Electromagnetism

Goldstein H 1950 *Classical Mechanics* (Reading, MA: Addison-Wesley) (B)

Jackson J D 1962 *Classical Electrodynamics* (New York: Wiley) (B)

Leech J W 1958 *Classical Mechanics* (London: Methuen) (B)

Lorrain P and Corson D R 1970 *Electromagnetic Fields and Waves* (New York: Freeman) (B)

Geometry, Relativity, Gravitation and Cosmology

Abbott L F and Pi S-Y (eds) 1986 *Inflationary Cosmology* (Singapore: World Scientific) (A)

Brandenberger R H 1985 Quantum field theory methods and inflationary universe models *Rev. Mod. Phys.* **57** 1

Foster J and Nightingale J D 1979 *A Short Course in General Relativity* (London: Longman)

Hawking S W and Ellis G F R 1973 *The Large Scale Structure of Space-time* (Cambridge: Cambridge University Press) (A)

Lightman A P, Press W H, Price R H and Teukolsky S A 1975 *Problem Book in Relativity and Gravitation* (Princeton, NJ: Princeton University Press)

Misner C W, Thorne K S and Wheeler J A 1973 *Gravitation* (San Francisco: Freeman)

Peebles P J E 1971 *Physical Cosmology* (Princeton, NJ: Princeton University Press)

Ryan M P and Shepley L C 1975 *Homogeneous Relativistic Cosmologies* (Princeton, NJ: Princeton University Press) (A)

Schutz B F 1980 *Geometrical Methods of Mathematical Physics* (Cambridge: Cambridge University Press)

—— 1985 *A First Course in General Relativity* (Cambridge: Cambridge University Press)

Silk J 1980 *The Big Bang* (San Francisco: Freeman) (B)

Weinberg S 1972 *Gravitation and Cosmology* (New York: Wiley)

—— 1977 *The First Three Minutes* (Glasgow: Collins) (B)

Quantum Theory, Quantum Field Theory and Elementary Particles

Aitchinson I J R and Hey A J G 1989 *Gauge Theories in Particle Physics* 2nd edn (Bristol: Adam Hilger)

Bailin D 1982 *Weak Interactions* (Bristol: Adam Hilger)

Birrell N D and Davies P C W 1982 *Quantum Fields in Curved Space* (Cambridge: Cambridge University Press)

Cheng T-P and Li L-F 1984 *Gauge Theory of Elementary Particle Physics* (Oxford: Oxford University Press)

Collins J C 1984 *Renormalization* (Cambridge: Cambridge University Press) (A)

Halzen F and Martin A D 1984 *Quarks and Leptons: an Introductory Course in Modern Particle Physics* (New York: Wiley)

Itzykson C and Zuber J-B 1980 *Quantum Field Theory* (New York: McGraw-Hill) (A)

Perkins D H 1982 *Introduction to High Energy Physics* (Reading, MA: Addison-Wesley)

Ramond P 1981 *Field Theory: A Modern Primer* (Reading, MA: Benjamin-Cummings)

Schiff L I 1955 *Quantum Mechanics* (New York: McGraw-Hill) (B)

Sudbery A 1986 *Quantum Mechanics and the Particles of Nature* (Cambridge: Cambridge University Press)

Taylor J C 1976 *Gauge Theories of Weak Interactions* (Cambridge: Cambridge University Press)

Wess J and Bagger J 1983 *Supersymmetry and Supergravity* (Princeton, NJ: Princeton University Press) (A)

West P (ed.) 1986 *Supersymmetry: A Decade of Development* (Bristol: Adam Hilger) (A)

Ziman J M 1969 *Elements of Advanced Quantum Theory* (Cambridge: Cambridge University Press)

Thermodynamics, Statistical Mechanics and Phase Transitions

Amit D J 1978 *Field Theory, The Renormalization Group and Critical Phenomena* (New York: McGraw-Hill)

Fetter A L and Walecka J D 1971 *Quantum Theory of Many-particle Systems* (New York: McGraw-Hill)

Ma S-K 1976 *Modern Theory of Critical Phenomena* (Reading, MA: Benjamin)

Pippard A B 1966 *Elements of Classical Thermodynamics* (Cambridge: Cambridge University Press) (B)

Reichl L E 1980 *A Modern Course in Statistical Physics* (University of Texas Press and London: Edward Arnold)

Stanley H E 1971 *Introduction to Phase Transitions and Critical Phenomena* (Oxford: Oxford University Press)

References

Aarnio P *et al* 1989 *Phys. Lett.* **B231** 539
Abe F *et al* 1989 *Phys. Rev. Lett.* **63** 720
Abrams G F *et al* 1989 *Phys. Rev. Lett.* **63** 724
Adeva B *et al* 1989 *Phys. Lett.* **B231** 509
Akrawy M Z *et al* 1989 *Phys. Lett.* **B231** 530
Anderson C D 1933 *Phys. Rev.* **43** 491
Baier R and Satz H (eds) 1985 *Proc. Phase Transitions in the Very Early Universe, Nucl. Phys.* **B252**
Barrow J D 1983 *Fundam. Cosmic Phys.* **8** 83
Bernstein J, Brown L S and Feinberg G 1989 *Rev. Mod. Phys.* **61** 25
Decamp D *et al* 1989 *Phys. Lett.* **B231** 519
Dirac P A M 1928 *Proc. R. Soc.* **A117** 610
—— 1929 *Proc. R. Soc.* **A126** 360
Dolgov A D and Zeldovich Ya B 1981 *Rev. Mod. Phys.* **53** 1
Domb C and Green M S (eds) 1976 *Phase Transitions and Critical Phenomena* vol. 6 (London: Academic Press)
Eddington A S 1929 *Space, Time and Gravitation* (Cambridge: Cambridge University Press)
Einstein A 1905 *Ann. Phys., Lpz* **17** 891, **18** 639
Georgi H and Glashow S L 1974 *Phys. Rev. Lett.* **32** 438
Glashow S L 1961 *Nucl. Phys.* **22** 579
Guth A 1981 *Phys. Rev.* **D23** 347
Hawking S W 1974 *Nature* **248** 30
t'Hooft G 1971 *Nucl. Phys.* **B33** 173, **B35** 167
Kaluza T 1921 *Sitzungsber. Preuss. Acad. Wiss.* 966
Kirzhnits D A and Linde A D 1972 *Phys. Lett.* **42B** 471
Klein O 1926 *Z. Phys.* **37** 895
Landsberg P T (ed.) 1982 *The Enigma of Time* (Bristol: Adam Hilger)
Lawrie I D 1988 *Nucl. Phys.* **B301** 685
Lawrie I D and Lowe M J 1981 *J. Phys. A: Math. Gen.* **14** 981
Lawrie I D and Sarbach S 1984 in *Phase Transitions and Critical Phenomena* vol. 9, ed. C Domb and J-L Lebowitz (London: Academic Press)
Lorentz H A 1904 *Proc. Acad. Sci. Amsterdam* **6** 809
Lucas J R 1973 *A Treatise on Space and Time* (London: Methuen)
Maxwell J C 1864 *Phil. Trans. R. Soc.* **155** 459

—— 1873 *A Treatise on Electricity and Magnetism* (Oxford: Clarendon Press) reprinted in 1954 by Dover, New York

Michelson A A and Morley E W 1887 *Am. J. Sci.* **34** 333 and *Phil. Mag.* **24** 449

Minkowski H 1908 Address to the 80th Assembly of German Natural Scientists and Physicians; translation in *The Principle of Relativity* (Methuen 1923, reprinted by Dover 1952)

Newton I 1686 *Philosophiae Naturalis Principia Mathematica* English trans. by A Motte 1927. Revised trans. ed. F Cajori 1966 (Berkeley, CA and Los Angeles: University of California Press)

Onsager L 1944 *Phys. Rev.* **65** 117

Ornstein R E 1969 *On the Experience of Time* (Harmondsworth: Penguin)

Penzias A A and Wilson R W 1965 *Astrophys. J.* **142** 419

Pound R V and Rebka G A 1960 *Phys. Rev. Lett.* **4** 337

Prigogine I 1980 *From Being to Becoming* (San Fransisco: Freeman)

Salam A 1968 in *Elementary Particle Physics* (Nobel Symposium No 8) ed. N Svartholm (Stockholm: Almqvist and Wilsell)

Salam A and Ward J C 1964 *Phys. Lett.* **13** 168

Schwarzschild K 1916 *Sitzungsber. Preuss. Acad. Wiss.* 189

Simmons G F 1963 *Introduction to Topology and Modern Analysis* (Tokyo: McGraw-Hill)

Smart J J C (ed.) 1964 *Problems of Space and Time* (New York: Macmillan)

Strawson P F 1959 *Individuals* (London: Methuen)

Weinberg S 1967 *Phys. Rev. Lett.* **19** 1264

Whitrow G J 1972 *What is Time?* (Thames and Hudson: London) reprinted as *The Nature of Time* by Pelican Books, Harmondsworth, 1975

Wilson K G and Fisher M E 1972 *Phys. Rev. Lett.* **28** 240

Yang C N 1952 *Phys. Rev.* **85** 809

Yang C N and Mills R L 1954 *Phys. Rev.* **96** 191

Index